THE NEUTRON

A Tool and an Object in Nuclear and Particle Physics

THE NEUTRON

A Tool and an Object in Nuclear and Particle Physics

Hans G Börner
Institut Laue Langevin, France

Friedrich Gönnenwein
University of Tübingen, Germany

NEW JERSEY • LONDON • SINGAPORE • BEIJING • SHANGHAI • HONG KONG • TAIPEI • CHENNAI

Published by

World Scientific Publishing Co. Pte. Ltd.
5 Toh Tuck Link, Singapore 596224
USA office: 27 Warren Street, Suite 401-402, Hackensack, NJ 07601
UK office: 57 Shelton Street, Covent Garden, London WC2H 9HE

British Library Cataloguing-in-Publication Data
A catalogue record for this book is available from the British Library.

THE NEUTRON
A Tool and an Object in Nuclear and Particle Physics

ISBN-13 978-981-4273-08-4
ISBN-10 981-4273-08-2

Printed in Singapore.

Preface

The Institut Laue-Langevin, or ILL, is an internationally-financed scientific facility, situated in Grenoble, France. It is one of the world centres for research using neutrons. Founded in 1967 and honoring the physicists Max von Laue and Paul Langevin. The ILL currently provides the most intense neutron source in the world, at 1.5×10^{15} neutrons/cm ^{2}s, reliably delivering intense neutron beams to 40 unique scientific instruments. The Institute welcomes 1700 visiting scientists per year to carry out world class research in solid state physics, crystallography, soft matter, biology, chemistry and fundamental physics. Funded primarily by its three founder members: France, Germany and the United Kingdom, the ILL has also signed scientific collaboration agreements with twelve additional European countries.

The high-flux research-reactor produces neutrons through fission in a specially designed, compact-core fuel element. Neutron moderators cool the neutrons to useful wavelengths, which are then directed at a suite of instruments and used to probe the structure and behavior of many forms of matter by elastic and inelastic neutron scattering, and to investigate the fundamental physical properties of the neutron. Nothing goes to waste: Fission products and gamma rays produced by nuclear reactions in the reactor core are analyzed by specialized instruments, which forms an important part of the instrument suite.

The Ill's neutron source is based on a single element 58.3 MW nuclear reactor designed for high brightness. The main moderator is the ambient D_2O coolant surrounding the core which delivers intense beams of thermal neutrons to 11 beam lines and to four neutron guides. A graphite hot source operating at 2400 ^{0}K delivers hot neutrons to 3 beam lines with energies up to 1 eV and wave lengths down to 0.3 Å. A renewal project has resulted in the installation of a new hot source and beam tubes during 2003. Two liquid deuterium cold sources at 25 ^{0}K deliver cold neutrons to 9 neutron guides with energies down to 200 meV and wavelengths beyond 20 Å. An ultra cold neutron source fed from the top of one of the cold sources delivers neutrons vertically through the reactor pool to five instruments on the operational floor of the reactor. In all there are more than 40 neutron delivering measuring stations at the ILL. Those form the basis for a program of research covering a wide variety of fields. The majority of instruments are dedicated to the study of problems in solid-state physics, materials science, chemistry, the bio-sciences and the earth sciences, but there is also a small group, dedicated to nuclear and low energy particle physics. A summary of the main activities of this group, carried out over the last 35 years, is the basic subject of this book where in the following the properties of the neutron as a particle and nuclear reactions induced by neutrons with energies not higher than a few eV are discussed.

Today the Nuclear and Particle Physics group – which is one out of five ILL instrumental groups – pursues two main lines in nuclear physics studies. They use neutrons as a tool with which nuclei can be excited. In the first one thermal-neutron capture uses the binding energy of a neutron to bring the nucleus into an excited state. Subsequently emitted gamma rays are measured with high precision. In the second one, neutron-induced fission allows one not only to study the fission process itself, but also the structure of the very neutron-rich isotopes produced. Most current measurements are performed using in-pile target arrangements, but some work is also carried out with neutrons at positions fed by neutron guides. ILL's nuclear physics instruments belong to the first generation instruments at the ILL. Although they are unique in the world in their combination of instrumentation and high neutron flux, the science performed on them is well embedded in a world-wide context. Over the years these instruments were either completely rebuilt or substantially up graded. Common to them is that they use in-pile target arrangements, exploiting the maximum thermal neutron flux of close to 10^{15} neutrons/cm^{2}s. The experiments performed on these nuclear physics instruments will constitute the main part of this book.

Low energy particle physics research at ILL uses the neutron as an object for the study of the fundamental forces by carrying out very precise experiments at very low energies. These studies may be divided into four subdivisions: Measurements of the neutron's static properties (such as the electric charge and the magnetic moment), its decay properties (lifetime, directional correlations of the decay products), the interaction between neutrons and nuclei (spin rotation, parity non conserving capture reactions) and neutron optics (interference experiments, inertial to gravitational mass equivalence, neutron gravitational levels). So called scheduled instruments (instruments which are open to the user community), dedicated to these studies, came to life as late as in the 90s after a long reactor shut down. Since, the scientific activity at the ILL in this area has increased steadily. Within the last decade the sensitivity of slow neutron experiments to small changes in momentum or energy has been increased by orders of magnitude. Therefore neutrons have become of prime interest in high precision studies of fundamental interactions. As already several quite exhaustive reviews have been published on this subject (see for instance [DUB91, PEN93, ABE08, ABE09, DUB11] we will here only present a summary to cover this part of activities in the Nuclear and Particle Physics Group of the ILL (and limit ourselves in this part to experiments carried out after 1990 when scheduled instruments became available at ILL in this area).

In the following first a short introduction to reactor neutron production and neutron beam techniques will be given. Then the basic studies of the neutron properties in low energy particle experiments will be summarized very briefly. just to give a feeling for the importance of such experiments in elementary particle physics. Finally, the main part of this contribution will start with an introduction to the neutron capture process followed by rather detailed reports in the fields of nuclear structure and nuclear fission studies by means of neutron induced reactions. We will cover a series of experiments carried out at the ILL over a period from about 1973 to 2010, using mainly the GAMS and LOHENGRIN spectrometers, respectively.

Bibliography

[ABE08] H. Abele, Prog. Part. Nucl. Phys 60 (2008) 1-81
[ABE09] H. Abele., Nucl. Instr. Meth. A 611 (2009) 193-197
[DUB91] D. Dubbers, Progr. Part. Nucl. Phys. 26 (1991) 173
[DUB11] D. Dubbers and M.G. Schmidt, Rev.Mod. Phys. in print arXiv: 1105.3694v1 [hep-ph]
[PEN93] J.M. Pendlebury, Annu. Rev. Nucl. Part. Sci. 43 (1993) 687

Acknowledgements

It is not possible to give credit to all staff scientists and technicians of the Institut Laue-Langevin for the ingenious construction of instruments and the continuous effort to further improve them and push their performance to the limit of technology. Visitors from all over the world have been attracted by the unique features of instruments installed at a powerful and reliable research reactor. In turn the "users", as the ILL's scientific visitors are called, contributed ideas for investigating either specific properties of the neutron as an object or for studying in thermal neutron induced reactions novel models or theories in nuclear physics. In these studies, very often it was just the performance and the precision of the instruments allowing for experiments which could not have been performed elsewhere. Experiments at the border lines of feasibility became in many cases decisive for the validation of new vistas on nuclear structure or fission phenomena. In addition the quality of cold and ultra-cold neutron beams has triggered the search for "new physics", complementary to High Energy Physics.

This book owes much to many people, but it would have been impossible without those scientists who thought of the first generations of our instruments and who built them. We would like to mention especially: P. Ageron, P. Armbruster, R.D. Deslattes, M.S. Dewey, D. Dubbers, T. v. Egidy, D. Heck, E.G. Kessler, H.R. Koch, B.P.K. Maier, H. Maier Leibnitz, W. Mampe, E. Moll, N. Ramsay, K. Schreckenbach, O.W.B. Schult, A. Steyerl and H. Wollnik.

Moreover there was quite a number of colleagues (the majority of ILL's scientists had – and still has – limited term contracts) having been responsible for the operation, maintaining, refurbishment and improvement of those instruments having contributed to the physics discussed hereafter: G. Barreau, J.P. Bocquet, R. Brissot, J. Butterworth, G. Colvin, W. Davidson, W. Drexel, H. Faust, P. Geltenbort, G. Fioni, F. Hoyler, S. Ivanov, M. Jentschel, J. Jolie, S.A. Kerr, U. Köster, B. Krusche, H. Lehmann, T. Materna, P. Mutti, S. Neumair, V. Nesvizhevsky, C. Plonka, J.A. Pinston, S. Robinson, P. Schillebeeckx, G. Siegert, H. Schrader, G. Simpson, T. Soldner, I. Tsekhanovich, D.D. Warner and O. Zimmer.

As to the visitors it is with gratitude that we acknowledge colleagues and friends having proposed pioneering ideas opening up new fields of research and the perseverance with which these ideas were pursued. The following list is by no means exhaustive and we apologize for colleagues having not been mentioned (for instance the about 100 bright students who carried out their thesis works using ILL's Nuclear and Particle Physics instrumentation): A. Abrahamian, H. Abele, P. Armbruster, M. Asghar, P. von Brentano, R.F. Casten, H.-G. Clerc, G. Danilyan, H.-O. Denschlag, R.D. Deslattes, D. Dubbers, A. Gagarski, K. Heyde, R.W. Hoff, F. Käppeler, K.P. Lieb, M. Mutterer, A. Oed, M. Pendlebury, G. Petrov, C. Signarbieux, C. Wagemans and N.V. Zamfir.

We are very happy to acknowledge the Institut Laue Langevin for supporting this project and for the infrastructure that made it possible.

Finally, we would like to thank World Scientific for asking us to write this book and for their patience during several delays and lapsed deadline.

Contents

Chapter 1

Low Energy Neutron Beams

1.1 Production of Slow Neutrons

In a nuclear reactor the fission process (of ^{235}U at the ILL) is utilized for the production of neutrons with energies in the order of MeV. The fuel element is surrounded by a moderator material of low atomic mass ("heavy water", D_2O at ILL) and the slowing down process of neutrons proceeds by multiple scattering of neutrons at the moderator atoms. In this process neutrons pass from energies in the range of keV to eV and finally come to thermal equilibrium with the moderator temperature ("thermal neutrons"). As neutrons behave very similar to a gaz, they can be extracted out of the moderator by evacuated beam tubes and be

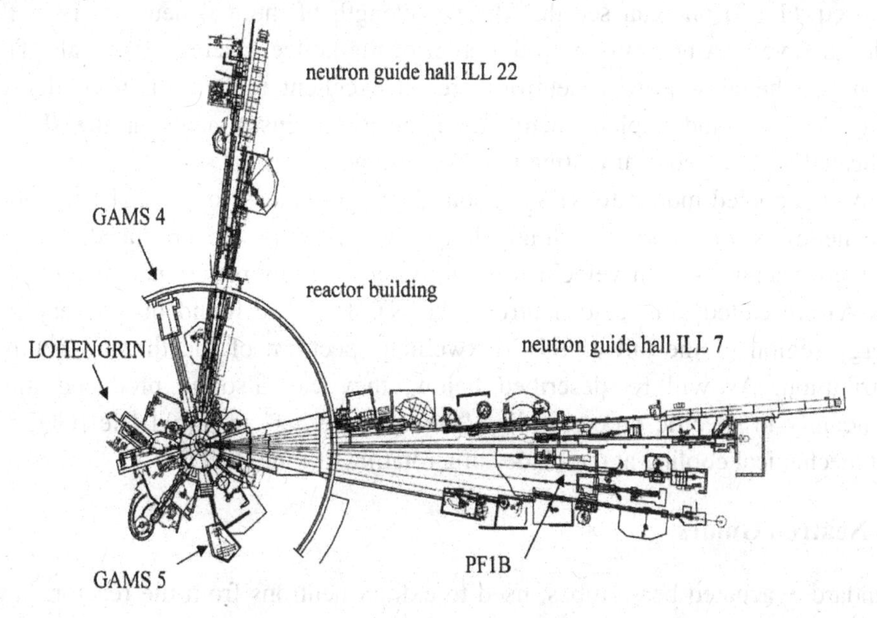

Fig. 1.1.1. Layout of beam tubes, neutron guides and experimental set-ups at the ILL reactor.

transported over quite some distances from the reactor core (Fig. 1.1.1.). At the ILL reactor two neutron guide halls (called ILL7 and ILL22, respectively) extend the experimental hall of the reactor building. This provides space for quite some additional instruments and experiments.

The spectrum of thermal neutrons is defined by a Maxwellian velocity distribution with the temperature of the moderator (typically around room temperature). This corresponds to a velocity which is typically about $v_0 = 2200$ m/s in the maximum of the distribution. Their de Broglie wavelength is given by

$$\lambda = h/mv_0 \tag{1.1.1}$$

with h Planck's constant and m the rest mass of the neutron.

At the most probable velocity $v_0 = (2kT/m)^{1/2}$ the kinetic energy of the neutrons is given by

$$E_n = \tfrac{1}{2} m v_0^2 = kT, \text{ typically around 0.025 eV for thermal neutrons} \tag{1.1.2}$$

As the wavelength of the neutrons is proportional to $1/v$ or $\sqrt{1/E_n}$ handy relations for the wavelength are

$$\lambda[\text{Å}] = 3970/v[m/s] = 0.287/\sqrt{E_n}\ [eV] \tag{1.1.3}$$

With eq. (1.1.3) one can see that the wavelength of thermal neutrons is in the order of several Ångstrom i.e. in the order of the lattice spacing of crystals. This is one of the reasons why neutrons are an excellent tool to study condensed matter physics and explains why the majority of instruments at the ILL is dedicated to this kind of investigations (see preface).

With a cooled moderator (D_2 at about 20° K in the case of the ILL) so called cold neutrons with correspondingly longer wavelengths are produced. Finally, very slow neutrons with velocities in the order of a few meters per second ($\lambda >$ 100 Å) are called ultra cold neutrons (UCN). They are found in the very low energy region of the tail of the Maxwellian spectrum of the thermal neutron distribution. As will be described below, they can also be produced more abundantly in specific processes like "down scattering" in liquid ^{4}He [GOL83] and mechanical cooling at the blades of a rotating turbine [STE75].

1.2 Neutron Guides

Standard evacuated beam tubes, used to extract neutrons from the reactor, have the disadvantage that it is difficult to reduce background by collimation and at large distances losses are substantial. In consequence measuring times are long.

These disadvantages can be avoided by using so-called neutron guides. Neutron guides are specially prepared rectangular tubes in which slow neutrons can be transported over large distances (even more than 100 m) by reflection at the inner walls of the tubes. The reflection can either be achieved by total reflection on a thin coating layer of metallic Nickel (best is isotopically pure ^{58}Ni), or by interferences at complex multilayer surfaces (super mirrors). The reflecting layers are generally deposited on float glass plates and are assembled as rectangular tubes with cross sections of several cm^2. Thermal and cold neutrons can be reflected when impinging at grazing incidence whilst background from the source is out scattered at the guide walls. Therefore the signal to background ratio becomes better at increasing distances. By bending the guides (radii of curvature are typically in the range of kilometres) one may shift the exit of the guide out of the direct line of sight to the source thereby suppressing completely the background coming from the source. This concerns especially the high energy part of the neutron beam. The guides start close to the neutron sources (thermal or cold) and direct the neutrons from the sources into neutron guide halls where they can feed a variety of experiments.

Analogous to electromagnetic waves, the wave vector of neutrons inside a material can be described as *nk*, with *k* the wave vector in vacuum and *n*, as usual, the index of refraction. In analogy to X-rays n can be written as:

$$n = 1 - 2\pi\rho<b>/k^2 = 1 - \delta \tag{1.2.1}$$

with ρ the number density of the nuclei in the reflecting material, and $<b>$ the average scattering length of the material. For natural nickel (68% ^{58}Ni, 26% ^{60}Ni, 1% ^{61}Ni, 4% ^{62}Ni, and 1% ^{64}Ni) $\rho = 9 \cdot 10^{28}$ m^{-3} and $<b> = 10.3 \cdot 10^{-15}$ m. This yields $\delta \approx 3.7 \cdot 10^{-5}$ for 5Å neutrons. The index of refraction for neutrons is thus comparable to the index for X-rays. By contrast, for some isotopes, such as natural Ti or pure ^{62}Ni, the scattering length is negative, so that n > 1. However, most materials have a refractive index smaller than 1. Hence neutrons flying along a neutron guide are totally reflected up to a critical grazing angle $\theta_{c,}$ which depends on the material and also on the neutron wavelength λ. For natural Nickel the critical grazing angle is $\theta_{c,Ni} \approx 0.1^{0.}\lambda$ with λ *in* Å. At the critical angle the intensity of the reflected beam drops rapidly. The momentum transfer *p* perpendicular to the surface is:

$$p = 4\pi/\lambda \sin\theta \tag{1.2.2}$$

Reflections of neutrons can be described in terms of the average Fermi pseudo potential P_F of the reflecting material. Compared to vacuum it can be positive or negative. Neutrons are totally reflected for $P_F > 0$ provided the critical momentum component p_c of the neutron perpendicular to the reflecting surface is smaller than $\sqrt{2mP_F}$. For natural Ni the critical momentum transfer is $p_{c,Ni} = 0.217\ nm^{-1}$. From that follows that for a natural Ni Mirror, the reflectivity is low if $\sin\theta \sim \theta > \lambda(nm) \cdot 0.217(nm^{-1})/4\pi$ [ICE10]. Here λ is the neutron wavelength in nm, θ is the incident angle of the beam with respect to the mirror surface in radians and the momentum transfer constant $0.217\ (nm^{-1})$ is a physical property of Ni. The corresponding Fermi potential is $P_F=2.2\cdot10^{-7}\ eV$. This gives for the critical angle a value of $\theta_c = 1.7\cdot10^{-3}\ \lambda$[Å] *radian* which for thermal neutrons of λ = 1.8 Å becomes $\theta_c = 0.17^0$. Since the critical angle decreases with increasing neutron energy, the reflection works better the colder the neutrons are.

In a magnetic environment with field strength B the neutron sees a potential $\mu_n\cdot B$ due to its magnetic moment μ_n. This potential adds to the Fermi potential P_F:

$$P = P_F + \mu_n\cdot B \tag{1.2.3}$$

The sign of $\mu_n\cdot B$ depends on the orientation of neutron spin and magnetic field, respectively. If for a given orientation $P < 0$, then the neutrons are not reflected. However, those with the opposite orientation are. This can be used to obtain polarized neutrons. The degree of polarization obtained depends on the reflecting material. Nowadays more and more so-called super mirrors are used for coating neutron guides. They exhibit larger critical angles. They consist of multilayers of elements such as Ni and Ti which are chemically similar, but whose indices of refraction are very different. Multilayers correspond to an artificial one-dimensional lattice and Bragg reflection occurs at an appropriate momentum transfer similar to the Bragg reflections from the lattice planes of a crystal. Super mirrors are essentially characterized by their reflectivity and the m value. The m value defines the range of the super mirror regime in multiples of $p_{c,Ni}$. From that follows, that neutrons are efficiently reflected when:

$$\sin\theta \sim \theta < \lambda\,(nm)\cdot m/57.9 \text{ [ICE10].}$$

Bibliography

[GOL83] R.Golub et al., Z. Phys. B51 (1983) 187
[STE75] A. Steyerl, Nucl. Instr. 125 (1975) 461
[ICE10] G.E. Ice, J.Y. Choi, P.A. Takacs, A. Kkhounsary, Y. Puzyrev, J.J. Molaison, C.A. Tulk, K.H. Anderson, T. Bigault, Appl. Phys. A 99 (2010) 635

Chapter 2

Particle Properties of the Neutron: A Summary

2.1 Mass of the Neutron

The mass of the neutron can be measured using the thermal neutron capture reaction:

$$p + n_{th} \rightarrow d + E_\gamma + E_{recoil}$$

Following neutron capture (in Hydrogen) a deuteron d is formed and the excess in binding energy is released by the emission of a γ-ray which causes a recoil of the deuteron with an energy of

$$E_{recoil} = E_\gamma^2/2m_d$$

The rest mass of both, the proton and the deuteron, respectively, were determined directly by charged particle mass spectroscopy. The gamma-ray energy was determined with high precision gamma-ray spectroscopy at ILL. Details are described in chapter 3. The mass of the neutron was obtained to be 939.565360(81) MeV/c^2 and exceeds that of the proton by 1.2933317(5) MeV [PDG06].

2.2 Studies of Neutron Properties with Cold Neutrons

Routinely scheduled experiments with cold neutrons started at ILL with the creation of a beam position called PF1 at the H53 cold neutron guide situated in the guide hall ILL22 (see also Fig. 1.1.1). This instrument was completely reconstructed when moving from the guide H53 in ILL22 to the new guide H113 in the guide hall ILL7. H113 was constructed with considerable manpower and financial contributions from the University of Heidelberg [HAE02, ABE06]. The guide cross section is 6×20 cm^2 and the neutron flux density is 1.4×10^{10} neutrons cm^{-2}s^{-1}. The big advantage of H113 with respect to H53 is that, besides the significant increase in flux, there are no other experimental installations upstream. Especially important for all experiments at PF1 was the continuous improvement of neutron polarization techniques [KRE05]. Carrying out experiments at these facilities is different from all other scheduled ILL instruments: the ILL provides beams of neutrons with optimized parameters like intensity or polarization plus necessary equipment. Typically the users then install at these beams – in collaboration with the ILL – their own dedicated equipment to carry out the experiments. The duration for such experiments is typically one to several reactor cycles of about 55 days each.

At first glance, the weak force seems not to play an important part in our everyday life. Nevertheless, it is actually vital to our existence. 50 years ago the discovery of parity violation in beta decay was a sensation. Today evidence that parity is not maximally violated would again be good for a sensation and this would point to left-right symmetric Grand Unification. Beyond-the-Standard-Model physics is a very active field. At PF1 great efforts were made to determine limits on right-handed current admixtures. Whereas nowadays most neutron lifetime measurements are performed with neutrons trapped in material bottles (see below), one of the first experiments on PF1 – when it was still installed at the H53 beam port – concerned an up to date most precise beam measurement of the lifetime τ_n [BYR90]. Quite a number of neutron decay studies were carried out by using several generations of dedicated instruments (for instance PERKEO III [MAE09]), developed by the University of Heidelberg. The methods employed here were precision measurements of various asymmetry coefficients observed in the decay of free polarized neutrons. Neutrons decay into electrons, protons, and electron-anti-neutrinos. Observables are the parameters describing the correlation between the spin of the decaying neutron and the momentum of one of the decay particles: the observable A describes the asymmetry of the electron [REI00], [KUZ00], B that of the neutrino [KRE05a, KRE05b, SCH07], and C that of the proton [SCH08]. Additional correlations of observables are also of interest. One of the first ones to be studied at PF1 was the parameter a which in the neutron decay links the outgoing neutrino momentum to that of the electron [BYR94]. PF1 experiments have (together with neutron lifetime measurements) provided stringent limits on the unitarity for the first row of the quark mixing Cabibbo-Kobayashi-Maskawa (CKM) matrix [ABE02, ABE00, ABE08, DUB99], which determines the strength of the quark mixing (the Standard Model requires the mixing to yield unity for the sum). The ILL findings have triggered in high energy physics new analyses, new measurements, and new investigations on the theoretical uncertainties, respectively.

A particular topic was the study of Time Reversal Invariance T which is broken only to a very small degree in the Standard Model. Triple correlation experiments, testing T-violation in neutron decay have steadily been improved at PF1 [SOL04, SOL03]. Searching for the weak neutral current in nucleon-nucleon interactions, several nuclear reactions induced by polarized neutrons were studied in view of testing Parity Non Conservation. Asymmetries in reactions like $^{6}Li(n,\alpha)t$ and $^{10}B(n,\alpha)^{7}Li$ were investigated [VES05, VES11]. Very challenging experiments dealt with various correlations in fission (see chapter 4.9).

2.3 Studies of Neutron Properties with Ultra Cold Neutrons

At the time of writing the present contribution the ultra-cold neutron facility PF2 is still the strongest source of UCN in the world. It was built by A. Steyerl from the TU Munich [STE89] and installed at the operational level D of the reactor in collaboration with the ILL. A system of Ni-coated vertical (TGV) and curved (TGC) guides transmits neutrons against the gravitational field from one of the cold sources of the reactor to the so-called Steyerl turbine. Reflected from the receding blades of the rotating turbine very cold neutrons (VCN, with speed around 50 m/s) are Doppler-shifted to the UCN-velocity region of around 5 m/s. At such velocities the de Broglie wave length of the neutrons is of the order of 1000Å and, therefore, when a neutron interacts with matter, it does not see single atoms but instead an ensemble of them. This interaction can be described by the Fermi potential. The UCNs are then conducted by horizontal guides to several experimental positions. This possibility of installing several experiments simultaneously was always one of the big assets of PF2. At the experimental positions ultra cold neutrons with densities of up to $30/cm^3$ and with speeds less than 8 m/s are available. There is also an output for very cold neutrons (VCN) with a wavelength of 100Å. A specific advantage of UCN physics is the option for storing neutrons in material and/or magnetic bottles. Consequently the majority of measurements carried out at PF2 make use of this feature.

The very first ILL storage measurement started with an experiment where neutrons were captured in a ring trap by magnetic and inertial forces [ANT89]. However, in most cases neutron lifetime measurements at PF2 were carried out with neutrons stored in material bottles, using the fact that ultra cold neutrons are totally reflected (under all angles) from the surface of materials with the appropriate Fermi potential. The first of these experiments was performed at the ILL by W. Mampe, P. Ageron and collaborators [MAM89, SCH92].

In contrast to neutrons being stable when bound in nuclei, free neutrons decay with a half life of about a quarter of an hour. Besides the neutron decay parameters (see above) also the lifetime of free neutrons depends on the structure and the strength of the weak force. The precise knowledge of the neutron's lifetime is hence important because it is related to the CKM matrix (see above). Under the hypothesis that three generations of quarks exist, this matrix has to be unitary. This imposes for the elements of the first row of this matrix (which describe the coupling of the up quark (u) to the d, s and b quarks, respectively:

$$|V_{ud}|^2 + |V_{us}|^2 + |V_{ub}|^2 = 1$$

This relation has to be verified experimentally by measuring the 3 matrix elements separately. The first one, V_{ud}, brings the main contribution and governs the decay of the neutron (which consists of one up (u) and two down (d) quarks) into a proton (two u and one d quark), an electron and an anti neutrino. It is directly linked to the neutron lifetime τ_n via relations as cited in [PDG06]:

$$|V_{ud}|^2 = (4908.7 \pm 1.9)s / \tau_n(1 + 3\lambda^2)$$

where $\lambda = g_A/g_V$ is the ratio of the axial vector and vector coupling constants of the weak interaction for the case when only left handed currents exist. This ratio is determined by neutron decay asymmetry measurements, as summarized above. The numerical values in the expression reflect the status of the values and their experimental uncertainties averaged over different experiments of measurements of the neutron's decay parameters and lifetime, respectively (see also [PIC10, ARZ00, SER05, SER08].

These values are also significant in cosmological theories seeking to explain the evolution of the Universe after the Big Bang. Their knowledge determines the precision with which one can predict the abundances of the very light elements (^{2}H, ^{3}He, ^{4}He, ^{7}Li).

An intriguing issue has been the search for an electric dipole moment of the neutron. A finite electrical dipole moment of the neutron (EDM) would not only imply parity (P) violation, but also time (T) reversal violation. This is equivalent to CP (charge parity) violation if one accepts the conservation of CPT. Today one would like to observe a new source of CP violation beyond the Standard model. This has already been observed for the decay of the K^0 meson and, more recently, the decay of the B meson. One can also expect the existence of such sources in the baryon sector. This was proposed by Sakharov [SAK67] to explain the matter-antimatter asymmetry observed in the universe. Consequently T violation turns out to be necessary to explain the survival of matter at the expense of antimatter after the Big Bang. With ultra cold neutrons, the search for these hypothetical new channels of T-violation were therefore flagship experiments at the ILL reactor. A whole generation of EDM measurements at the ILL was started by N. Ramsey (who had already carried out the very first EDM measurement at all [SMI57]) and collaborators and has now been terminated. An upper limit for the EDM approaching 10^{-26} *ecm* [BAK06] was obtained. Also the Standard Model predicts an EDM, but only at the level of $< 10^{-32}$ ecm. Any result above this value would be an indicator for "new physics" beyond the Standard Model.

Neutrons have a mass and therefore experience gravitational forces. They are ideal particles for studying gravity at the microscopic level on the earth: they are electrically neutral, easy to detect, have a relatively long lifetime, and reflect well off mirrors. Recently at the ILL, quantum states of the neutron in the Earth's gravitational field were successfully observed for the first time [NES02]. This success has triggered rather intense and rapidly growing research. Many related studies were conducted in the meantime at the ILL. Based on the experience gained a new gravitational spectrometer was constructed in a common effort of several groups. The spectrometer is expected to come into operation before long.

Similarly to PF1 certain scientific questions addressed with UCN at the facility PF2 are also challenged in high-energy physics.

Finally, the facility is also unique for testing techniques and components for future UCN facilities. Techniques in UCN physics investigated are polarization methods, detector development and improvement of neutron storage times in traps. New possibilities of UCN production by means of down scattering of a cold neutron beam in super fluid ^{4}He was and still is under study at the ILL and other laboratories. Advanced tests by a Sussex/ RAL/ILL collaboration in the frame of a new EDM measurement have already demonstrated that one can create high UCN densities by this technique. This principle is also applied for the construction of a new UCN source at ILL [ZIM11] which should deliver UCNs with higher density than presently available.

Bibliography

[ABE00] H. Abele, Nucl. Instr. Meth. A 440 (2000) 499

[ABE02] H. Abele, M. Astruc Hoffmann, S. Baeßler, D. Dubbers, F. Glück, U. Müller, V. Nesvizhevsky, J. Reich, O. Zimmer, Phys. Rev. Lett. 88 (2002) 211801

[ABE06] H. Abele, D. Dubbers, H. Häse, M. Klein, A. Knöpfler, M. Kreuz, T. Lauer, B. Märkisch, D. Mund, V. Nesvizhevsky, A. Petoukhov, C. Schmidt, M. Schumann, T. Soldner, Nucl. Instr. Meth. A 562 (2006) 407

[ABE08] H. Abele, Prog. Part. Nucl. Phys. 60 (2008) 1-81

[ABE09] H. Abele., Nucl. Instr. Meth. A 611 (2009) 193-197

[ANT89] F. Anton, et al., Z. Phys. C45 (1989) 101

[ARZ00] S. Arzumanov, L. Bondarenko, S. Chernyavsky, W. Drexel *et al.*, Phys. Lett. B 483 (2000) 15

[BAK06] C.A. Baker, D.D. Doyle, P. Geltenbort, K. Green, M.G.D. van der Grinten, P.G. Harris, P. Iaydjiev, S.N. Ivanov, D.J.R. May, J.M. Pendlebury, J.D. Richardson, D. Shiers, K.F. Smith, Phys. Tev. Lett. 97 (2006) 131801

[BYR90] J. Byrne, et al. Phys. Rev. Lett. 65 (1990) 289

[BYR94] Nucl. Instr. Meth. A 349 (1994) 454

[DUB91] D. Dubbers, Progr. Part. Nucl. Phys. 26 (1991) 173

[DUB99] D. Dubbers, Nucl. Phys. A 654 (1999) C297

[GOL83] R.Golub et al., Z. Phys. B51 (1983) 187

[HAE02] H. Häse, A. Knöpfler, K. Fiederer, U. Schmidt, D. Dubbers, W. Kaiser Nucl. Instr. Meth. 485 (2002) 453

[ICE10] G.E. Ice, J.Y. Choi, P.A. Takacs, A. Kkhounsary, Y. Puzyrev, J.J. Molaison, C.A. Tulk, K.H. Anderson, T. Bigault, Appl. Phys. A 99 (2010) 635

[KRE05a] M. Kreuz, V. Nesvizhevsky, A. Petoukhov, T. Soldner, Nucl. Instr. Meth. A 547 (2005) 583-591

[KRE05b] M. Kreuz, T. Soldner, S. Baeßler, B. Brand, F. Glück, U. Mayer, D. Mund, V. Nesvizhevsky, A. Petoukhov, C. Plonka, J. Reich, C. Vogel, H. Abele, Phys. Lett. B 619 (2005) 263-270

[KUZ00] I.A. Kuznetsov, Yu.P. Rudnev, A.P. Serebrov, V.A. Solovei, I.V. Stepanenko, A.V. Vasiliev, Yu.A. Mostovoy, O. Zimmer, B.G. Yerozolimsky, M.S. Dewey, and F. Wietfeldt, Nucl. Instr. Meth. A 440 (2000) 539

[MAE09] B. Märkisch, H. Abele, D. Dubbers, F. Friedl, A Kaplan, H. Mest, M. Schumann, T. Soldner and D. Wilkin, Nucl. Instr. Meth. A 611 (2009) 216-218

[MAM89] W. Mampe, P. Ageron, C. Bates, J.M. Pendlebury, A. Steyerl, Phys. Rev. Lett. 63 (1989) 593

[NES02] V. Nesvizhevsky, H. G. Börner, A. Petukhov, H. Abele, S. Baeßler, F. Rueß, T. Stöferle, A. Westphal, A. Gagarski, G. Petrov, V. Strelkov, Nature 415 (2002) 299

[PDG06] Particle Data Group's Review, J. Phys. G33 (2006) 1

[PIC10] A. Pichlmaier, V. Varlamov, K. Schreckenbach, P. Geltenbort, Phys. Lett. B693 (2010) 221

[REI00] J. Reich, H. Abele, M. Astruc Hoffmann, S. Baeßler, P. v. Bülow, D. Dubbers, V. Nesvizhevsky, U. Peschke, O. Zimmer, Nucl. Instr. Meth. A 440 (2000) 535

[SCH92] K. Schreckenbach and W. Mampe, J. Phys. G18 (1992) 1

[SCH08] M. Schumann, M. Kreuz, M. Deissenroth, F. Glück, J. Krempel, B. Märkisch, D. Mund, A. Petoukhov, T. Soldner, H. Abele, Phys. Rev. Lett. 100 (2008) 151801

[SCH07] M. Schumann, T. Soldner, M. Deissenroth, F. Glück, J. Krempel, M. Kreuz, B. Märkisch, D. Mund, A. Petoukhov, H. Abele, Phys. Rev. Lett. 99 (2007) 191803

[SMI57] J.H. Smith, EM. Purcell, N.F. Ramsey, Phys. Rev. 108 (1957) 120

[SAK67] A. D. Sakharov, JETP Lett. 5 (1967) 24

[SER05] A. Serebrov, V. Varlamov, A. Kharitonov, A. Fomin *et al.*, Phys. Lett. B 605 (2005) 72

[SER08] A.P. Serebrov, V.E. Varlamov, A.G. Karitonov, A.K. Fomin, Y.N. Pokotilovski, P. Geltenbort, I.A. Krasnoschekova, M.S. Lasakov, R.R. Taldaev, A.V. Vasiljev, O.M. Zherebtsov, Phys. Rev. C78 (2008) 035505-1

[SOL03] T. Soldner, L. Beck, C. Plonka, K. Schreckenbach, O. Zimmer, Nucl. Phys. A 721 (2003) C469

[SOL04] T. Soldner, L. Beck, C. Plonka, K. Schreckenbach, O. Zimmer, Phys. Lett. B 581 (2004) 49

[STE 75] A. Steyerl, Nucl. Instr. 125 (1975) 461

[STE89] A. Steyerl, S.S. Malik, Nucl. Instr. Meth. A284 (1989) 200

[VES05] V.A. Vesna, Yu.M. Gledenov, V.V. Nesvizhevsky, A.K. Petukhov, P.V. Sedyshev, T. Soldner, O. Zimmer, and E.V. Shulgina JETP Lett. 82 (2005) 463-466.

[VES11] V.A. Vesna, Yu.M. Gledenov, V.V. Nesvizhevsky, P.V. Sedyshev, E.V. Shulgina, Europ. Phys. J. A47 (2011) 43.

[ZIM11] O. Zimmer, F.M. Piegsa, S.N. Iwanov, Phys. Rev. Lett. 107 (2011) 134801

Chapter 3

High Resolution Spectroscopy

3.1 The Neutron Capture Reaction

As neutral particles neutrons can easily approach the atomic nucleus. In the range of the strong interaction (~10^{-13} cm) they can be captured by a nucleus A forming an excited nucleus (A+1)*.

$$A + n \rightarrow (A+1)^* \tag{3.1.1}$$

By adding a neutron to the nucleus a gain of typically 5 to 10 MeV in the binding energy of stable nuclei is obtained. The gain in total nuclear binding energy can be expressed in terms of mass differences:

$$\begin{gathered} S_n = (m_{A+1} - m_A - m_n)c^2 \\ E_x(A+1)^* = S_n + E_n \end{gathered} \tag{3.1.2}$$

S_n denotes the neutron separation energy and E_x the excitation energy of the nucleus (A+1). Thermal neutrons are captured as s waves with angular momentum transfer $l = 0$. For keV neutrons p-wave capture ($l = 1$) becomes important too. The capture state has spin I and parity π of

$$\begin{gathered} I_{A+1^*} = I_A \pm \tfrac{1}{2} \pm l \\ \pi_{A+1^*} = \pi_A \cdot (-1)^l \end{gathered} \tag{3.1.3}$$

The capture state can in general be described in terms of a compound nucleus model where the captured neutron is integrated in the nuclear matter and where the compound state has a relatively long lifetime of about 10^{-14} s compared to a much shorter timescale of direct reactions of about 10^{-22} s. The capture state then decays in the form of electro-magnetic radiation, particle emission (protons, α particles, etc.) or induces nuclear fission (see chapter 4). Thus a broad range of nuclear reactions can be observed – in laboratory experiments with neutron sources just like in the well known astrophysical processes where neutrons are produced by fusion and fission.

The outstanding importance of neutron capture in astrophysics is demonstrated by the stellar nucleosynthesis, the production process for the heavy elements. The fusion process in stars cannot produce elements beyond iron in sufficient quantities. Heavier elements are created essentially by neutron capture, mainly in the r process ("rapid") and s process ("slow"). Additionally particle induced reactions (p-process) are of some importance. The r process is supposed to occur in certain types of super-nova explosions. During very short exposure times to extremely high neutron fluxes neutron-rich nuclei are produced up to the neutron drip line (where no additional neutrons can be bound anymore). They then decay back to the region of stable nuclei by beta decay. The stellar material may then be reintegrated in stars (red giants for instance) where it is exposed to lower neutron fluxes with about 27 keV neutron energy (s process). Many nuclear and astrophysical parameters enter into the estimations of the r and s process, respectively, and different scenarios are discussed (see for instance [FOW84, THI79, KLA81]). An example for the s process is given in section 3.8.5.

The neutron capture reaction, exploited in ultra high resolution gamma spectroscopy, is a rather unique tool to investigate in a non selective way a large number of properties of nuclear structure. When a nucleus captures a neutron it is excited to the binding energy of the neutron. Then gamma rays cascade down from this initial state to the lower lying levels. Precise measurement of these gamma rays allows to determine the sequence of the excited states and to study their manner of de-excitation. This gives insight into the structure of the excited states which they connect. It is also possible to measure the lifetimes of the excited nuclear states using the GRID technique [BÖR88] which was pioneered at ILL [YSF91]. This method relies on the observation of Doppler shifted gamma rays emitted when the nucleus is in flight. The nuclear motion is here induced directly by the recoil of a nucleus following the emission of a gamma ray. Therefore the method is called Gamma Ray Induced Doppler broadening (GRID, section 3.7.1). The recoils induced are extremely small and they can only be detected by spectrometers with part per million resolution as it was possible with ILL's unique crystal spectrometers GAMS4/5 (Section 3.7). Conventional semi-conductor gamma ray detectors yield a resolution which is more than 3 orders of magnitude (!) less good. With crystal spectrometers one determines the wavelengths of gamma rays by precise measurement of their reflection angles at ideal single crystals. To obtain the highest possible resolution in a measurement has always been one of the major challenges for ILL's nuclear structure physicists because increased resolution always results in progress- sometimes even unexpected. During the now 30 years of existence of the ILL already five generations of high resolution crystal spectrometers have been developed. This allowed to steadily improve resolution and precision by about three orders of magnitude from the first to the currently last generation!

With the GAMS spectrometers one can study nuclei in the region of stability, or at least close to it. If we want to do gamma-ray spectroscopy far away from stability, we need an additional tool. Neutron induced fission allows to investigate the properties of very neutron rich nuclei like those relevant for astrophysical element synthesis paths. Extracting the fission fragments from the reactor makes ILL's mass separator LOHENGRIN (chapter 4) a unique facility. Very neutron rich nuclei are created by neutron induced fission of Uranium (or other actinides). The fission fragments pass through the mass separator in about 1 micro second. Electromagnetic fields are used to separate the ionized fragments according to their mass. An array of germanium detectors can then be used to efficiently detect gamma rays from these nuclei. As this can be done with coincidence set ups this is again equivalent to high energy resolution. Just to cite a few of the problems which can be addressed with this spectrometer: The boundaries which limit the nucleus in terms of its neutron number or total mass define the limits of existence of nuclear matter. As neutrons are successively added to a nucleus on the stability line, the binding energy of the last neutron decreases steadily until it is no longer bound. This defines the neutron drip line. The structure of nuclei is expected to change drastically as the limit of nuclear stability is approached. A direct consequence of the diminished binding of the last neutrons is that they spend more and more of their time far from the normal density core of the nucleus and form a 'neutron halo'. Only light halo nuclei have been identified to date and a lot of research has still to be done here. As in the LOHENGRIN mass separator fission products need about a micro second to fly from the target to the detector, the study of a big variety of micro second isomers is possible (see 3.8.6) and properties of the excited states lying below these isomeric states can be extracted. We can study nuclei important for astrophysics, we can look into the features of double magic neutron rich nuclei, or in contrary look into strongly deformed ones. All that gives a huge variety of investigations to fulfill one of the main objectives of nuclear structure: To gain insight into the nature of strongly interacting many-body systems, the effective interaction between nucleons within a nucleus and the variety of highly coherent many-particle excitations arising from them.

In the following chapters we will present some basic characteristics of the capture process of slow neutrons followed by the emission of γ-rays or internal conversion electrons. In parallel, a number of selected nuclear structure studies carried out at the ILL instruments will be discussed.

3.2 Gamma-ray Emission Following Neutron Capture

Photons carry spin 1 and therefore the minimum order of the multipole radiation field is $l = 1$. That means E1 or M1 multipole radiation. Compared to the processes which we know to take place in the atomic shell, multipole radiation of higher order is much more common in nuclear matter due to the extended many body system of the nucleus. Quite often E2 transition probabilities are of similar magnitude as M1, for instance. For mixed M1/E2 transitions a mixing parameter $\delta^2 = I_\gamma(E2)/ I_\gamma(M1)$ is defined. Spin and parity selection rules for γ transitions between the initial state i and the final state f are:

$$E_\gamma = E_f - E_i \tag{3.2.1}$$

$$I_f = I_i + l\,,\ \ |\,l\,| \geq 1$$

$$|\,I_i + I_f\,| > l > |\,I_i - I_f\,| \tag{3.2.2}$$

$$\pi_f = \pi_i\,(-1)^l \text{ electric multipolarity}$$

$$\pi_f = \pi_i\,(-1)^{l+1} \text{ magnetic multipolarity} \tag{3.2.3}$$

$$m_f = m_i + m$$

m: magnetic quantum numbers

The γ-ray transition probabilities are given by

$$W_{i,f}(\sigma,l) \sim E_\gamma^{\,2l+1} \cdot Bi{,}f\,((\sigma,l)$$

$$B_{i,f}(\sigma,l) = 1/(2I_i + 1)\sum_{mimf}|<I_f,m_f|M^\sigma{}_{l,m}|I_i,m_i>|^2 \tag{3.2.4}$$

The expression in < > denotes the nuclear transition matrix element for a multipole type σ,l. $B_{i,f}$ is the “reduced tansition probability” since the energy factor $E_\gamma^{\,2l+1}$ is removed. In general the transition probabilities decrease strongly with increasing multipole order.

3.3 Primary Gamma Rays

The most common deexcitation modes of the neutron capture state are the emission of primary gamma rays accompanied by the emission of neutrons. The total energy width Γ of a neutron resonance is therefore given by $\Gamma_{tot} = \Gamma_\gamma + \Gamma_n$, where Γ_γ and Γ_n denote the partial width for the gamma and neutron decay branch, respectively. Typical values for Γ_{tot} are in the order of 0.1 eV for eV neutrons. The uncertainty principle yields then a lifetime of the capture state of $\tau = \hbar/\Gamma_{tot} = 6.6\cdot10^{-16}[\text{eVs}]/\Gamma_{tot} \sim 10^{-14}$ s = 10 fs, a typical value for the compound state.

The energy of primary gamma rays is given by

$$E_\gamma = S_n + E_n - E_x - E_{recoil} \tag{3.3.1}$$

with a nucleus of mass M experiencing a recoil energy of $E_{recoil} = E_\gamma^2/2Mc^2$. Knowing S_n and E_n the primary gamma rays with energies E_γ yield therefore directly the excitation energies E_x of the states populated by the primary gamma rays.

Quite large fluctuations in intensity can be observed in the decay by primary gamma rays following thermal and/or resonance neutron capture, respectively. This holds even for cases where the primary transitions populate different nuclear states with the same spin and parity values. This is due to statistical fluctuations (so-called Porter Thomas fluctuations) depending on the number of entrance channels possible for capture into specific resonances. A detailed description of this behavior may be found in [POR56]. The consequence is that only limited information can be deduced from the intensities of those gamma rays. This applies directly for thermal neutron capture. The problem can be overcome by averaging over many entrance channels (many resonances). Only in average E1 transitions are favored over M1 transitions by about an order of magnitude. Primary gamma rays of higher multipolarity are generally extremely weak.

3.3.1 *The PN4 Facility at ILL*

Figure 3.3.1 shows an example of a primary gamma ray spectrum recorded at the former PN4 facility [KER81] at the ILL high flux reactor. The target was placed at the in-pile beam position (in common with the GAMS spectrometers described below), 55 cm distant from the reactor core.

The thermal neutron flux at the target position was $5.5 \cdot 10^{14}$ n/cm^{2}s. Primary gamma rays were measured with a Ge(Li) pair spectrometer situated at a distance of about 15 m from the target. The pair spectrometer (Fig. 3.3.2) consisted of a planar Ge(Li) detector sandwiched between two NaI scintillation crystals. Primary gamma rays of typically 4 – 10 MeV are mainly absorbed in the Ge(Li) by e^-e^+ pair production. The corresponding electrons and positrons are then stopped in the detector by depositing an energy of $E_\gamma - 2m_ec^2 = E_\gamma - 1.022$ MeV. The e^+ annihilation creates two 511 keV gamma rays emitted in opposite directions and analysed in the two NaI crystals. Triple coincidence events (e^+e^- in the Ge(Li), 511 keV in each of the two NaIs) are then recorded.

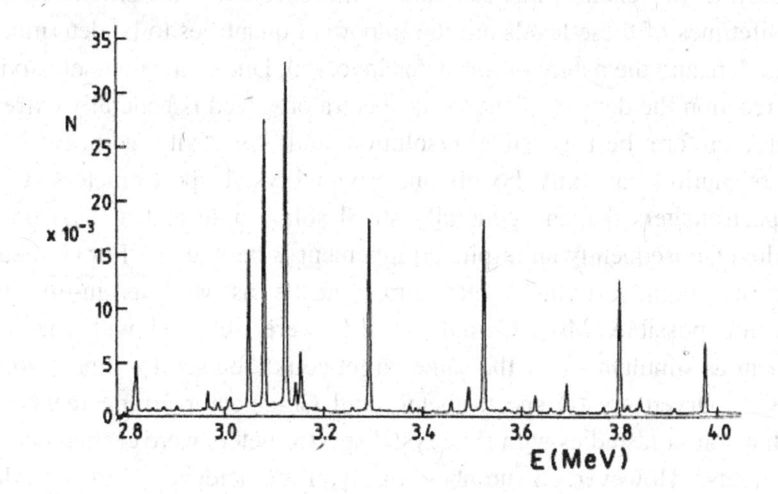

Fig. 3.3.1. Part of a pair spectrum, measured with the PN4 facility in 28Al [SCH82].

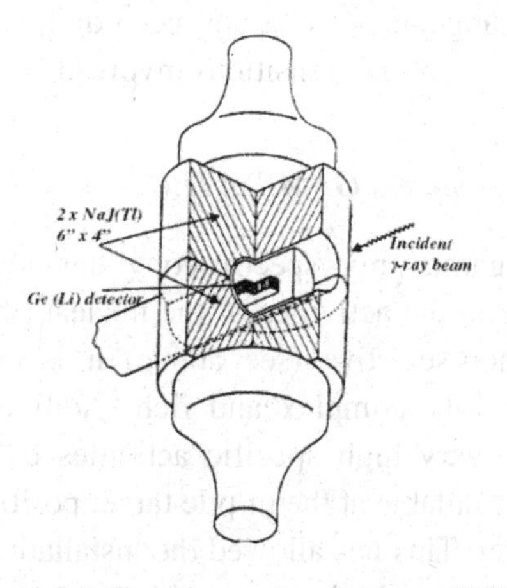

Fig. 3.3.2. The pair spectrometer set-up PN4.

3.4 Secondary Gamma Rays Following Thermal Neutron Capture

For most nuclei the measurement of secondary γ-rays is restricted to transitions between the lower lying states, since the level density strongly increases with the excitation energy. Only for the lighter elements (around ^{42}K [KRU85] and/or or lighter) rather complete (n,γ) studies were reported. The big advantage of the (n,γ) reaction is its non-selectivity which allows the study of excitation modes of very different nature. Energies, intensities and multipolarities of γ-transitions interconnecting the excited nuclear states (the secondary transitions) together with the lifetimes of these levels are the important quantities to be determined in order to understand the nature of the states involved. Due to the non-selectivity of the (n,γ) reaction the density of the γ-ray spectra observed is generally extremely high and therefore best possible resolution and sensitivity is needed. Best possible resolution can only be obtained with crystal spectrometers. Curved crystal spectrometers demand generally small solid angle and highest possible neutron flux. Consequently an in-pile arrangement is best suited. For Ge-detector measurements both, external target arrangements as well as in-pile target geometry are possible. Most Ge-data at ILL were obtained with the in-pile arrangement as simultaneously the same target could be used for high resolution studies with curved crystal spectrometers and Ge-detector arrangements. Also ultra-high resolution studies with flat crystal spectrometers were carried out using in-pile targets. However, a number of (γ,γ) coincidence and correlation experiments were also performed with external targets and Ge spectrometers at neutron guides. In correlation experiments one takes advantage of the fact that γ-rays in cascade show angular correlations according to the spin sequence of levels and the multipolarity of the transitions involved.

3.4.1 *The Crystal Spectrometers GAMS at ILL*

Ultra-high resolution gamma-ray spectroscopy has always played, and still plays, a significant role in the activities of the nuclear physics group at ILL. As the (n,γ) reaction is non-selective (see above) it is evident that the spectra obtained are generally very complex and rich. Additionally, at the high-flux reactor one can obtain very high specific activities by exploiting the flux of 5×10^{14} neutrons/cm^{2}s available at the in-pile target position which was also used for the PN4 spectrometer. This has allowed the installation of additional gamma-ray spectrometers with the highest possible energy resolution. This was accomplished using Bragg reflection on ideal crystals, where the wavelength can be deduced by precise measurements of the angle of reflection of gamma rays impinging on a crystal:

$$n\lambda_\gamma = 2d\ sin\theta \qquad (3.4.1)$$

with n the order of reflection, λ the wavelength, d the lattice spacing of the crystal (which is known to a precision of about 10 ppb) and finally θ, the Bragg angle measured. The energy resolution ΔE_γ follows then directly as:

$$\Delta E_\gamma \propto E^2_\gamma (d\Delta\theta/n) cos\theta d\theta \qquad (3.4.2)$$

Currently an angular resolution ($\Delta\theta$) in the milliarcsec range can be obtained, which corresponds to an energy resolution in the ppm range. The crystal spectrometers GAMS have been extensively used to address a large variety of different fields, with an emphasis to nuclear-structure studies. Originally they (GAMS1 and GAMS2/3 [KOC80], respectively) were mainly used to construct level schemes with the aim to do this as completely as possible. A well known example is ^{168}Er [DAV81] (see also section 3.8.3) where – within the spin window which can be reached in neutron capture – a complete level scheme was established up to an excitation energy of about 2 MeV. Today one uses (with GAMS4 [KES01] and GAMS5 [DOL00], respectively) the ppm resolving power mainly to observe the tiny Doppler shifts obtained when a nucleus recoils (section 3.7.1).

3.4.2 *The Curved Crystal Spectrometers GAMS 1 and GAMS2/3*

Focusing spectrometers are either of the Cauchois geometry [CAU32, CAU34, KAZ60] or the DuMond geometry [DUM47]. The latter arrangement requires a narrow source and thus a relatively small amount of target material. This allows the use of enriched isotopes and simultaneously reduces nuclear heating problems at the high neutron flux. The total amount of radioactivity of the source of an in-pile DuMond spectrometer is about three orders of magnitude smaller than that of a corresponding Cauchois spectrometer. This makes the target handling procedure easier. On the other hand the orientation of the source and its stability during the measurement require very careful mounting and control of the target position. For the ILL spectrometers the DuMond type geometry was chosen, in order to be able to study also very rare isotopes. The spectrometers were used for the measurement of low energy γ-rays up to about 1500 keV emitted by the in-pile target after neutron capture. The targets were located in the centre of the tangential through tube H6/H7 were the flux of thermal neutrons is $5.5 \cdot 10^{14}$ /cm^{2}s. Two separate spectrometer systems, one of focal length 5.76 m (GAMS1, using a quartz crystal with a thickness of 4 mm) and the other of focal length 24.0 m (GAMS2/3, using quartz crystals with a thickness of 13 mm) viewed the source from the opposite ends of the beam tube. The GAMS1 spectrometer was designed to operate in the γ-ray energy interval $20 \leq E_\gamma \leq 400$ keV. An angular resolution as small as 1.1 arcsec was achieved with this instrument. The GAMS2/3 system consisted of two spectrometers on top of each other, operating simultaneously at $+\theta_{Bragg}$ and $-\theta_{Bragg}$, respectively. The GAMS2/3 facility was designed to cover the energy interval $200 \leq E_\gamma \leq 1500$ keV. Minimum line widths of ~0.4 arcsec have been achieved. In routine measurements the detection sensitivity for neutron capture γ-rays was ~1 mb, for GAMS1 between 80 keV and 300 keV and for GAMS2/3 from 300 keV to 800 keV. Figure 3.4.1 shows the installation of the two spectrometers on both sides of the through-going beam tube H6-H7. The fact of using two set-ups with crystals of different thicknesses and focal lengths allowed to obtain rather uniform sensitivity over a quite large energy interval. More details about the advantage of combining spectrometers of different focal length may be found in [KOC80]. Up to about 1 MeV, even routine resolution obtained with these spectrometers was largely superior to what could be obtained with the best thick Ge(Li) detectors as can be seen from Figs. 3.4.2 and 3.4.3. The reflection angles in the spectrometers were measured with "first generation" Michelson-type angle interferometers by performing point by point angle measurements of a spectrum. Details may be found elsewhere [KOC78, KOC80]. The axis of rotation of the respective crystal was located between two totally reflecting isosceles 900 prisms (crystal and prisms being fixed on a common rotation table). The number of interference periods $(N\text{-}N_0)$ in an angular step φ is given by $N\text{-}N_0 = (4D/\lambda_L) \sin\varphi$ with D the distance between the optical points of the prisms (see [KOC80,KOC78]) and λ_L the wavelength of the He-Ne laser beam used for the interferometers. When the diffraction crystal is adjusted such that the zero points of φ and the Bragg angle θ, respectively, coincide, one obtains, together with (3.4.1) a linear relation between the wavelength λ_γ and the number of interference steps N.

$$\lambda_\gamma = (2d/n)\,(\lambda_L/\,4D)(N\text{-}N_0) \tag{3.4.3}$$

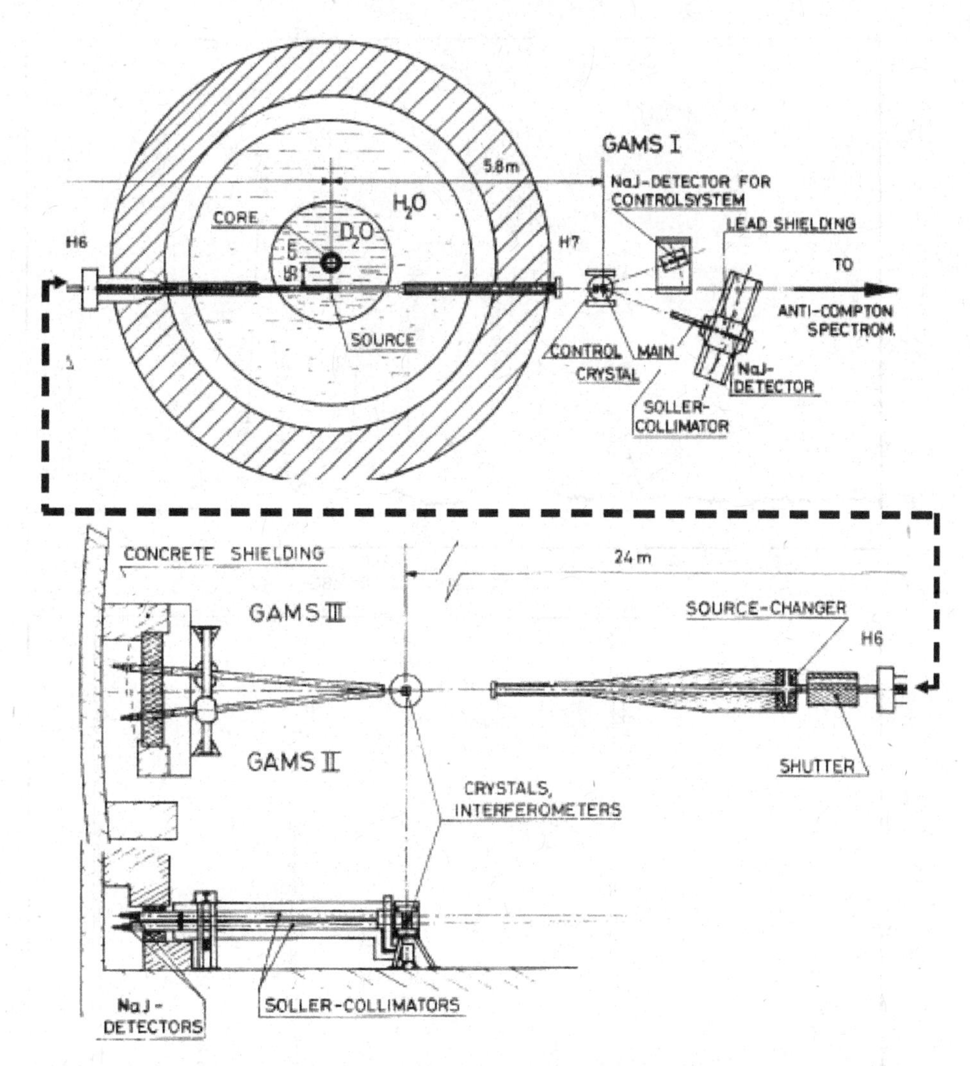

Fig. 3.4.1. Schematic arrangement of the curved crystal γ-ray spectrometers at the tangential through tube H6-H7. The components of the left and right side, respectively, are put on top of each other for better display of details. This is indicated by the arrow, connecting the two parts. The figure is partly compressed in the horizontal direction.

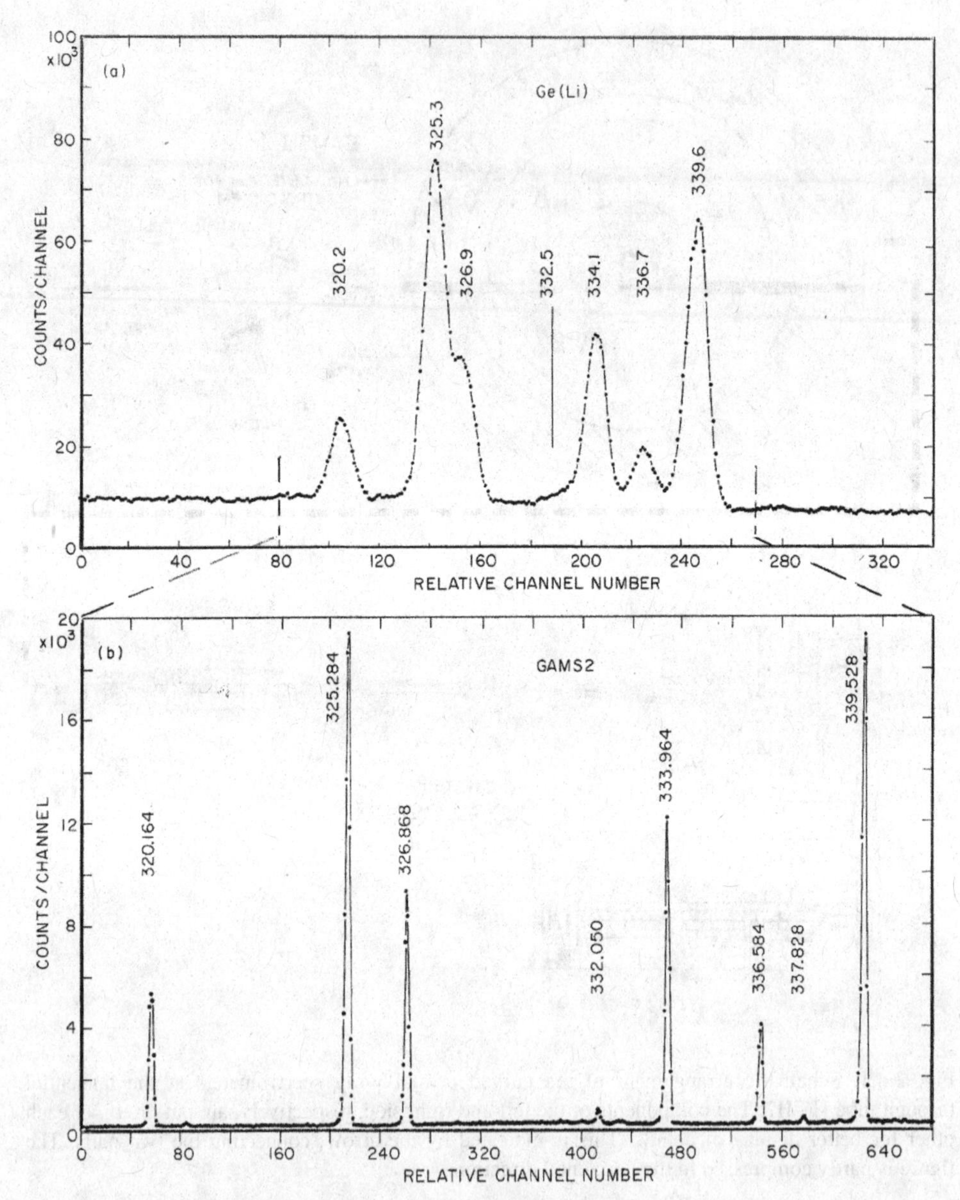

Fig. 3.4.2. Portion of a ^{108}Pd (n,γ) spectrum, comparing a measurement with a Ge(Li) detector (upper part) and GAMS2 (lower part). It was taken in 1977 in 2nd order of reflection [CAS80]. The comparison shows the much better resolution obtained with the crystal spectrometer, but also the much better sensitivity (peak to background ratio) observed. Energies are given in keV.

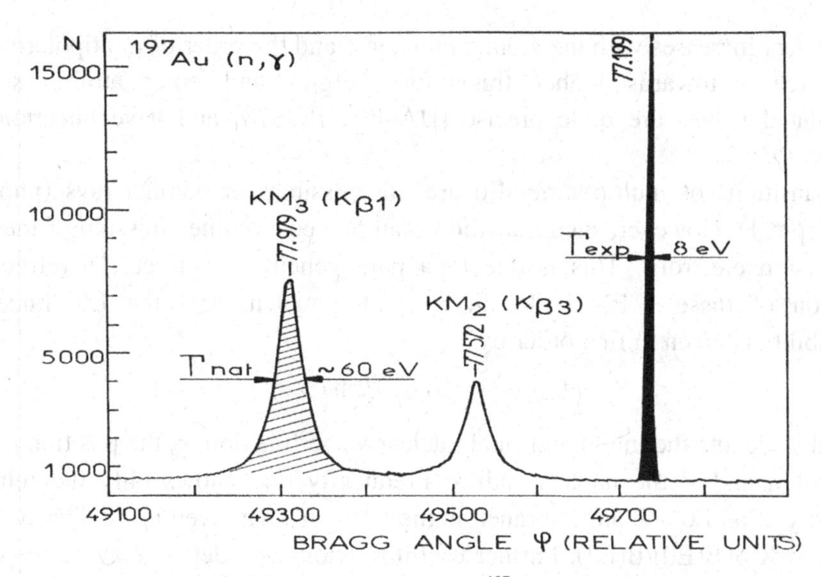

Fig. 3.4.3. Photon spectrum following neutron capture in ^{197}Au, measured with GAMS1 in 1977. In this measurement gold KX-rays were observed close to a γ-transition. The width, ΔE = 8 eV, of the γ-line corresponds to the instrumental resolution. The more pronounced width of the X-rays is due to their natural line width, caused by the short lifetime of atomic levels emitting these X-rays.

3.5 Internal Conversion Electrons

The emission of internal conversion electrons is a process which competes with the emission of secondary gamma rays. In this process the nuclear transition energy E_{trans} is directly transferred to an electron in the atomic shell. The conversion electrons are emitted with energies of

$$E_e^{(i)} = E_{trans} - E_b^{(i)} \tag{3.5.1}$$

With E_{trans} the nuclear transition energy and $E_b^{(i)}$ the binding energy of a given electron in the atomic shell i. After the internal conversion process took place the so created atomic shell vacancy is filled from outer shells. This leads to the emission of X-rays and/or Auger electrons. Both, gamma rays and internal conversion electrons, respectively, are due to the electromagnetic interaction and their intensity ratio

$$\alpha_i = I_e^{(i)}/I_\gamma \tag{3.5.2}$$

is in most cases independent of the nuclear structure. The value α_i for an electron from an atomic shell i is called the internal conversion coefficient (ICC). This coefficient increases with the atomic number Z and the order of multipolarity but it decreases towards higher transition energies and outer atomic shells. Calculated values are quite precise [HAG68, RÖS78] and have uncertainties smaller than %.

Transitions of multipolarity E0 are not possible for gamma rays (photons carry spin 1). However, such transitions can proceed via the emission of internal conversion electrons. This is due to a pure penetration effect. Therefore the creation of these ICE's is localized at the nucleus and the E0 transition probability is given in first order by

$$\rho(E0) = \sum_p \langle \psi_i | r_p^2 / R^2 | \psi_f \rangle \tag{3.5.3}$$

ψ_i and ψ_f denote the initial and final nuclear wave function, r_p the position of the pth proton and R the nuclear radius. Frequently one knows only the relative intensity of an E0 and an E2 branch in the decay from a given level. This defines the ratio X = B(E0)/B(E2). Further useful relations and details may be found in [COL85] and references therein.

3.5.1 *The Conversion Electron Spectrometer BILL*

For not too light nuclei, the spectra of gamma rays and their corresponding internal conversion electron lines are of comparable density. The high resolution consequently needed is commonly achieved by magnetic beta spectrometers. At the ILL again an internal target arrangement was chosen (especially favourable when only small amounts of target material are available as is the case for isotopically enriched ones). The conversion electron spectrometer BILL [MAM78] was operational from 1973 to 1990. A schematic view of this spectrometer is shown in Fig. 3.5.1. It was composed of two double focussing iron core magnets. Double focussing was achieved by a combination of homogeneous and 1/r fields where the first magnet produced a 1:10 reduced image of the target. This image was viewed by the second magnet with high momentum dispersion. An energy range from 15 keV to 10 MeV could be covered and a best resolution of $8 \cdot 10^{-5}$ δp/p had been measured at an electron energy of 300 keV. Spectra were recorded point by point by increasing the magnetic field in small steps. At 500 keV a sensitivity of 100 μb was achieved. In Fig. 3.5.2 a typical spectrum is displayed.

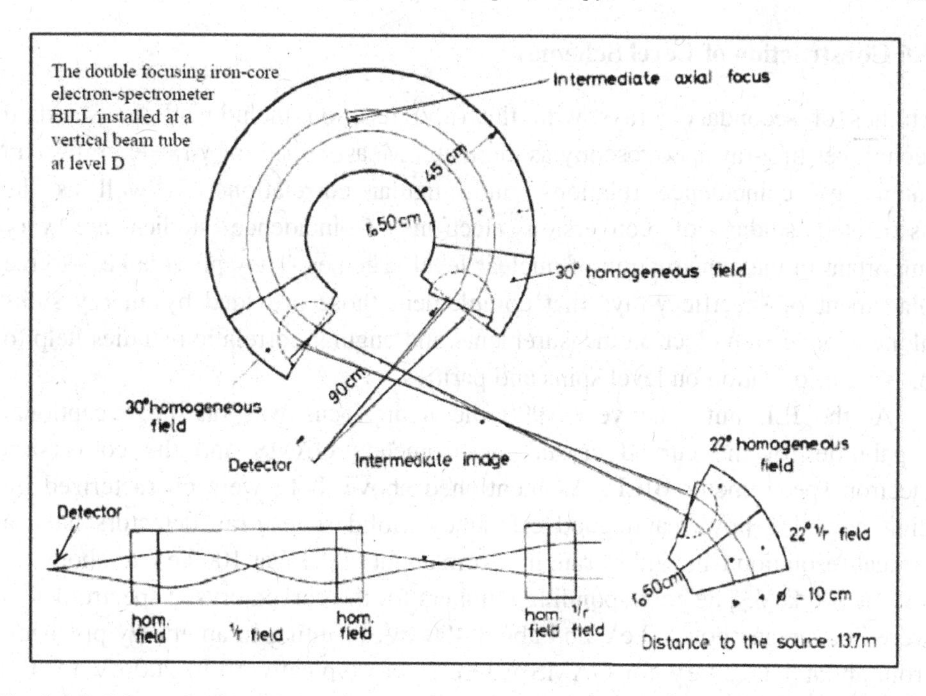

Fig. 3.5.1. Schematic view of the spectrometer Bill [MAM78], demonstrating the arrangement of the magnets and showing radial and axial electron trajectories in the two magnets.

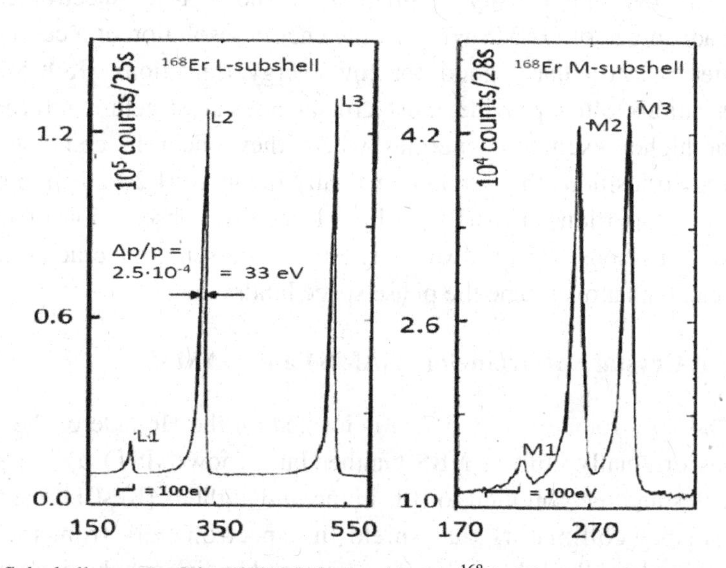

Fig. 3.5.2. Sub shell groups of the 79.998 keV transition in ^{168}Er. The energy scales for the two sub shells L and M, respectively, differ by about a factor of two (see channel numbers).

3.6 Construction of Level Schemes

Studies of secondary γ-rays with the (n,γ) reaction include all the standard techniques of γ-ray spectroscopy as there are: Measurement of γ-ray energies and intensities, coincidence relations, and angular correlations, as well as the associated studies of conversion electrons. Coincidence studies are very important in the construction of nuclear level schemes. They provide keys to the placement of specific γ-rays that complement those provided by energy sums alone. Conversion electron measurements and angular correlation studies help to provide information on level spins and parities.

At the ILL until the year 1990 the main focus was on the exceptional capabilities of the curved crystal spectrometers GAMS and the conversion electron spectrometer BILL. As mentioned above, both were characterized by high resolving power and good efficiency. Solid state γ-ray detectors have a typical resolution (linewidth) ranging from about 800 eV at 100 keV to about 1.5 keV at 800 keV. The corresponding numbers for the curved crystal spectrometers were in average about 50 eV and about 400 eV, resulting in an energy precision from about 1 to 20 eV for GAMS where it was typically 50 to 200 eV for the solid state detectors. This led to a tremendous difference in the probability for chance (incorrect) Ritz combinations: One to two orders of magnitude – depending on the line density – smaller for the GAMS spectrometers. An additional advantage of GAMS was that the energy resolution of a curved crystal spectrometer is particularly good for low-energy transitions [KOC80]. Low-energy transitions often provide most crucial nuclear structure information at medium or higher excitation energies where they occur in competition with higher energy transitions that are kinematically favoured (E2 transition rates, for instance, are proportional to B(E2) x E_γ^5). Hence the observation of highlying low-energy transitions nearly always signals the presence of crucial collective matrix elements that overcome the phase space hindrance.

3.7 The Flat Crystal Spectrometers GAMS4 and GAMS5

The GAMS4 spectrometer (Fig. 3.7.1) is located on the H6 side of the through-tube. It was originally built at NBS Gaithersburg (now NIST) by the group of R.D. Deslattes and then brought to ILL in the mid eighties [Kes01].The through-tube has special collimators that shield the spectrometers from the intense radiation emitted by the tube walls. The source changing mechanism is located on the H6 side of the through-tube between the reactor biological shield and the

GAMS4 environmental chamber. Details concerning the through-tube and the source changer are available in [KOC80]. The GAMS4 environmental chamber is constructed out of concrete blocks lined with acoustical insulation. Inside the concrete blockhouse is a thermal enclosure that contains the vibration isolation platform, the spectrometer with its two interferometers, and the collimators. Figure 3.7.2 shows two common crystal arrangements for the spectrometer and introduces the important concept of non-dispersive and dispersive geometries. Radiation from the source strikes the first crystal and all wavelengths, which satisfy the Bragg condition, are diffracted. The wavelength region that can satisfy the first crystal Bragg condition is determined by the angular spread of the incoming beam. For the GAMS4 facility the angular spread of the beam striking the first crystal is on the order of $1.3 \cdot 10^{-4}$ rad. In the non-dispersive geometry, shown on the left, the planes of the two crystals are parallel so that all wavelengths that are diffracted by the first crystal simultaneously satisfy the Bragg condition at the second crystal. By rocking the second crystal around the Bragg angle, a profile that is an accurate representation of the instrument response function is recorded. For a true non-dispersive position (same lattice spacing and orders for both crystals) the recorded profiles are insensitive to the

Fig. 3.7.1. View of the GAMS4 double flat crystal spectrometer. The orientation of two perfect Si crystals is controlled by optical angle interferometers. The absolute calibration of the interferometers is carried out using an optical polygon.

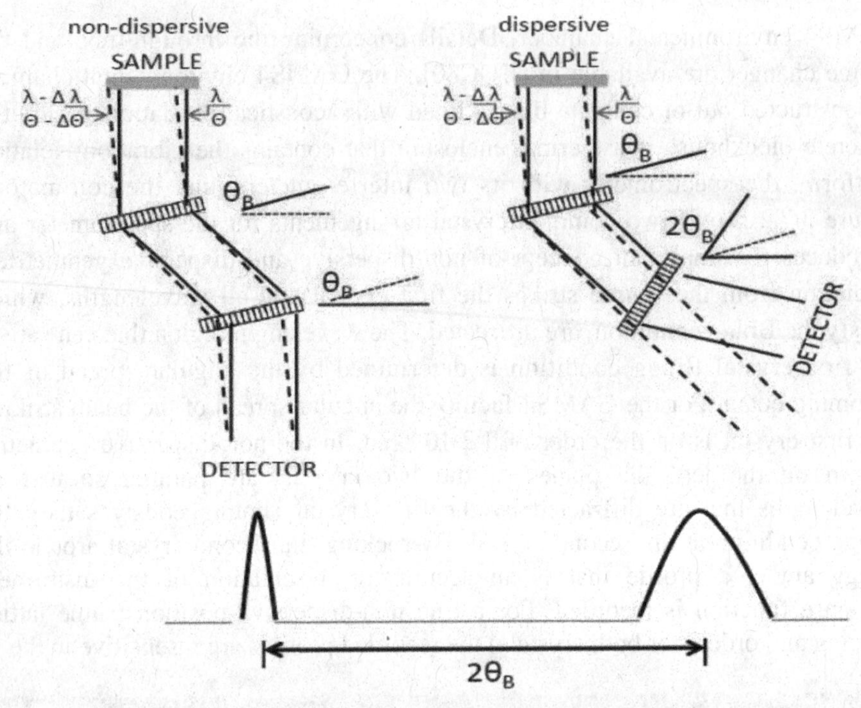

Fig. 3.7.2. Schematic drawing of Bragg angle measurements using a two crystal spectrometer. The instrument response function is recorded using the non-dispersive geometry on the left and the gamma-ray instrument convolution function is recorded using the dispersive function on the right.

spread in wavelength of the incoming radiation. As the crystals are nearly perfect specimen of Si and Ge their performance closely follows dynamical diffraction theory which yields for the intrinsic reflectivity of a single parallel crystal slab used in transmission [PET1984]:

$$P(t,y) = \{sin^2[A(y^2 + 1)^{1/2}]\}/(y^2 + 1) \tag{3.7.1}$$

where

$$A = \pi t/t_0$$
$$y = (\theta - \theta_B)sin2\theta_B$$
$$t_0 = \pi V cos\ \theta_B)/\lambda r_e F$$

θ_B is the Bragg angle, t is the crystal thickness, t_0 is the extinction length, V is the volume of the unit cell, r_e is the classical electron radius and F is the structure factor.

The ability to directly record the instrument response function (an example is shown in Fig. 3.7.3) and precisely measure the angular positions at which the planes of the two crystals are parallel makes the two crystal arrangement very useful for certain types of measurements. For example, measurements that depend on the width and shape of the dispersive profile need an accurate estimate of the instrument response function in order to properly analyze the data. Wavelength measurements that depend on the determination of the Bragg angle, θ_{Bragg}, use the position at which the crystal planes are parallel as an angular reference. The spectrometer is transformed from the non dispersive to the dispersive geometry shown on the right by rotating the second crystal through $2\theta_{Bragg}$ while keeping the first crystal fixed. In the dispersive geometry all wavelengths diffracted by the first crystal are not all simultaneously diffracted by the second crystal. As the second crystal is rocked about the Bragg angle, the Bragg condition is satisfied sequentially for the wavelengths diffracted by the first crystal. Thus, the recorded profile is a convolution of the instrument response function and the wavelength spread of the incoming gamma-ray. For characteristic X-ray lines, the spread in wavelength is so large that the width of the dispersive profile is usually an order of magnitude (or more) larger than the width of the non-dispersive profile. For gamma-rays, the spread in wavelength is generally quite small so that the non-dispersive and dispersive profiles are often nearly identical. After removing the instrument response function from the dispersive profiles, a true measure of the wavelength distribution of the emitted gamma-ray is obtained. From this information the lifetimes of nuclear levels and the motion of the atoms in the source can be deduced. For wavelength measurements, the angular interval between the dispersive profile and the non-dispersive profile is twice the Bragg angle, θ_{Bragg}. If the crystal lattice spacing, d, is known, then wavelengths directly follow from the Bragg equation, $\lambda = 2d \sin \theta_{Bragg}$.

In its basic conception the GAMS5 two axes crystal spectrometer (Fig. 3.7.5) followed closely its predecessor at ILL, the GAMS4 spectrometer. The γ-rays are also diffracted by a two axis crystal (either flat or bent) spectrometer used in transmission. Nearly ideal crystals (Si or Ge) are rigidly connected to the rotational axes whose angular rotations are again measured with Michelson type angle interferometers. The main difference to the GAMS4 spectrometer consists in the use of glass ceramics (ZERODUR instead of INVAR and/or iron) for all major building blocks of the interferometers. Due to the very low thermal expansion coefficient of ZERODUR ($\Delta d/d \approx 10^{-8}/^{o}K$ at 20° C) the effect of thermal gradients on the instrument are minimized. Nevertheless the temperature in the spectrometer housing is stabilized to better than 0.1 °C/day. Additionally,

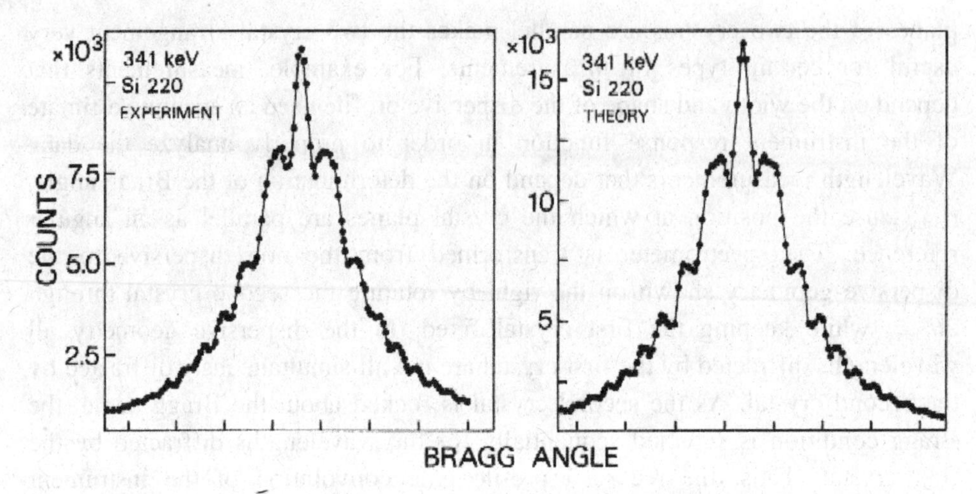

Fig. 3.7.3. Experimental (left) and theoretical (right) response functions obtained for a 341 keV transition (in ^{49}Ti) in non-dispersive mode. The examples were obtained for (n,m) = (1,1) first order reflections using the [220] planes in 1.6 mm thick crystals mounted on the GAMS4 spectrometer. The Bragg angle is $\theta_{Bragg} = 0.57°$ and the displayed angular range is about 1 arcsec.

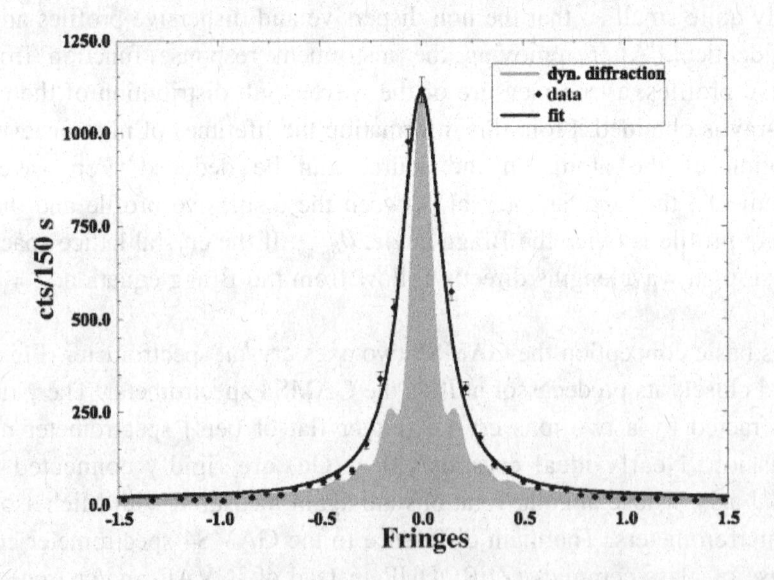

Fig 3.7.4. The instrumental resolution of the GAMS5 spectrometer, measured with flat 2.7 mm thick Si crystals. The curve shown is for a 816 kev line in ^{168}Er, recorded using the Si (660) reflection. The shaded area shows the shape calculated from dynamical diffraction theory. The data points are fitted to a curve which is the dynamical diffraction theory folded with a Gaussian (instrumental broadening) of 0.006". This instrumental broadening of 0.006" corresponds for this energy to an energy resolution of $\Delta E/E = 1.2 \cdot 10^{-6}$. On the abscissa scale 1 fringe = 0.040".

Fig. 3.7.5. View of the GAMS5 double crystal spectrometer. Two perfect crystals of Si or Ge are mounted on the rotating arms of the two angle interferometers. The mirrors on the front side are used for control of the orientation of the optical elements and the crystals themselves. On the left upper side one can see the first crystal mounted on a rotating interferometer arm. The second crystal is mounted opposite, on the right side – in front of a long collimator which is no more shown on the photograph. In the space between the crystal one see the interferometers. The mounts which support the optical elements of the two interferometers are made from glass ceramics.

relevant temperatures are precisely measured and used for corrections. This concerns i) the air temperature to follow the laser wavelength, e.g., "length" of an interference fringe, ii) the crystal temperatures for the temperature dependence of the grating constant and iii) the interferometer arm to account for the interference fringe/angle ratio. As the laser wavelength depends also on further environmental parameters like air pressure and humidity, these are also monitored and enter into corrections. GAMS5 is installed at the H7 side of the through going beam tube, opposite to GAMS4, and has replaced the GAMS1 spectrometer. Figure 3.7.4 shows an example for a non-dispersive measurement of an 816 keV transition, measured in 3rd order of reflection with both crystals, respectively. Figure 3.7.5 shows a photograph of the GAMS5 spectrometer [DOL00].

3.7.1 *Lifetime Measurements Using the GRID Technique*

Various lifetime measurement methods have been developed in the past [NOL79]. For the γ-decays the range of observed lifetimes span from ~ 10^{-18}

seconds up to years. Correspondingly many methods of measurement – each appropriate for a particular range – have been applied. We will discuss here a method, developed at ILL and based on the observation of the Doppler shift, which gives access to the critical range below several pico seconds where the big majority of lifetimes of nuclear excited states is observed. This method makes use of the outstanding resolution, in the ppm range, which can be obtained with two axes flat crystal spectrometers.

In the classical Doppler techniques accelerated ions interact with the target nuclei in a solid medium and the reaction products recoil with a velocity typical several percent of the velocity of light. In contrast to that the gamma ray induced Doppler broadening (GRID)-technique [BÖR88, BÖR93], developed at the ILL, is not based on any external acceleration but the recoil induced is that by the nuclear decay itself. As already mentioned earlier, thermal neutron capture is used to excite nuclei to the neutron binding energy. The capture state will then decay preferably by the emission of γ-rays. In the GRID technique the lifetime of a nuclear state is obtained by an analysis of the Doppler broadening of a de-excitation γ-ray transition (see Fig. 3.7.6): Excited nuclei, produced by thermal neutron-capture, de-excite to lower energy-states via emission of γ-quanta which induce a recoil to the nuclei. The excited atoms move in the bulk of the target material and are slowed down by collisions with the surrounding atoms. Meanwhile the still excited atomic nuclei de-excite by secondary γ-emission. The probability of these emissions depends on the ratios (t/τ) where t is counted from the emission of γs inducing the recoil, and τ is the lifetime of the intermediate state. If the nucleus is in motion when it de-excites ($t \approx \tau$) the γ-ray energy will be Doppler-shifted. If it is stopped, the normal transition energy will be measured. The Doppler effects are on the order of eV. Thus the γ-ray (energy typically ~ 1 MeV) must be detected with a resolution of $\Delta E/E \sim 10^{-6}$ – three orders of magnitude better than obtainable with the usual Ge semi-conductor detectors. Fortunately this is feasible for the GAMS4 and GAMS5 crystal-spectrometers. Because there is no preferred direction in the emission of the primary γ-rays one will observe a Doppler broadening, rather than a Doppler shift (the situation for cascade feeding is more complex; details may be found in [BÖR93]). The measured line shape originates from a convolution of the instrumental response function measured experimentally and the Doppler profile. This Doppler profile is for the ILL spectrometers with their small solid angle described by:

$$I_D(E)\, dE = C \int_0^\infty exp(-t/\tau)\, dt\, Y \tag{3.7.2}$$

with $Y = [arctan\,(2/\Gamma)\,\{E - E_\gamma(1 - v(t)/c)\} - arctan\,(2/\Gamma)\,\{E - E_\gamma(1 - v(t)/c)\}]$

C is a normalization constant, Γ represents the natural width, τ the lifetime of the depopulated nuclear state, E_γ the energy of the depopulating transition and $v(t)$ the average velocity of the recoiling atoms. At the low recoil velocities the slowing down time is quite short and this limits the determination of lifetimes to below about 10 picoseconds. However, there is in principle no lower limit as below about 1 femto second one can measure experimentally the natural width $\Gamma[\text{eV}] \approx 6.6{\cdot}10^{-16}/\tau[\text{s}]$ of the corresponding transitions.

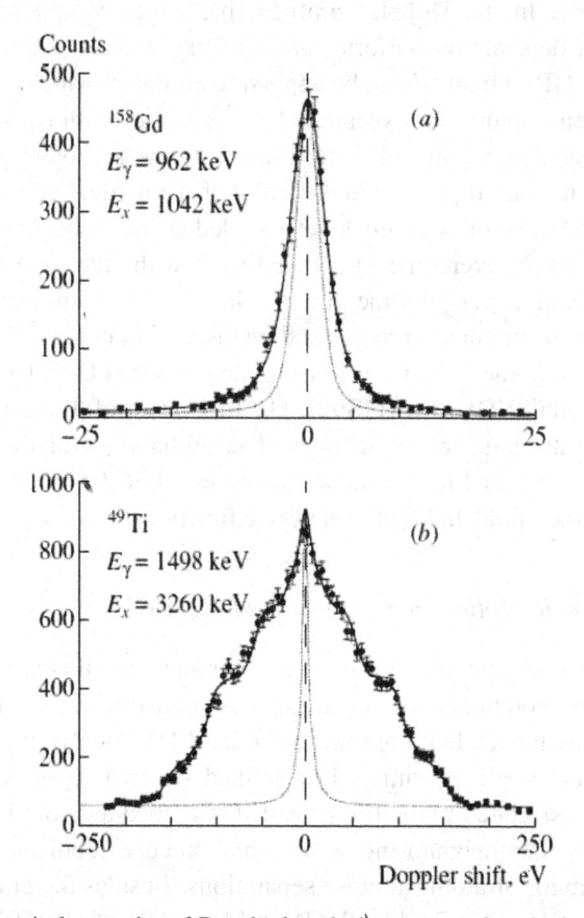

Fig. 3.7.6. Two typical examples of Doppler broadening compared to the instrumental resolution (dashed lines). Mind the change in scale of the abscissa in the two parts. In the heavy nucleus Gd the Doppler broadening is small because the recoil is mainly caused by cascade feeding. In the "light" nucleus Ti the Doppler broadening is stronger because of the smaller mass and because the feeding is mainly due to primary – higher energetic – gamma rays. Structure appears in the Ti profile because a single crystal was used as target (see below).

Calculations of *v(t)* have first been carried out with the so called Mean Free Path Approach [JOL97] but the slowing down process has also been studied in detail by using molecular dynamics (MD) simulations [KEI91]. This has proven to be especially powerful when applied to the measurement of Doppler profiles of γ-rays emitted by atoms from oriented single crystals [JEN96, STR99a]. Due to the regular arrangement of atoms in a single crystal the rate of slowing down depends on the recoil direction (channeling and/or blocking caused by the surrounding atoms). In the Doppler profiles fine structure appears (Fig. 3.7.6 lower part) which depends on the orientation of the crystals with respect to the spectrometer. The GRID method can be applied to all nuclei which can be reached by thermal neutron capture. In exceptional cases nuclei with up to 2 neutrons beyond stability have been studied. Due to the extreme resolving power natural target materials can be used. The main difficulty of the method concerns the often sparsely known feeding of a given level (needed to describe the initial recoil velocities). This can be overcome by simulating it with statistical models or by extracting upper and lower lifetime limits which depend on extreme feeding assumptions about the population routes and intensities. In practice, lifetimes were obtained by comparing the MD simulations to the measured Doppler profiles via a fitting routine called GRIDDLE [ROB92]. The lifetimes τ of the excited states are directly related to absolute transition rates: The probability for the decay of an excited state is proportional to the square of the matrix element of the transition operator between the initial and final state wave functions.

3.7.2 *Study of Atomic Motion in Single-Crystal Metals*

Numerous interatomic potentials exist in the literature and there are many ways to elaborate them. Potentials for metals can be constructed from equilibrium properties using the embedded atom method (EAM [DAW93]), where they give fairly good results. These potentials are defined by two terms: an embedded function which is correlated to the force needed to embed an atom in the electron density caused by the neighboring atoms, and a core repulsive term which describes the behavior of atoms at close separations. Besides the embedded atom method potentials other theories like the Born Mayer (BM [ABR69]), defined by an exponential function, or the Ziegler, Biersack, Littmark (ZBL [ZIE85]), based on a coulomb screened function, are calculated for very small interatomic separation. A more detailed summary of many of the potentials may be found in [STR99b], however not all of them are suited for the slowing down in the energy range of a few hundreds eV to a few eV.

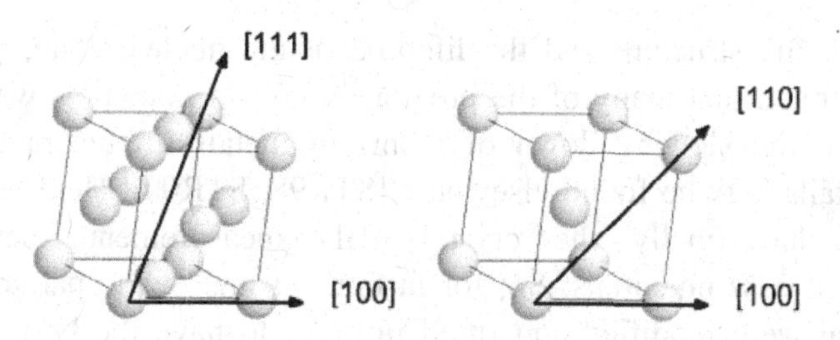

Fig. 3.7.7. Direction of observation for the 2 crystalline structures used for metals. Nickel has a fcc structure shown on the left, Fe and Cr have a bcc structure.

The Crystal-GRID technique fills in the missing energy gap. Interatomic potentials, which govern the slowing down, can be tested in the recoil energy domain by comparing the line shape measured with the double-flat crystal-spectrometer and the one simulated with different inter atomic potentials. The Doppler line-structure depends on three terms: i) the lifetime of the nuclear state populated, which gives the probability emission of the second gamma ray, ii) the orientation of the crystal, because blocking and channelling of the recoiling atom due the ordered structure of the crystal influence the trajectories and velocities, and finally iii) the inter atomic potential which determines the slowing down and the motion of the atom recoiling inside the crystal.

At ILL for each transition metal, almost 10 different potentials were tested with 2 different crystal orientations. Figure 3.7.7 shows the crystalline structure of the metals and the directions of observation of the second gamma-rays. The slowing down theories are always simulated by molecular-dynamics simulation which calculate the trajectories and velocities of the recoiling atom in a molecular-dynamics cell of about 1000 atoms by solving the Newtonian equations of motion for each individual atom. From the stored value of the trajectories and velocities, a line shape was reconstructed, assuming a given lifetime value. The simulated line structure was then fitted to the measured one leaving the nuclear state lifetime as a free parameter. Figure 3.7.8 demonstrates that blocking and/or channelling – due to the ordered atom positions in crystals give rise to different line profiles. In a bcc crystal, as is the case for Chromium, the atom is more free to move in the [110] direction. This is because the closest neighbor is further away than in the [100] direction. Therefore, as can be seen from Fig. 3.7.8, the line profile will in this case be more broad for the [110] direction. When proceeding with the fitting of the data, a perfect match with the theory – depending on a given potential – should return a χ^2 value of 1. It was found that among the known potentials, very few were able to reproduce both, the measured line structure and the lifetime of the nuclear state, respectively. This demonstrated that many of the previously known potentials were not good candidates for the slowing down of atoms in metals in this range of recoil energies. Details may be found elsewhere [STR98, STR00]. Here we only want to mention that finally the crystal-GRID measurements permitted the construction of new potentials: So, for instance, some of the parameters of the ZBL potential were modified and fitted in order to have the best χ^2 values by comparing simulated and measured line shapes [STR00]. The experiments presented opened a way to a more systematic study of interatomic potentials in alloys following a building up principle [STR98, STR00].

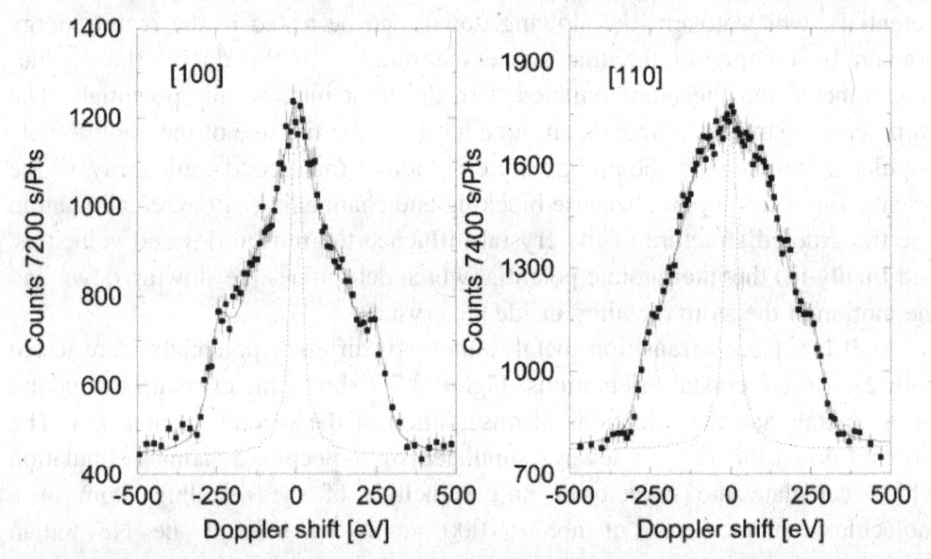

Fig. 3.7.8. The gamma-ray line shape of the 2239.07 keV transition for Chromium, measured and fitted in the [100] and [110] direction, respectively. The simulated Doppler-broadened line-shape was obtained with an EAM potential for the two orientations and for a lifetime of 14.72 fs. The dashed profile is the instrumental response function as calculated by dynamical diffraction theory.

3.7.3 *Neutrino Induced Recoil*

Also atomic electron capture can provide the kinetic energy leading to Doppler broadened line profiles. It is a process where the nucleus captures an electron from its shell and then emits a neutrino which, in first approximation, is the only particle which leaves the nucleus. Because the neutrino has a well defined energy this leads to a well defined recoil energy. It should be noted, that pure electron capture does only occur when the mass difference between initial and final atom does not exceed 1022 keV, as in this case the competing β^+ decay is forbidden.

Consequently beta decay associated with electron capture takes place via the following nuclear reaction:

$$A + e \rightarrow (A+1) + \nu$$

If the Q value is lower than 1.022 MeV then the sharp recoil energies that can be obtained are only as low as a few eV for heavy nuclei. The recoil velocity of the newly formed atom follows then from the conservation of momentum (for zero-rest mass neutrinos and using non relativistic kinematics) by:

$$v_r / c = (Q_{ec} - B_e)/(A_r(X)u)c^2 \tag{3.7.3}$$

The Q_{ec} value for the reaction is defined as:

$$Q_{ec} = (A_r(X)u - A_r(Y)u)c^2 \tag{3.7.4}$$

In eqs. (3.7.3) and (3.7.4) $A_r(X)u\, c^2$ and $A_r(Y)u)\, c^2$ denote the atomic rest masses in the initial and final states (expressed in unitless relative atomic masses A_r and the atomic mass unit u) and B_e is the binding energy of the captured electron.

For the neutrino induced recoil (NID) the recoil energy is generally smaller than the binding energies of the atoms (typically around 10 eV) and therefore the recoiling atoms are not able to definitively leave the equilibrium positions. In order to analyze the measured line shapes different approaches were developed at ILL. The first, analytic, approach is based on a phonon creation model [JOL97]. The second, numerical, approach relies on molecular dynamics (MD) simulations of the slowing-down process at ultra-low recoil energies [KUR95, STR97, STR98]. Both approaches allow to obtain information on the lifetime of the nuclear state fed by the electron capture process, and/or to study the effect of ultra-low recoil energies on atoms located in a lattice of bulk matter.

All NID experiments carried out at ILL used electron capture in the isomeric 0^- state of ^{152}Eu. In order to populate this state, which has a half life of 9.3 h, natural europium was used. Placed at the GAMS4 in-pile target position, the isotope ^{151}Eu, which has a 48% natural abundance, captures a thermal neutron and forms ^{152}Eu. The cross section for this reaction is 9204 barns, making the targets "black" for thermal neutrons. Thus the neutron capture rate per cm^2 of target area is limited to $2.5 \cdot 10^{14}\ s^{-1}$. Following the electron capture a 1^- level at 963 keV in ^{152}Sm is populated. This 1^- state can either decay to the nuclear ground state by emitting a single 963.4 keV gamma ray, or to a 121.8 keV 2^+ level via a 841.4 keV gamma transition. Different kinds of targets have been used in quite a number of experiments. In the earlier measurements powder targets of different chemical composition (Eu_2O_3, EuF_3, EuF_2 and $EuCl_3$) were used

[JOL97]. In later experiments oriented single crystals of EuO were observed for different orientations with respect to the spectrometer [STR97, STR98]. EuO crystals have a fcc cubic structure and the NID measurements were performed once with the [100] direction towards the spectrometer and once with the [110] direction. Details on this experiment are given elsewhere [STR98].

Using Molecular Dynamics (MD) simulations and Monte Carlo simulations the description of the slowing down can be done in a very detailed manner [KUR95]. By solving the Newtonian equations for atoms recoiling in random directions it is possible to study the spread of velocities and to treat in detail the effects of the thermal motion, x-ray and Auger-electron emission. For the MD simulations the main input are the interatomic potential which are treated as sets of pair potentials dependent on the distances in between the atoms. Knowing the interatomic potentials, the lifetime can be fitted and compared to those available in the literature. The experimental line shapes were analyzed using the computer code GRIDDLE [ROB92], and Fig. 3.7.9 illustrates the very good quality of the fit of the line shape for the EuO measurement [STR97].

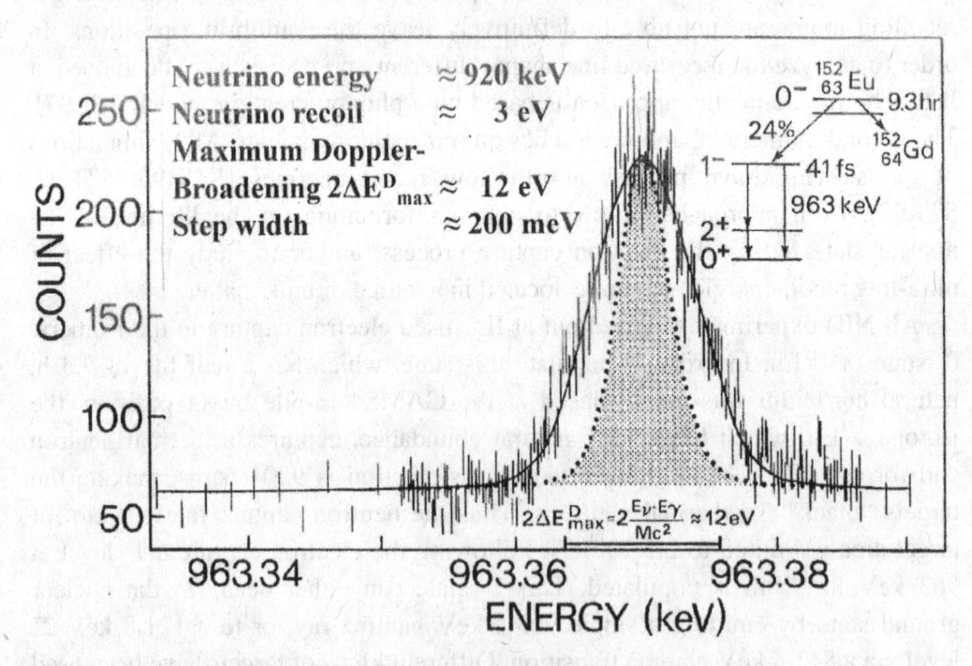

Fig. 3.7.9. Comparison between fitted and measured line shape of the 963 keV 1^- to 0^+ ground state transition in ^{152}Sm. The shaded area indicates the instrumental resolution.

3.7.4 *Precision Measurement of High Energy Gamma-Rays*

The precise measurement of high energy gamma-ray photons was one of the original reasons for establishing a collaboration between the ILL and NIST. This finally led to the precision gamma-ray metrology facilities, GAMS4 and GAMS5. One of the first gamma-ray photons that was measured during the early days (1986, Fig. 3.7.10) of the first flat crystal spectrometer installed at ILL, GAMS4, was the 2.2 MeV gamma-ray produced in the reaction, n + ^{1}H = ^{2}H + γ (2.2 MeV). This measurement can be combined with mass spectroscopic measurement of the ^{2}H-^{1}H atomic mass difference (see below) to obtain a value for the neutron mass. At that time, the neutron mass uncertainty was dominated by the mass spectroscopy measurements. By the mid nineties, new mass spectroscopy techniques had been developed. Those were more accurate than previous atomic mass determinations. When these new mass measurements were combined with the 1986 2.2 MeV gamma-ray photon mass, the neutron mass uncertainty was dominated by the uncertainties in the GAMS measurements. However, also the GAMS facilities had been continually improved towards the goal of providing high energy gamma-rays with a relative uncertainty of 10^{-7}. NIST and ILL have shared the responsibility for these improvements which include precisely positioned vibration isolation platforms for the crystal spectrometers, improved angle interferometers with reduced errors, angle calibration facilities, state-of-the-art gamma-ray detection facilities which reduce the background, temperature stabilization of the spectrometers, improved spectrometer control electronics, and more perfect Si and Ge crystals whose lattice spacings have been precisely measured.

The new mass spectroscopic techniques that had been developed, measure atomic masses by comparing the cyclotron frequencies of two different trapped ions. As already mentioned, the results obtained were more than an order of magnitude more accurate than previous atomic mass determinations [NAT93, DIF94]. This dramatic reduction in the uncertainty of atomic mass measurements has renewed interest in the connection between atomic masses and precisely measured wavelength intervals that can lead to precise values for the neutron mass and the molar Planck constant.

One atomic mass unit u is defined as 1/12 times the mass of a ^{12}C atom. The related Avogadro constant N_A is defined as the number of atoms present in 12 g of ^{12}C.

In a typical neutron capture reaction n + A → (A+1)* = (A+1) + gammas, the energy conservation applied yields:

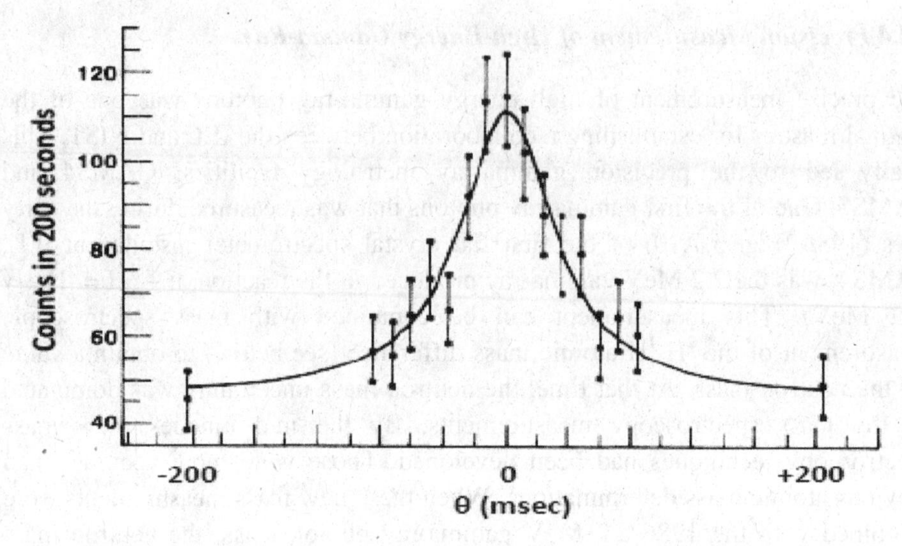

Fig. 3.7.10. One of the first 2.2 MeV n-p capture γ-ray measurements using the GAMS4 spectrometer in 1986. He diffraction angle is about 0.12^0; total range displayed is ~ 0.4 arc sec. Here the theoretical curve is a simple Lorentzian. The total data accumulation time was ~ 1h.

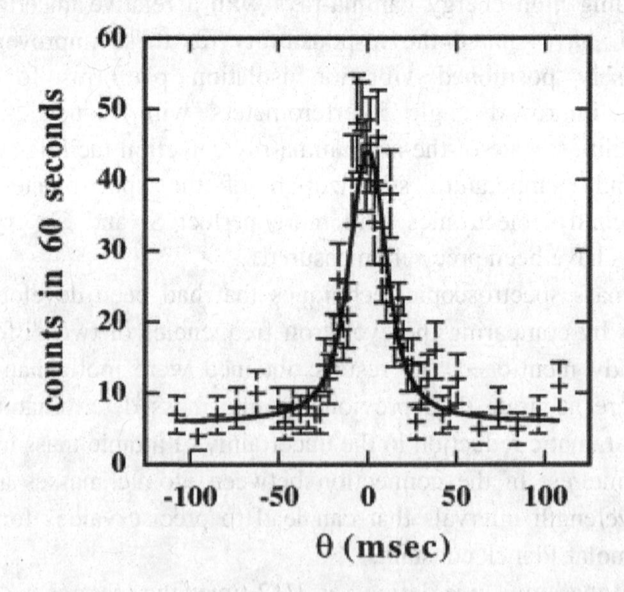

Fig. 3.7.11. The 2.2 MeV gamma-ray profile [KES99] using 2.5 mm thick Si crystals in the (220,440) configuration in 1999 measurements; total range displayed is ~ 0.2 arc sec. The improvements had lead to a significant increase in the quality of data available from this facility. Compared to the 1986 measurement [GRE86] the intensity has increased and the angular width and background have been reduced, all of which contributed to more accurate angle measurements.

$$[A_r(n)u]c^2 + [A_r(A)u]c^2 = [A_r(A+1)u]c^2 + hc/\lambda^*_{(A+1)} \quad (3.7.5)$$

where A_r is the unitless relative atomic mass ($A_r(^{12}C)=12$), u is the mass of an atomic mass unit, h is the Planck constant, c is the speed of light, and $hc/\lambda^*_{(A+1)}$ represents the binding energy $S_n(A+1)$ of the nucleus (A+1) expressed as the sum of reciprocal wavelengths over those gamma-rays which connect the (A+1) capture state to the ground state (typically 3 gamma-rays for GAMS measurements). The atomic mass unit u is given via the Avogadro constant by $u = \{10^{-3}/N_A\}_{SI}\,kg$ where $\{\}_{SI}$ indicates its numerical value when expressed in SI units. $\lambda^*_{(A+1)}$ includes a correction for the (A+1) recoil energy. Expressing u via the Avogadro constant and then rearranging eq. (3.7.5) yields:

$$A_r(n) + A_r(A) - A_r(A+1) = \{10^3\, N_A h/c\}_{SI}\{1/\lambda^*_{(A+1)}\}_{SI} \quad (3.7.6)$$

This relation connects mass measurements in atomic mass units to wavelength measurements in meters. The corresponding conversion factor is the molar Planck constant $N_A h$ divided by the speed of light c (see also [COH87], c being an exactly defined quantity). This factor can also be derived from another relation:

$$\{10^{-3}\, N_A h/c\}_{SI} = \{A_r(e)\, \alpha^2/2R_\infty\}_{SI} \quad (3.7.7)$$

with $A_r(e)$ the relative atomic mass of the electron [FAR95], α the fine structure constant and R_∞ the Rydberg constant [UDE97]. This led to a conversion factor of:

$$1m^{-1} = 1.331\,025\,045(11)\,10^{-15}\,u \quad (3.7.8)$$

Before the year 2007 the dominant source of the uncertainty of about $8\cdot 10^{-9}$ came from the fine structure constant. However, to date new measurements of the magnetic moment of the electron together with QED calculations [GAB07] yield a much smaller error of $7\cdot 10^{-10}$!

Due to the improvements (evident by comparing Figs. 3.7.10/11) undertaken at ILL's two axes crystal spectrometers, those are to date capable of measuring $\lambda^*_{(A+1)}$ (in favorable cases) with a relative standard deviation of around 10^{-7}. The best mass measurements are done with uncertainties close to 10^{-10}. As masses of light nuclei are in the order of some 10 GeV whereas their neutron binding energies are around 10 MeV, 3 orders of magnitude smaller, this results in an overall precision of about 10^{-7} for the determination of the atomic mass unit u (see below) and/or the molar Planck constant. Evidently there can be different reasons why it is interesting to determine the ingredients of equation (3.7.7). Some of them are discussed in the following. Due to the neutrons neutrality its mass cannot be directly measured with a trap but it can be determined from neutron capture on hydrogen, complemented by mass difference measurements in traps:

$$A_r(n) + A_r(^1H) - A_r(D) = \{10^{-3} N_A h/c\}_{SI}\{1/\lambda^*_{(D)}\}_{SI} \quad (3.7.9)$$

Expressed in the classic way that means: The neutron mass $m(n)$ is obtained by expressing the nuclear reaction n + p = d + γ in atomic mass units, $m(n) = m(^2H) - m(^1H) + S(d)$, and combining the deuteron binding energy with precision atomic mass difference measurements.

The wavelength of the 2.2 MeV gamma-ray was determined by combining the measured Bragg angle (obtained from averaging over more than 100 scans in each case) and the crystal lattice spacing using the Bragg equation for diffraction, $\lambda = 2d \cdot sin\theta$. The wavelength was such measured to be $\lambda = 5.576\ 712\ 99(99) \cdot 10^{-13}$ m. This measured wavelength has to be corrected for the recoil given to the deuterium atom, when the gamma ray is emitted to obtain the wavelength, λ^*, that corresponds to the binding energy of the deuteron, $S(d)$. The value for λ^* is then $\lambda^* = 5.73\ 409\ 78(99) \cdot 10^{-13}$ m. $S(d)$ can then be expressed in atomic mass units and in eV by using the inverse meter to atomic mass unit conversion factor (equation (3.7.8)) and the inverse meter to eV conversion factor, respectively. The results are: $S(d) = 2.388\ 170\ 07(42) \cdot 10^{-3}\ u$ and $S(d) = 2\ 224\ 566.14(41)$ eV. Remember: $1\ u \approx 1.6605 \cdot 10^{-27}$ kg. This gamma-ray result was combined with the most precise mass spectroscopic measurement of the $m(^2H)\text{-}m(^1H)$ atomic mass difference [DIF94] to obtain an improved value for the neutron mass. The equation for the neutron mass follows directly from the reaction given above: $m(n) = m(^2H) - m(^1H) + S(d)$ with $m(^2H)\text{-}m(^1H) = 1.006\ 276\ 746\ 30(71)\ u$ and the new binding energy measurement one obtained $m(n) = 1.008\ 664\ 916\ 37(82)\ u$ [KES99].

In the 1995 Atomic Mass Evaluation [AUD95] the mass of the neutron is reported as $m_n = 1.008\ 664\ 9232(22)\ u$. The improved 2.2 MeV gamma-ray mass reported here lead to a neutron mass which was $\approx 7 \cdot 10^{-9}\ u$ less than the value in the 1995 Atomic Mass Evaluation and with an uncertainty which was reduced by a factor of 2.5. The contributions to the neutron mass uncertainty by the gamma-ray and mass spectroscopic measurements cited above were $0.42 \cdot 10^{-9}\ u$ and $0.71 \cdot 10^{-9}\ u$, respectively.

Based on this more precise value for the mass of the neutron, it was logic to try to obtain also more precise values for heavier nuclei. Therefore the binding energies of ^{29}Si, ^{33}S, and ^{36}Cl have been measured [DEW06]. They were determined with a relative uncertainty $< 6 \cdot 10^{-7}$ using the GAMS4 spectrometer.

The binding energies of these three nuclei are in the 8.5 to 8.6 MeV range and are obtained by measuring lower energy lines that form a cascade scheme connecting the capture and ground states (Fig. 3.7.12). For all three nuclei, the cascade scheme with the most intense transitions includes a gamma-ray with energy > 4.9 MeV. Such high energies present a significant measurement challenge for gamma-ray spectroscopy because the Bragg angles are < 0.1^0 and the diffracted intensity is rather small (a few counts/s or less). The binding energies determined from these gamma-ray measurements are consistent with highly accurate atomic mass measurements within a relative uncertainty of 4.3×10^{-7}, in agreement with earlier precision gamma-ray measurements. The results for the binding energies obtained are summarized in table 3.7.1. As soon as more precise atomic mass measurements in these nuclei become available, then comparison with those can be used (equation (3.7.6)) to determine a competitive value for the molar Planck constant $N_A h$. Currently the most precise value for $N_A h$ comes from the determination of the fundamental constants $A_r(e)$, α, and R_∞ combined with $N_A h$ as shown in eq. (3.7.7).

Table 3.7.1. Measured binding energies in meters (m), atomic mass units (u) and electr. volts (eV)

	$\lambda_{be} \times 10^{12}$ (m)	$S_n \times 10^3$ (u)	S_n eV	wavelength rel. uncertainty
^{29}Si	0.146318275(86)	9.0967793(53)	8473595.7(5.0)	0.59
^{33}S	0.143472991(54)	9.2771820(35)	8641639.8(3.3)	0.38
^{36}Cl	0.144507180(80)	9.2107883(51)	8579794.5(4.8)	0.55
^{2}H	0.557341007(98)	2.38816996(42)	2224566.10(44)	0.18

There is yet another point of view from which one can consider these results. Clearly, the mass of a photon is zero, but as the photon has a momentum we can relate its "mass", m_γ, to its energy, E, or wavelength, λ, through the Einstein relation, $E = hc/\lambda = m_\gamma c^2$. The so determined "mass" of a photon in the visible part of the spectrum is very small, $4.42 \cdot 10^{-36}$ kg, much smaller than the masses of fundamental particles and atoms.

However, a million times more massive photons are available in the 1 to 10 MeV gamma-ray region of the electromagnetic spectrum ($2 \cdot 10^{-30}$ to $2 \cdot 10^{-29}$ kg) available from neutron capture. One could see from the results described earlier that, if they are measured with sufficient accuracy, they can provide a significant contribution to our knowledge of the masses of fundamental particles and atoms.

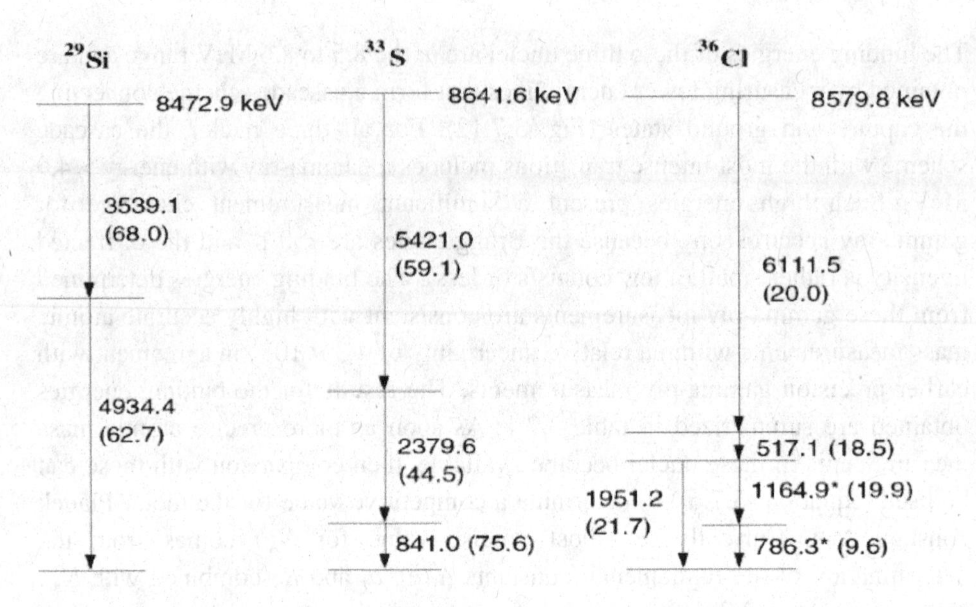

Fig. 3.7.12. Partial decay schemes for ^{29}Si, ^{33}S, and ^{36}Cl showing the transitions that were measured in this study. The numbers in parentheses are the number of gammas per 100 neutron captures. The transitions marked with asterisks in Cl were not used to deduce binding energies.

In that spirit – combining the precise atomic mass and gamma-ray wavelength measurements involving isotopes of Silicon and Sulphur as described above – two tests were obtained [RAI05] that separately confirmed Einstein's relationship. This was done by considering balance in a nuclear reaction, which is initiated by particles with a minimum of kinetic energy. This is the case in the (n,γ) reaction, when a nucleus with mass number A captures a neutron: The mass of the resulting isotope, with mass number A+1, ought to differ from that of the original nucleus (plus unbound neutron) by the neutron binding energy $S_n(A+1)$. Using eq. (3.7.5) Einstein's equation can be rewritten as:

$$(A_r(A)+A_r(n)-A_r(A+1))c^2 = (1/u)S_n(A+1)=10^3 N_A\, hc/\lambda^*_{(A+1)} \qquad (3.7.10)$$

As described before, the mass of the neutron can be eliminated from equation (3.7.10) by introducing the masses of Hydrogen ^{1}H and Deuterium ^{2}D combined with the wavelength λ^*_D corresponding to the deuterium binding energy:

$$(A_r(A)+A_r(2D)-A_r(1H)-A_r(A+1))=10^3\, N_A h/c(1/\lambda^*_{(A+1)} + 1/\lambda^*_D) \qquad (3.7.11)$$

For this exercise the molar Planck constant is taken as known independently $N_A h = 3.990\ 312\ 716(27) \cdot 10^{-10}$ J s mol^{-1} (u/kg) with a precision at the $5{\cdot}10^{-8}$ level (through its relationship with the fine structure constant [MOH05]). The values of $S_n(^{29}Si)$, $S_n(^{33}S)$ and $S_n(^2H)$, obtained from the experiments are displayed in table 3.7.1. These numbers combined yield relative uncertainties of $5.1 \cdot 10^{-7}$ (^{33}S) and $8.0 \cdot 10^{-7}$ (^{29}Si) for the right-hand side of equation (3.7.10). The mass difference was determined at the Massachusetts Institute of Technology using a technique to directly compare the cyclotron frequencies of two different ions simultaneously confined in a Penning trap [RAI04]. Mass ratios of $A_r(^{32}S)+A_r(H)-A_r(^{33}S) = 0.00843729682(30)u$ and $A_r(^{28}Si)+A_r(H)-A_r(^{29}Si) = 0.00825690198(24)\,u$ were obtained. By adding $A_r(^2H)-2A_r(H) = -\,0.001\,548\,286\,29\,(40)\,u$ to each one, one obtains the mass differences with a relative uncertainty of $7 \cdot 10^{-8}$ for both. The comparison of the measured energies and masses leads to two independent tests for a deviation from $(1-E/mc^2)$ of $(2.1 +/- 5.2) \cdot 10^{-7}$ and $(-9.7 +/- 8.0) \cdot 10^{-7}$ with sulphur and silicon isotopes, respectively, and a combined value of $(-1.4 +/- 4.4) \cdot 10^{-7}$. This test is 55 times more accurate than the previous best direct test of $E = mc^2$, performed by comparing the electron and positron masses to the annihilation energy [GRE91]. The error on this comparison is currently dominated by the uncertainty on the gamma-ray measurements.

3.8 Nuclear Physics Studies

Nuclear physics is a dynamic area in science with many features which remain to be discovered and explained. As a consequence of the three, in the subatomic region mainly involved interactions with very complicated properties – the strong, the weak, and the electromagnetic interaction – there is an extraordinary big variety of phenomena observed and the history is full of unexpected discoveries. In order to show this we give a short historical outline: The starting signal for nuclear physics was given in 1895 by W.C. Röntgen who discovered the existence of X-rays. In 1909 E. Rutherford finished the characterization of alpha-, beta-, and gamma radiation and in 1911 he proposed that atoms might consist of a nucleus surrounded by a cloud of electrons [RUT11]. In 1913 then N. Bohr presented indeed a model with electrons orbiting around a nucleus [BOH13, BOH14]. Lise Meitner discovered in 1924 that gamma radiation follows beta and alpha decay in the de-excitation of daughter nuclei [MEI24]. In 1930 W. Pauli postulated a new elementary particle, the neutrino (called at that time "neutron" by Pauli in a letter to L. Meitner and H. Geiger) to explain the beta decay and afterwards he formulated the exclusion principle. J. Chadwick discovered the neutron in 1932 [CHA32]. A couple of years later E. Fermi presented the first theory of the beta decay [FER34a, FER34b] and he also was the first to show that with capture of slow neutrons one can indeed produce new nuclei. C.F. von

Weizsäcker presented his famous droplet model in 1935 in order to explain the masses of nuclei [WEI35]. In 1939 O. Hahn, F. Strassmann and in the same year L. Meitner and O.R. Frisch discovered nuclear fission and offered a first explanation of fission with neutrons, respectively [HAH39, MEI39]. In 1942 E. Fermi started the first nuclear reactor in Chicago [FER46]. O. Haxel, J.H.D. Jensen H.E. Suess and M. Goeppert-Mayer in 1949 further developed the shell model for the nuclear core [HAX49, GÖP49] by introducing a spin-orbit coupling of the nuclear forces. This was followed in 1950 by the introduction of a nuclear model by J. Rainwater [RAI50], A. Bohr and B. Mottelson [BOH52, BOH53], which combined collective and particle aspects of the nuclear interaction. S.G. Nilsson introduced in 1956 his collective nuclear model [NIL55] in single particle approximation for deformed nuclei. E.P.J. Wigner obtained in 1963 the Nobel prize for his contributions to nuclear theory, especially for the development of fundamental symmetry principles (see for instance [WIG27], introduction of the D-matrix). When the first high resolution measurements started in 1973 at the ILL high flux reactor, the basic idea was to corroborate the earlier findings summarized above and to try to elucidate some "less understood" details, mainly from nuclei in transitional regions and in the actinides. But then dynamic symmetries were introduced in nuclear physics in 1974 by F. Iachello and A. Arima [IAC74, ARI75] and this had direct impact on the direction of further nuclear structure studies at ILL. The Interacting Boson Model (IBM), they presented, has three dynamic symmetries: U(5), SU(3), and O(6). Experimental examples of all three types have been studied at ILL. Dynamical symmetries are not only a tool for the nuclear physicist but, although started in this field, they find today wide applications in systems ranging from nuclei to molecules and even polymers. For O(6) the first experimental case at all was discovered at ILL in 1978 [CIZ78, CIZ79] by J. Cizewsky, R.F. Casten and collaborators in studying the structure of ^{196}Pt, using high resolution gamma ray spectroscopy. This was an important experimental confirmation of the theoretical predictions formulated in the IBM and was followed at ILL by many studies to disentangle more and more details in a wide variety of nuclei in order to probe the validity of the theoretical predictions. Very recently, new concepts, the critical point symmetries have been introduced by F. Iachello in 2000 and 2001 [IAC00, IAC01]. Critical point symmetries extend the concept of dynamic symmetry to systems at the critical point of a phase transition. ILL has contributed to the experimental verification of these new ideas by contributing with studies of lifetimes of excited states in nuclei in the Samarium region [BÖR06, ZAM02].

We have displayed above many facets of nuclei by mentioning them in a kind of historical enumeration. It might sound surprisingly, but many of the problems addressed in earlier days are still relevant today. Such an example are GRID measurements (see below) in ^{168}Er – carried out at ILL in 1990 – which demonstrated for the first time that an excited state in a heavy deformed nucleus could be associated with a 2-phonon excitation [BÖR91]. In solids, multiple vibrations can be piled on top of one another in many ways. In nuclei, however, the Pauli exclusion principle has long been thought to impede significantly such modes. The ^{168}Er experiment ended a discussion which was going on for more than three decades and clearly demonstrated that multiphonon vibrational excitations could indeed exist in nuclei that are non spherical. The atomic nucleus is not a single object but a collection of species. Over the last 35 years at ILL nuclei ranging from hydrogen to the actinides have been studied and it was shown that they display an extremely rich and fascinating variety of phenomena. The nucleus is very small, about 10^{-12} to 10^{-13} cm in diameter, and can contain up to a couple of hundred individual protons and neutrons that orbit relative to each other and interact primarily via the nuclear and Coulomb forces. At least the heavier nuclei are such complex systems that we have still not yet reached a full understanding of their structure! On the other hand, as seen from the examples above, we know already a huge number of facts about the nucleus, theoretical and experimental ones, and we understand often in detail what the individual nucleons do in nuclei and how specific phenomena change from nucleus to nucleus. If we look back over the last 30 years of nuclear structure research at ILL we see that important contributions have been made in quite different directions: It helped to elucidate the nature of the nucleus as a many-body quantal object and to look into the role of the Pauli Principle in this fermionic system, to find many of the basic symmetries which are also important for particles physics, to trace the evolution of the universe by adding knowledge to element synthesis processes which might take place in so different systems like a red giant or in super novae explosions, to identify the signature of quantum chaos in complex nuclei, or to study very neutron rich nuclei close to the neutron drip line.

3.8.1 *Studies of Light Mass Nuclei*

Quite a number of spectroscopic studies of light mass nuclei were performed with the bent crystal spectrometers at ILL. The aim was to establish their respective level schemes as completely as possible. Especially nuclei in the sd shell, but also the heavier fd shell nuclei are in regions of the nuclear chart where the shell model can give good predictions [TAL93], without having to rely on large truncations of the model space.

It was already mentioned that the nuclear shell model was "born" in 1949 following independent work by several physicists. The shell model uses the Pauli exclusion principle to describe the nuclear structure. Evidently shells exist for both protons and neutrons individually and one speaks of "magic nuclei" where one nucleon type is at a magic number of nucleons (2, 8, 20, 28...). The shell model potential represents the average interaction of all the other nucleons exerted on any one of the nucleons in the nucleus. Since the nucleon-nucleon interaction is short range one uses a potential in the quantum-mechanical calculations that mimics the mass distribution of the nucleus. Therefore nuclear shell model calculations frequently use the Wood Saxon form of the potential [WOO54]. It is then necessary to compare the theoretical predictions to experimental data.

Due to its non-selectivity the (n,γ) reaction is extremely well suited for establishing complete level schemes in the regions of spins which can be populated by neutron capture. For very light nuclei one can combine thermal neutron capture and high resolution spectroscopy. For heavier nuclei one can also complement this by adding average resonance capture (ARC) techniques [POR56]. At ILL extensive high resolution studies of light and medium light nuclei were carried out for ^{20}F [HUN83a], ^{24}Na [HUN83b], ^{28}Al [SCH82], ^{36}Cl [KRU82], ^{40}K [EGI84], ^{41}K [KRU84], and ^{42}K [KRU85]. Many of the level schemes obtained for these nuclei were also used as input for statistical model predictions.

3.8.2 *Empirical Manifestation of New Symmetries for Nuclei*

As mentioned above, research at the ILL has played a major role in pioneering studies of dynamical symmetries. In the IBA model, developed by Iachello and Arima [IAC74, ARI79], the Hamiltonian is written in terms of interactions between bosons which can occupy L=0 and L=2 (s and d) states. This model was phrased in the group theoretical language of SU(6). For this group three natural limits arise for which analytical solutions can be obtained. These limits correspond to three subgroups of SU(6), namely SU(5), SU(3), and O(6). SU(5) corresponds to a (anharmonic) vibrator, SU(3) to the quadrupole deformed rotor. Many examples of nuclei close to these two limits were well known before the introduction of the IBA. Following an extensive set of high resolution γ-ray measurements, carried out at ILL and at the Brookhaven National Laboratory (BNL), the third limit, O(6), could for the first time be associated empirically with the structure of low spin positive parity states in ^{196}Pt [CIZ78, CIZ79]. The experiments performed in this study involved an essentially complete set of (n,γ) techniques: Singles spectra of secondary and primary γ transitions at thermal, eV, and keV neutron energies have been recorded with Ge(Li) detectors at BNL. These data were complemented by high-resolution spectra of secondary transitions taken at the ILL with the GAMS1 and GAMS23 bent-crystal spectrometers. Especially the 2 keV average resonance neutron beam measurements allowed to make sure that all 0^+, 1^+, and 2^+ levels below 2.5 MeV have been identified. Thanks to the high sensitivity and precision of the GAMS spectrometers an important number of very crucial low-energy transitions could be placed which, though weak in intensity, represent the largest deexcitation B(E2) values from their respective levels.

^{196}Pt belongs to a class of nuclei, situated towards the ends of major shells, for which none of the "classical limits", vibrator and rotor, respectively, could be applied with good success. In the O(6) limit the energies of collective states are given by:

$$E(\sigma,\tau,J) = (1/4)A(N - \sigma)(N + \sigma + 4) + B\tau(\tau + 3) + CJ(J + 1) \qquad (3.8.1)$$

Where N is the number of bosons, defined as ½ of the sum of the number of protons plus the number of neutrons away from the nearest respective closed shells (for ^{196}Pt, $N=6$); $\sigma = N, N\text{-}2, N - 4, \ldots, 0$, and $\tau = 0,1,\ldots\ \sigma$. J takes on the values $2\lambda, 2\lambda - 2, 2\lambda - 3, \ldots\ \lambda + 1, \lambda$, where λ is a nonnegative integer defined by $\lambda = \tau - 3\upsilon_\Delta$ for $\upsilon_\Delta = 0,1,2,\ldots$.Each level can be uniquely defined by the quantum numbers J^π $(\sigma,\tau,\upsilon_\Delta)$. A characteristic feature of the O(6) scheme is a recurring 0^+ - 2^+ - 2^+ pattern of levels with E2 selection rules predicting strong cascade γ-ray transitions within the sequence. Some of the results of the (n,γ) studies are summarized in Fig. 3.8.1.

As already mentioned above, in ^{196}Pt states have been identified which belong to families of levels which are labeled by the quantum number σ with values $\sigma = N, N\text{-}2, N\text{-}4$ and within each family levels are labeled by the quantum number τ. In ^{196}Pt, all positive-parity states below ~ 1.8 MeV are assigned O(6) character and all allowed transitions were observed, all forbidden ones are weak or unobserved, and the measured E2 branching ratios of allowed to forbidden transitions are large. Although this did only leave little doubt as to the structure of ^{196}Pt there was an additional conclusive test, namely the measurement of

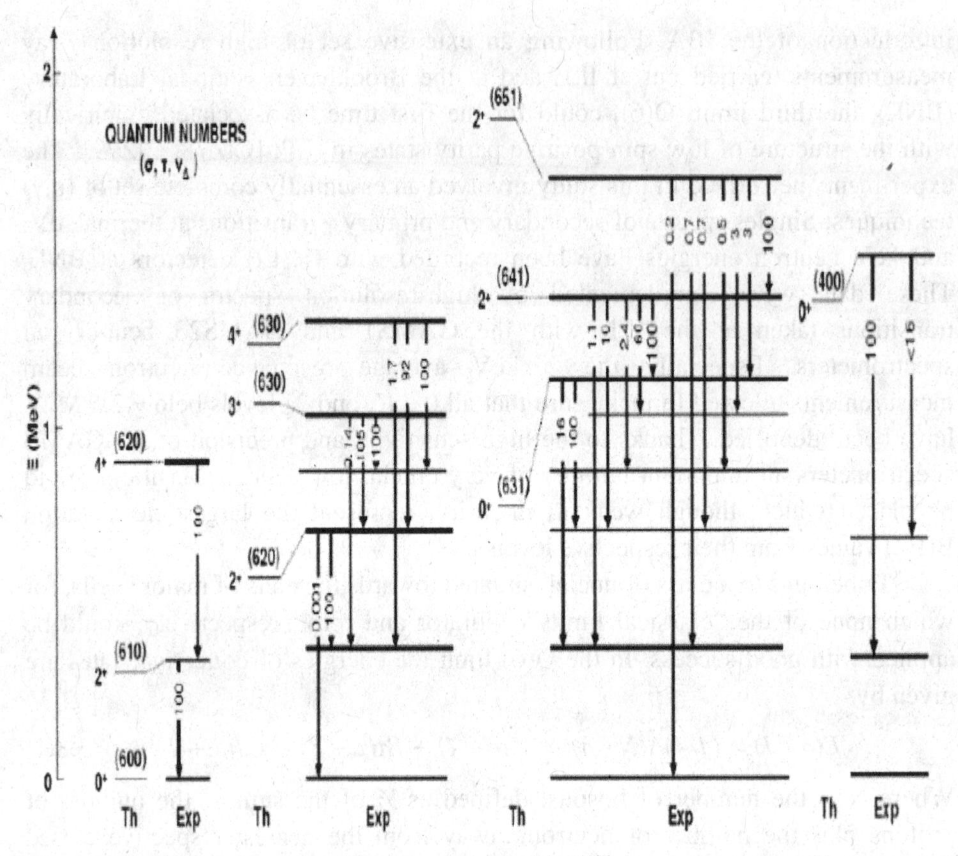

Fig. 3.8.1. Partial level scheme for positive parity states in ^{196}Pt. In parentheses of the theoretical levels are the quantum numbers (σ,τ,υ$_Δ$). The number on the transition arrows is the measured relative B(E2) value. To the very right one finds the 0^+_3 state with quantum numbers (400), used to test the goodness of the σ quantum number.

absolute transition rates (or limits) for the decay of states from one family to another one. Consequently – in a first application of the GRID method to a heavy mass nucleus – the first absolute transition rate from a $\sigma < N$ (σ_{max}) state in an O(6) nucleus was determined by measuring the lifetime of the 0^+_3 state at 1402 keV [BÖR90]. This corresponded to a first quantitative test of the goodness of the σ quantum number, which labels the fundamental irreducible representations of the O(6) group. In O(6), the E2 selection rules are $\Delta\sigma = 0$, $\Delta\tau = +/- 1$. Therefore the band head of each excited σ family should not decay. Clearly, eventually they will decay, generally by transitions preserving the τ rule, to a $\tau = 1$, $J^{\pi} = 2^+$ state. In ^{196}Pt, the most crucial transition was therefore the 1047 kev transition from the 1402 keV ($\sigma = N\text{-}2 = 4$) state to the 2^+_1 level at 355 keV. The lifetime measured allowed to set a limit on the B(E2) value that is about an order of magnitude weaker than allowed B(E2) values. This supported the goodness of the sigma quantum number at the same level.

O(6) describes the properties of even-even nuclei in this region. A close lying question was then how to understand the neighboring odd-even nuclei! It was mentioned above that S.G. Nilsson introduced in 1956 his collective nuclear model in single particle approximation for the interpretation of deformed odd even nuclei. However, in this region of Pt nuclei the equilibrium nuclear shape is changing rapidly and it became evident that a simple Nilsson model description could not be applied to the odd-even isotope ^{195}Pt, the neighbor of ^{196}Pt. A very complete gamma spectroscopic study of ^{195}Pt [WAR82] became an opportunity to investigate a – at that time – newly proposed boson-fermion formalism (IBFA) [IAC81,BAL82] in a case where the core structure (^{196}Pt) could be described analytically. The fact that the (n,γ) reaction is the experimental technique most capable of providing a guarantee of completeness was again demonstrated for ^{195}Pt [WAR82] where the studies allowed to establish a complete set of J^{π} =1/2, 3/2$^-$ states below 1500 keV excitation energy. Specifically, primary γ-rays had been measured using the ARC technique (at Brookhaven National Laboratory), while secondary γ-rays had been studied with the GAMS spectrometers at ILL. A portion of the spectrum measured with the GAMS2 spectrometer is shown in Fig. 3.8.2. The most striking feature of the level scheme constructed with the high resolution data was the large number of the J^{π} =1/2$^-$,3/2$^-$ states observed already below 1 MeV, a total of 11, too many to be explained in a simple Nilsson model. It was first pointed out by D.D. Warner that a comparison of the empirical decay scheme of ^{195}Pt with the predictions of the IBFA could offer an ideal basis within which to study the structure of ^{195}Pt: In the case of bosons with O(6) symmetry (^{196}Pt core), coupled to fermions in j=1/2, 3/2 and 5/2 orbits, the collective and single particle degrees of freedom can be described together in a single analytic framework [BAL82] which resulted in the prediction of a super symmetry, applicable to both the even-even and the odd-even nuclei. This situation corresponded very well to the negative parity orbits available to the odd neutron nucleus ^{195}Pt ($p_{3/2}$, $f_{5/2}$, $p_{1/2}$). In fact, it turned out that it seemed to be possible to establish a one to one correspondence between the low lying negative parity states in theory and experiment. It was concluded that this very first comparison of a ^{195}Pt level scheme with the predictions of multi-j super symmetry showed extremely promising agreement, but should be further corroborated.

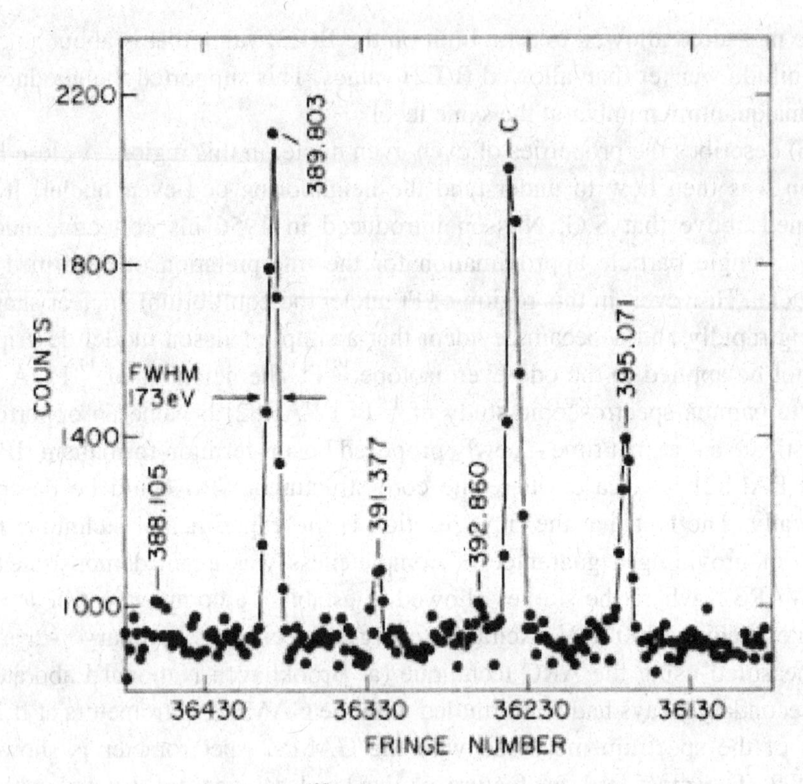

Fig. 3.8.2. Portion of the spectrum in ^{195}Pt observed with the curved crystal spectrometer GAMS2 in the second order of reflection. The peak labelled C is a contaminant from neutron capture into neighboring ^{196}Pt.

Since the first assignment of ^{196}Pt as an O(6) nucleus, the situation has further evolved: The Pt as well as the Os nuclei lie in a complex transition region which, in the past, has always been difficult to understand in a comprehensive way. Before the application of the IBA numerous degrees of freedom (prolate, oblate, hexadecapole, triaxial deformations, for example) had to be introduced in order to account for the diversity of phenomena observed. However, the evidence that ^{196}Pt corresponds to a new nuclear symmetry offered the possibility for a new understanding of this region in terms of small and progressive departures from that limiting symmetry. Also at ILL several high resolution measurements [CAS78a, CAS78b, WAR79] of gamma rays and conversion electrons in a series of Os nuclei, contributed to these new developments. The measurements in ^{192}Os and ^{194}Os are especially nice examples of how far one can push spectroscopy at high neutron flux. As both, ^{191}Os and ^{193}Os, respectively, are beta unstable, their neighboring isotopes can (in neutron capture) only be reached by successive capture of two neutrons. This so-called double neutron capture scales with the square of the neutron flux and has to compete with the beta decay time. As can be seen from the energy systematic in the Osmium isotopes with A < 194 [CAS78a, CAS78b, WAR79] become less and less deformed (decreasing $E_{4^+_1}/E_{2^+_1}$).

Starting from ^{196}Pt and decreasing in mass, the Pt-Os region was then viewed as undergoing a transition from the 0(6) limit toward (but not nearly reaching) the limit of the symmetric rotor [CAS78c]. This transition can be typified by a number of very characteristic changes which are empirically reflected in the level schemes of the even-even Os and Pt nuclei. The most striking of these changes concerns the decay modes of 0^+ states which, in Pt, de-excite to both the first and second 2^+ states whereas, in Os, they populate almost exclusively the second 2^+ level.

Another example where ILL was able to contribute to pioneering the study of symmetries in nuclei concerns lifetime measurements in nuclei which were important for the theoretical development of new "critical point" symmetries and the application of Landau Theory to the nuclear phase diagram. As already mentioned, the structure of atomic nuclei depends sensitively on the numbers of protons and neutrons in the outermost orbits (active nucleons) near the surface of the nucleus. It has been known for half a century that, in many regions of the Nuclear Chart, nuclei change structure and shape from spherical to deformed as the number of active nucleons increases. However, studies of absolute transition rates, especially in the nucleus ^{152}Sm, have radically revised our knowledge of this and similar nuclei with 90 neutrons and have shown that the structure changes abruptly at that point [CAS98]. These N = 90 nuclei can simultaneously take on both spherical and deformed shapes in different states. These observations suggest that finite nuclei can undergo phase transitional behavior as a function of the number of nucleonic constituents. To further map out the phase transition, GRID studies of lifetimes were carried out with the GAMS4 spectrometer in ^{152}Sm and in the spherical neighboring nuclei ^{148,150}Sm. In ^{150}Sm an extensive set of about 50 new transition rates were measured for the first time [BÖR06] and they provided additional knowledge for evaluating the structure in this phase/shape transition region. Figure 3.8.3 shows a typical example for a ^{150}Sm transition where the measured Doppler broadening allowed to deduce a lifetime.

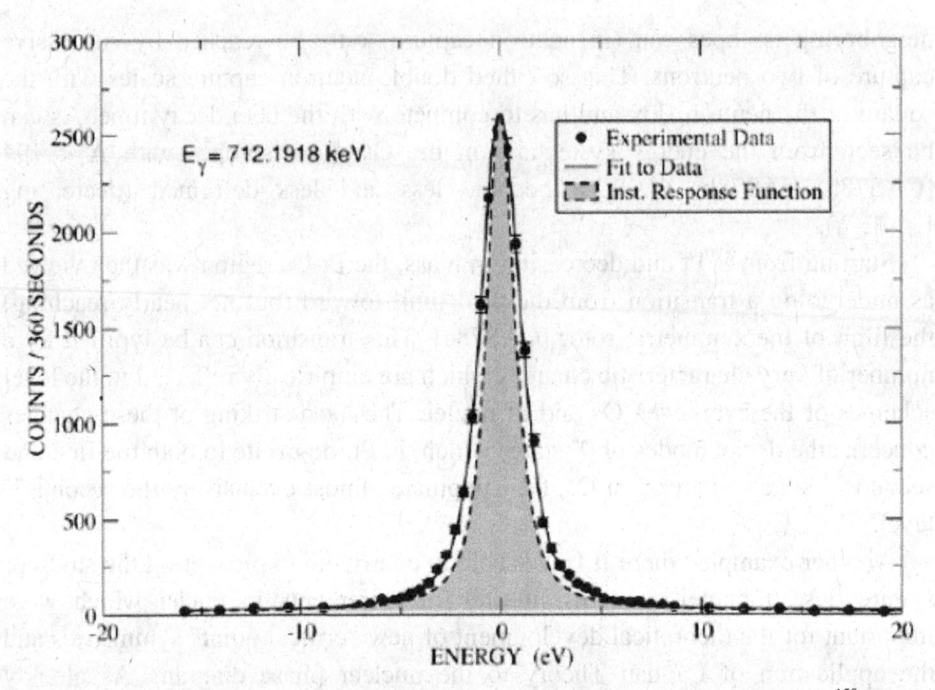

Fig. 3.8.3. The Doppler broadened line shape for the 712 keV $2_2^+ \rightarrow 2_1^+$ transition in ^{150}Sm. As usual, the dotted line shows the instrumental line width. The solid line is a fit to the data that incorporates Doppler broadening due to the finite lifetime obtained to be 1.2 ps < τ < 2.7 ps.

The new observations altered significantly the understanding of how collectivity and deformation develop in certain regions of nuclei. Instead of a gradual softening of the nucleus to deformation as valence nucleons are added, the nucleus remains essentially spherical in going away from closed shells, albeit developing greater anharmonicities, until enough correlations build up that the nucleus 'snaps' to a deformed structure [IAC98, CAS99]. Geometrically, the deformation variable β, which describes the magnitude of ellipsoidal deformation, is "bi-valued": either ~0 or fairly large, without a transitional region in between.

It has been proposed by F. Iachello that there are two possibilities for this new class of symmetry, one describing a transition from a vibrator to a deformed γ–unstable structure and the other a vibrator to axially symmetric transition region. These critical point symmetries are the first entirely new paradigms for the structure of atomic nuclei since the 1950's. Historically, nuclei in shape transitional regions have been the most difficult to treat, because they involve intense competition of different degrees of freedom. However, F. Iachello has developed a concept of symmetries, which, in contrast to previous numerical calculations, was essentially parameter-free. One of them, called X(5) [IAC01], describes first order spherical-deformed phase transitions. ^{152}Sm [ZAM02] emerges as the first empirical example [CAS98] of this symmetry. The second development is the application, by Jan Jolie, Pavel Cejnar and their colleagues [CEJ03], of the classic Landau theory of phase to the ground state structure of atomic nuclei. The resulting phase diagram transitions (illustrated in Fig. 3.8.4) shows that nuclei have three phases at low energy – spherical (denoted by $\beta = 0$), prolate deformed (American football shaped, $\beta > 0$), and oblate deformed (disc-like, $\beta < 0$) and that the first order phase transitions that separate these phases meet at the triple point, t, which is an isolated point of second order phase transition.

Whereas ^{152}Sm was the first nucleus identified to exhibit a structure close to the X(5) predictions (Fig. 3.8.5), at ILL special attention was devoted to study in detail the properties of neighboring ^{150}Sm. Precise gamma-ray energies were measured using the GAMS4 spectrometer and level lifetimes were determined using the GRID technique. Figure 3.8.6 shows a comparison of the decay rates measured for ^{150}Sm with those proposed by IBA calculations. A phonon-like structure is evident in both, the experimental spectrum and the IBA calculations.

The energies and decay strengths suggest that the 4^+_1, 2^+_2, and 0^+_2 states (Fig. 3.8.6 top) are members of the 2 quasi-phonon multiplet while the 6^+_1, 4^+_2, and 2^+_3 states (Fig. 3.8.6 bottom) are members of the 3 quasi-phonon multiplet. This establishes the ground state of ^{150}Sm as close to spherical vibrator. Taking into account the positions of the known 0^+ states it was suggested [BÖR06] that ^{150}Sm is located before the phase transitional point in the evolution from spherical to deformed structures. In conclusion we can say that ILL data contributed again to a new understanding of structural evolution in nuclear systems. These findings have raised the possibility that finite nuclei evolve in ways very close to those of systems showing real phase transitional behavior, with coexisting phases, critical points, and order parameters. The nature of collectivity and coherence in nuclei and their evolution with proton and neutron number is one of the most fundamental issues in nuclear structure. Although nuclei can change properties rapidly with changes in the number of their constituents, phase coexistence and phase transitions, as a function of nucleon number, in the condensed matter sense of these concepts, have generally been discounted in finite nuclei. The studies cited above marked a significant change in this situation: they have revealed that ^{152}Sm is a rare example of phase coexistence with a deformed ground state and a more spherical set of vibrator-like excited states. In contrary to that, the neighboring nucleus ^{150}Sm exhibits a ground state close to a spherical vibrator.

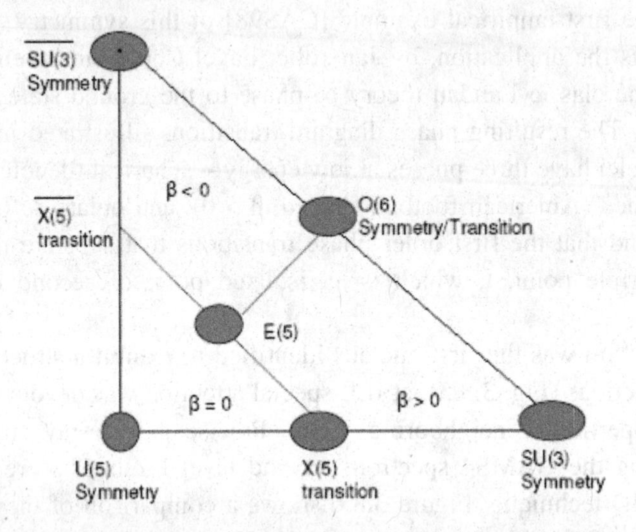

Fig. 3.8.4. Equilibrium phase diagram for nuclei. Nuclear models typically span the triangle using two variables which are analogous to pressure and temperature in Landau theory.

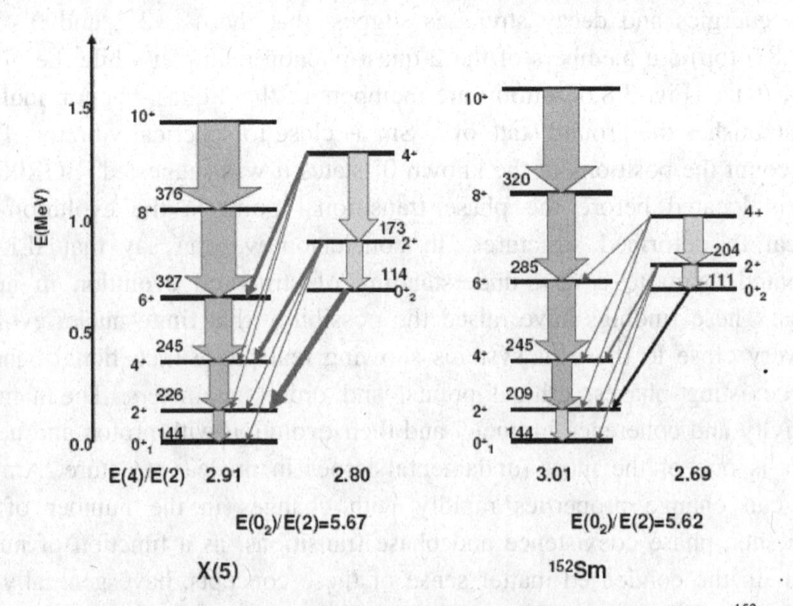

Fig. 3.8.5. A new class of symmetry: The critical point phase transitional nucleus ^{152}Sm. Partial decay scheme including transition strengths. The left side shows the predictions obtained from X(5). The right side shows the empirical results.

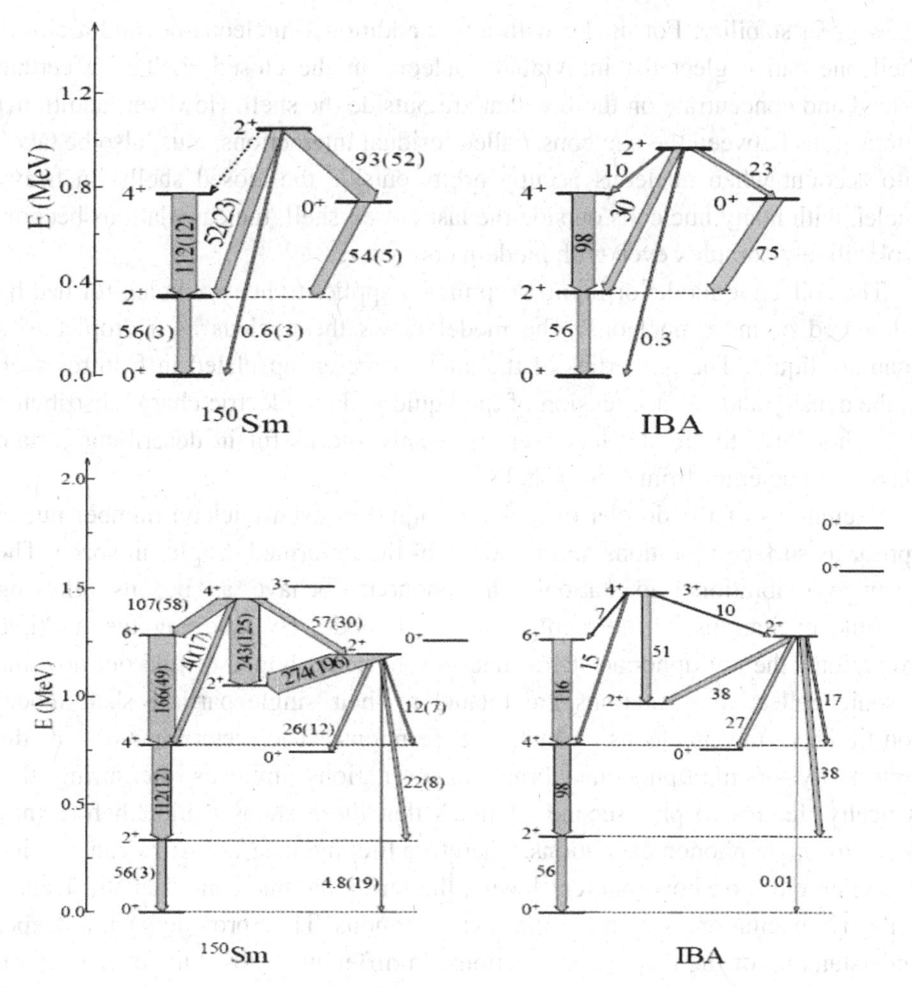

Fig. 3.8.6. Comparison of experimental (left) low-lying states in 150Sm with the IBA calculations (right).

3.8.3 *Vibrational Modes in Rare Earth Nuclei*

The shell model considers the atomic nucleus to be an ensemble of interacting neutrons and protons, where the nucleons can occupy various orbits – one for protons, one for neutrons. That this is possible in the dense atomic nucleus is due to the fermionic nature of the nucleons which, by the Pauli exclusion principle, forbids two neutrons or protons to occupy the same orbit. Nuclei with a closed shell of protons, or a closed shell of neutrons (and especially those with both) show great stability. For nuclei with a few additional nucleons beyond a closed shell one can neglect the individual nucleons in the closed shell to a certain extend and concentrate on the few that are outside the shell. However, additional interactions between the nucleons, called residual interactions, must also be taken into account when nucleons occupy orbits outside the closed shells. In heavy nuclei, with many nucleons outside the last closed shell, the calculations become prohibitively complex even with modern computers.

The collective model or liquid drop model applies to heavy nuclei, formed by a hundred or more nucleons. The model views the nucleus as a droplet of a quantum liquid. The properties of the nucleus are encapsulated in features such as the density and surface tension of the liquid and the electric charge distributed throughout it. This model has been extremely successful in describing certain classes of nuclei far from closed shells.

Excitations of the droplet of quantum liquid in even nucleon number nuclei appear as surface vibrations and rotations of the deformed droplet in space. The quantized vibrational excitations, the phonons, behave as bosons allowing multiple excitations of the same vibrational mode. By studying the multiple excitations, the multiphonon states, nuclear physicists hoped to find out how the bosonic collective excitations are related to their single particle shell model constituents, the nucleons, which are fermions. The determination of the collectivity of multiphonon vibrational excitations involves measuring the, typically, femto- to pico-second lifetimes that these states exhibit before they decay to single phonon excitations. Therefore lifetime measurements can provide important clues on how inherent loners, the fermionic nucleons, can still lead to collective excitations which are the jovial phonons. They provide such a deeper understanding of the residual interactions. In different words: One of the central challenges in nuclear physics has always been to understand the interplay between the concepts of each nucleon moving in individual quantized orbits and that of coherent excitations (e.g. vibrations or shape oscillations) of the nucleus as a whole. Despite years of studies, both experimental and theoretical, we do not yet fully understand these collective excitations. More accurately stated, the understanding we have long thought to be in hand now appears to need a major revision.

Aside from the "central" potential to which all the nucleons are subject, there are attractive residual interactions that primarily affect the outermost, or valence, nucleons. These interactions lead to correlated motions and, thereby, to collective excitation-modes of the nucleus. Macroscopically, these modes can be viewed as nuclear-shape oscillations. The best studied vibrational modes are those involving quadrupole (ellipsoidal) shapes. In spherical nuclei these are oscillations of quadrupole shape away from sphericity. In deformed nuclei, they are oscillations around the mean deformed quadrupole shape. These latter can be of two basic types, called β and γ. Beta-vibrations are oscillations that preserve the axial symmetry of the deformed ellipsoid but increase or decrease the degree of deformation. Gamma-vibrations are shape excursions perpendicular to this axis in which the nucleus momentarily takes on squashed axially asymmetric shapes. Figure 3.8.7 schematically illustrates these shape oscillations and shows the

As these nuclear level-schemes that result from them. (Note that, on top of each vibrational excitation in a deformed nucleus, there is a band of levels corresponding to an adiabatic rotation of the nucleus as a whole.) For the deformed case, there is the characteristic quantum number *K*. As β-vibrations represent oscillations in the degree of deformation they have no projection of angular momentum on the symmetry axis ($K^\pi = 0^+$), γ-vibrations are oscillations in which the nucleus takes on axially asymmetric shapes, and thus do have an angular momentum projection ($K^\pi = 2^+$).

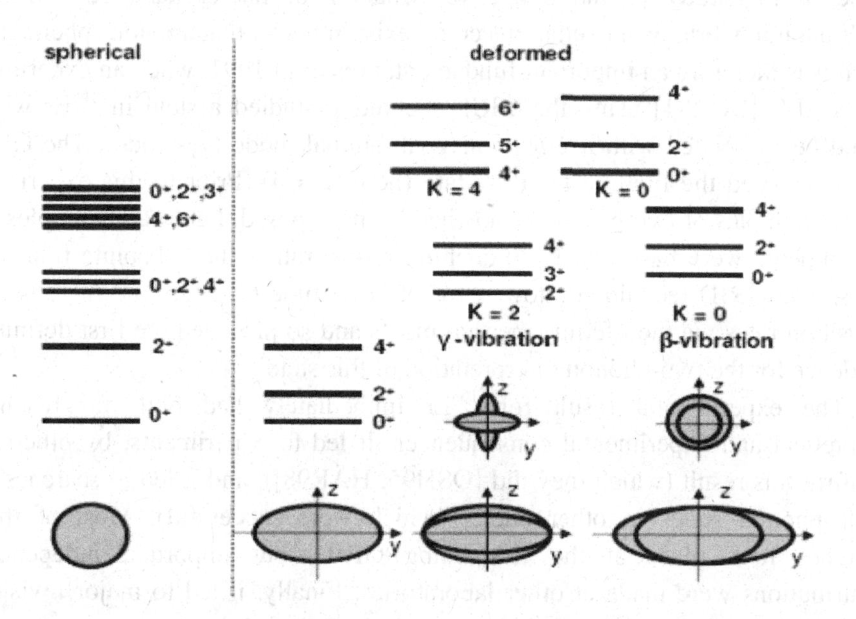

Fig. 3.8.7. Phonon and multi-phonon excitations in spherical and deformed nuclei.

As these vibrational phonons behave as bosons, one can envisage superpositions of them: Those would be the multi-phonon states. Projection quantum numbers add algebraically. Hence, for the γ-mode there are two 2-phonon excitations, with K = 0 and 4. The most telling experimental signature of these are the observation of de-excitation transitions to the single phonon band whose strength corresponds to matrix elements that are comparable to or greater than the de-excitation matrix elements from the γ-vibration to the ground-state band.

But there is a problem with this simple picture of multi-phonon states. Although the vibrations are boson-like, they are composed of fermions (protons and neutrons). Microscopically, the wave functions of these vibrations correspond to linear combinations of "particle-hole" excitations in which a particle in an occupied orbit is elevated to a vacancy in an unfilled orbit and the fermions out of which these vibrations are constructed occupy single-particle states with, in general, rather low angular momentum j. Due to the Pauli Principle, each orbit, which has 2j + 1 magnetic substrates, can only contain at most 2j + 1 identical nucleons. This has long been thought to impede significantly the creation of such modes. Indeed, for thirty years a long-standing issue in the study of the collective behavior of nuclei has been whether multiphonon vibrational excitations could exist in nuclei that are non spherical.

It was therefore an important fundamental result in 1991, when an experiment at the ILL [BÖR91] using the GRID technique studied a state in ^{168}Er which could be associated with a 2-phonon γ-vibrational mode (γγ-mode). The GRID study deduced the lifetime for this state (here K = 4). Prior to this experiment many examples of two phonon bands had been proposed, but in all examples the assignments were based on E2 branching ratios, rather than absolute transition rates. The GRID technique allowed for the first time to determine the absolute transition rates via the lifetime measurements and so provided the first definitive evidence for the two-phonon interpretation of this state.

The experimental result for ^{168}Er immediately had both far reaching theoretical and experimental consequences. It led to experiments, by others, to confirm this result (which they did [OSH95, HÄR98]), and it led to searches for multi-phonon states in other nuclei (which were successful). Most of these searches took place at the ILL using GRID, but important independent contributions were made at other laboratories. Finally, it led to major revisions and corrections to some of the theoretical models.

Experimental work, triggered by the observation of the 2-phonon state in ^{168}Er, has gone in three general directions. One has been to look for additional $\gamma\gamma$-modes. Remember: projection quantum numbers add algebraically and the first example found in ^{168}Er was of the γ-vibrational type where the *K*-quantum number (see introduction) added up to 4 (2^+_2). Such additional $\gamma\gamma$-modes with good collectivity have been found in ^{164}Dy [COR97] and ^{166}Er [FAL96, GAR97]. Secondly, there have been searches for the case where the K-quantum number for the γ-vibration adds up to 0. Without going into any detail one may say that the success in finding examples of these has perhaps had an even more profound impact. It turns out that for the lowest K = 0 excitation which was thought to be generally the β-vibration (see Fig. 3.8.7) there is evidence that in some cases it is instead this second partner of the $\gamma\gamma$-mode. This was again deduced from the pattern of the observed transition strengths – in ^{158}Gd for example –, determined via lifetime measurements. The third direction concerns the study of the β-vibration (where we have seen that some of the states at least, which previously were assigned to be β-vibrations, are instead $\gamma\gamma$-vibrations). New experiments - on ^{152}Sm [BUR02] and ^{166}Er [GAR97] by others and on ^{178}Hf [APR02] with GRID – tried to test good candidates for collective β-vibrations and ^{178}Hf revealed the first good candidate for a 2-phonon β-vibration. In other cases like ^{164}Dy and ^{168}Er the situation is more complex as several excitation modes (collective and single particle) seem to compete [LEH98]. Clearly, there is a richness in these vibrational modes heretofore unrealized. Whenever the appropriate, sufficiently sensitive, experiments have been carried out, many of them at ILL, they almost always reveal new examples of multi-phonon states. As a result of all this work most basic concepts of nuclear collective motions are undergoing radical and unexpected revision.

3.8.4 *Odd Mass Actinide Nuclei*

Studies of odd mass actinide nuclei were the first real challenges for the spectroscopy with ILL's curved crystal spectrometers. Prior to the measurements at ILL – which started around 1973 – not much was known about these nuclei. They exhibit very high level density, and the very high number of transitions arising from fission of the odd A nucleus (formed after neutron capture of the even-even target nucleus) adds furthermore to the complexity of the γ-ray spectrum. It is therefore necessary to use instruments of highest resolution and precision. Already in these first experiments the γ-ray energies measured with the

GAMS spectrometers had errors which were an order of magnitude smaller than those obtained in previous studies. Moreover, in most cases conversion electron data which are so helpful to determine spin and parities, had not at all been measured before. Additionally, isotopically enriched material is needed, and in many cases only very small amounts (at most some milligrams) are available. Therefore internal target geometry in conjunction with high sensitivity is a prerequisite for efficient gamma and conversion electron spectroscopy of actinides – exactly what had been made available at ILL. As stressed before, the non-selective (n,γ) reaction populates all low-lying levels with spins not too different from that of the capture state ($1/2^+$ in most cases for the odd A nuclei), regardless of their single-particle or collective character. Therefore the configurations of special interest in these early studies concerned those resulting from the coupling of one-phonon vibrational states of the core (see also section 3.8.3) to excited Nilsson single-particle states. These vibrational states are expected at energies close to those in the neighboring even-even nuclei, i.e. from about 700 keV to 1MeV. An outstanding result of these measurements were the first observations of E0 admixtures (characterized through the observed intensity patterns of the respective conversion coefficients measured) in transitions de-exciting specific levels in these nuclei. Most of these levels were then interpreted as $K^\pi = 0^+$ vibrational components coupled to the low lying Nilsson single-particle states, namely 1/2[631] and 5/2[622], respectively. Evidence for $K^\pi = 2^+$ phonons (via the observation of strong E2 admixtures) coupled to the same single particle states and additionally to 7/2[743] was found as well. Finally, octupole vibrational states with $K^\pi = 0^-$ phonons were assigned build on the 1/2[631], 3/2[631], 5/2[622] and 7/2[743] configurations, respectively. Most of these assignments were done for the first time in these nuclei. Extended studies of vibrational excitations in odd A actinides were carried out for ^{227}Ra [EGI81], ^{231}Th [WHI87], ^{233}Th [JEU79], ^{235}U [ALM79], ^{239}U [BÖR78], ^{241}Pu [WHI98], and ^{249}Cm [HOF82]. More details may also be found in [EGI79].

3.8.5 *High Resolution Measurements in ^{176}Lu*

As pointed out earlier, the main nucleosynthesis mechanisms for the heavy elements in the mass region A ≥ 56 are neutron capture reactions in both, the r and s processes, respectively. That means, that they are essentially produced by successive neutron captures, either during the Red Giant stage of stellar evolution (s process) or, in supernovae, when massive stars end their life in a gigantic explosion (at extremely high neutron fluxes = r process). Consequently, the reaction path of the r process is shifted to the region of very neutron rich nuclei close to the neutron drip line. After termination of the r process the reaction products decay back to the stability valley. The origin of ^{176}Lu (and of about 30 other nuclei) however, can be completely ascribed to Red Giants, where a modest neutron flux (slow = s) is produced in the He burning zones deep inside the star. For mass 176, the β-decay chain from the r-process region is terminated at stable ^{176}Yb, such shielding ^{176}Lu from population in the r process. The fact that ^{176}Lu is a naturally occurring unstable nucleus with a half-life of 36 billion years, and that its original *s* abundance yield can be quantitatively described, triggered several attempts to determine the age of the s elements, since the decay of ^{176}Lu can be considered as a perfect clock [AUD72, ARN73]. Such an analysis could, in principle, be performed by comparing the original *s* production of ^{176}Lu with its abundance today. It turned out, however, that these attempts encountered an obstacle. While a perfect clock under laboratory conditions, the decay of ^{176}Lu was suspected to change at the extremely high temperatures inside Red Giants. The reason for this exotic behaviour lies in the particular structure of this isotope which exhibits two families of nuclear states of very different type, one connected to the long-lived ground state, and a second related to an isomeric state which decays with a half-life of only 3.68 h. Under normal conditions, these families are completely separated due to their big differences in spin and K-values: The ground state has spin 7^-, K=7, the isomeric state has 1^-, K=0. Therefore direct electromagnetic transitions are inhibited by selection rules. The question that challenged ^{176}Lu as a suitable clock was whether thermal photons, which are very abundant and fairly energetic at the high s-process temperatures, may provide a link between the families, thus reducing the overall half-life of the ground state (see Fig. 3.8.8). Such additional feeding was shown to occur in the hot stellar photon bath [BEE81] via thermally induced transitions from the isomer to a mediating state at higher excitation with a decay branch to the ground state. The experimental evidence for such a state was found by using high resolution gamma spectroscopy, using the GAMS1 and GAMS2/3 spectrometers, which allowed to study the two families of states in ^{176}Lu with unsurpassed sensitivity [KLA91a, KLA91b]. And indeed, this study revealed a hitherto unknown link that is activated at 150 million degrees, well below the 300 million degrees at the *s*-process site. This discovery had drastic consequences: While the half-life of ^{176}Lu remained at its laboratory value up to 150 million degrees, it drops to about 1 year at its production site in Red Giants. This behaviour corresponds to a clock which works reliably for most of the time, but runs much faster in certain circumstances. It also turned out, that the coupling between isomer and ground state is not only determined by the position of the mediating state at 838 keV with $I^{\pi} = 5^{-}$, but also by the branching ratio and, in particular, by its half life. A GRID measurement, using the GAMS4 spectrometer allowed to set a lower limit for the lifetime of the 838 keV state of $\tau \geq 10$ ps [DOL99], providing a more stringent constraint for the thermal equilibration between the long-lived ground state and the short-lived K isomer at 123 keV, and, hence, for the resulting estimate of the temperature during shell helium burning in red giant stars. The present-day ^{176}Lu abundance compared to the fraction that survived the high temperature phases can now be used to obtain a realistic estimate of the temperature. In this way one finds that ^{176}Lu is produced at temperatures between 200 and 300 million degrees. This is an important information in itself because it provides a probe for the interior of Red Giant stars completely independent of the yet uncertain stellar models.

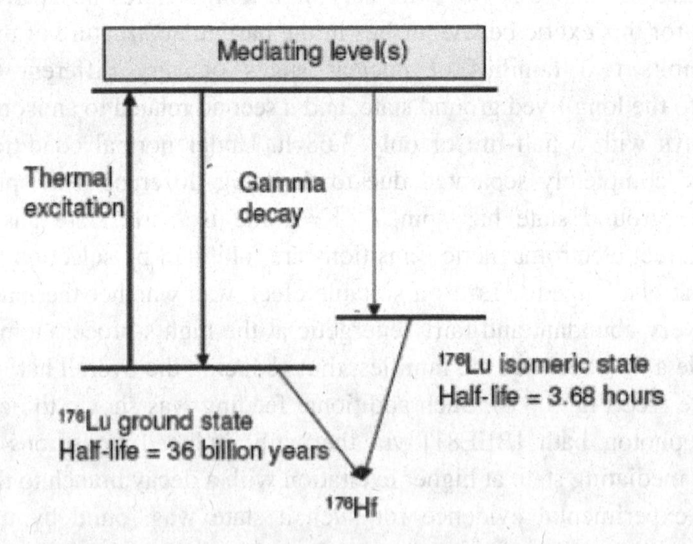

Fig. 3.8.8. The hot photon bath provides a link between the isomer and the ground state of ^{176}Lu.

3.8.6 *Spectroscopy of Fission Products*

The energy spectra of nuclei depend strongly on whether they have even or odd numbers of neutrons and protons. The reason is that each species likes to come in pairs. Nuclei which are even in both, neutrons and protons, have less excited states at low energies. In nuclei with one unpaired nucleon (either neutron or proton) one nucleon does not have a partner and that leads to more excited states. As in doubly odd nuclei, both, the neutron and the proton are unpaired, the situation is even more complicated. In some regions of the nuclear chart one encounters an accumulation of especially long lived excited states, the so called isomeric states. The reason for the existence of isomers is due to a big difference in spin of the initial and final states in a given nucleus. This is a hindrance for gamma transitions with low multipolarity.

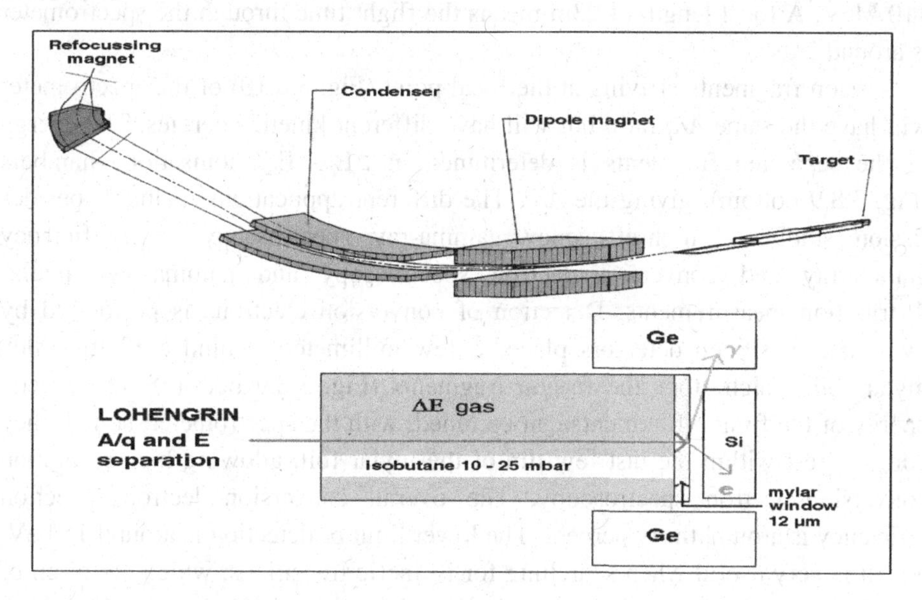

Fig. 3.8.9. Schematic drawing of flight path and detection system for studies of micro second isomers of neutron rich nuclei produced in neutron induced fission at the LOHENGRIN spectrometer.

The Lohengrin recoil spectrometer [MOL77] for unslowed fission products, whose construction was instigated by P. Armbruster, has not only contributed to research in fission (see chapter 4) but also to spectroscopy of very neutron-rich nuclei. The analysis of fission fragments on their flight path through magnetic and electric fields (Fig. 3.8.9 top) ensures high mass resolution with a clear separation between fragments with different mass numbers, or more precisely with different ratios of mass number *A* to ionic charge *q*. Fragments with the same ratio *A*/*q* are focussed along parabolas, with different kinetic energies. More details will be explained in chapter 4. Rather high fission count rates allow searches for very rare events. Fragment yields have been analysed down to the level of 10^{-8}/fission and have, for example, led to the discovery of new neutron-rich isotopes in Fe, Co and Ni. The useful range of atomic masses available at the separator for gamma-ray spectroscopy studies, is approximately70 to 170. These limits are set only by yield distributions from fission. Various fissile targets are available, from ^{229}Th to ^{251}Cf, and targets can be chosen to optimise the production rate for a particular nucleus of interest. The kinetic energies of the fragments passing through the spectrometer varies between about 50 and 110 MeV. A focal length of 23m means the flight time through the spectrometer is around 2 μs.

Fission fragments arriving at the focal point (Fig. 3.8.10) of the spectrometer will have the same *A/q* ratio but will have different kinetic energies. The energy of the separated fragments is determined in ΔE - E_{rest} ionisation chambers (Fig. 3.8.9 bottom), giving the *A/q*. The different applications include - besides fission studies – high-efficiency gamma-ray spectroscopy, high-efficiency gamma-ray and conversion-electron spectroscopy and gamma-ray angular distribution measurements. Detection of conversion electrons is performed by two adjacent silicon detectors placed a few millimeters behind a 12 μm thick mylar foil, which stops the fission fragments (Fig. 3.8.9 bottom). The kinetic energy of the fission fragments can be tuned, with the spectrometer, so that they come to rest within the last few μm of the mylar foil, allowing high-resolution conversion-electron spectroscopy. The overall conversion-electron detection efficiency is around thirty percent. The lower limit of detection is around 15 keV, which is very useful when searching for isomeric transitions, which are often of low energy. The gamma-ray multiplicity in experiments is generally low, allowing few detectors to be used. For isomers with μs lifetimes the correlation between the arrival time of the fission fragment in the ionisation chamber and the detection of a gamma ray, or conversion electron, is sufficient to measure the isomeric lifetime.

The best technique to obtain information on excited nuclear states of neutron-rich nuclei far from stability, at Lohengrin, is to observe the decay of states below μs isomers. The transit time of approximately 2 μs through the spectrometer currently allows isomeric states with lifetimes as short as about 0.5 μs to be observed. Isomeric states are present across the whole landscape of the nuclear chart resulting from both single particle effects (such as yrast spin traps) and collective motion (*K*-isomers).

Putting a time window of several tens of μs between the detection of the arrival of a fission fragment and the detection of a gamma ray, plus the ability to

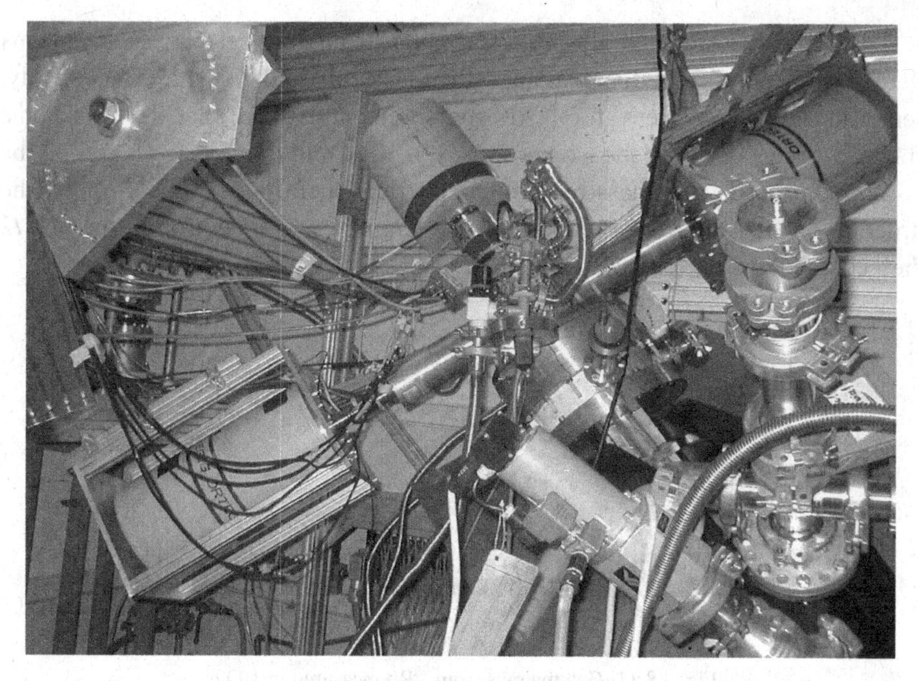

Fig. 3.8.10. A detector set-up at the LOHENGRIN focal point with two 60% Ge-detectors and one Si-detector in position.

select individual masses (or in favourable cases individual isotopes) gives very clean spectra. The cleanliness of these spectra allows transitions to be identified and assigned to nuclei with just tens of gamma-ray photo peak counts. The same is true for conversion electrons. The ratio of K- to γ-lines gives a measure of the multipolarity of the transition. Isomers are useful not only because they allow measurements of the properties of weakly produced nuclei but because these states have unique a character among many other excited states, giving valuable nuclear structure information. Isomers are generally at higher angular momentum than ground states, giving isomer spectroscopy of odd-odd nuclei a distinct advantage over beta-decay spectroscopy, where only low-spin states are populated, as intermediate-spin states cannot be accessed.

A very impressive series of gamma-ray spectroscopy experiments using isomeric states has been performed by the LPSC Grenoble group in collaboration with ILL. New isomers were discovered around the mass 100 region including ^{94}Y, ^{96}Rb [PIN05] and ^{106}Nb [GEN99]. In the case of ^{96}Rb this was the first measurement of any excited states in this very neutron-rich nucleus. Big progress in sensitivity was obtained after the year 2004 when new detector arrangements were installed at the focal point (Figs. 3.8.11 and 3.8.12, respectively, demonstrate the progress for ^{96}Rb measurements). The first level scheme obtained for the decay of the isomer in ^{96}Rb is shown in Fig. 3.8.13. Based on the assumption that in this mass region the spherical shell model should apply, the structure of the isomer was proposed to be composed of a neutron in the h 11/2 shell coupled to a proton hole in the f 5/2 shell.

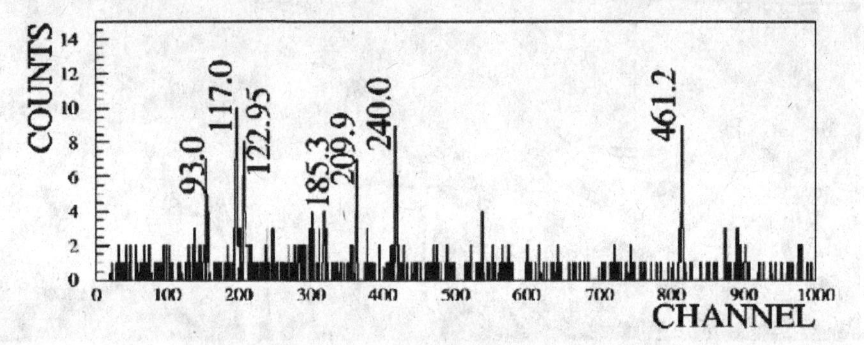

Fig. 3.8.11. Gamma rays from ^{96}Rb measured in 1999.

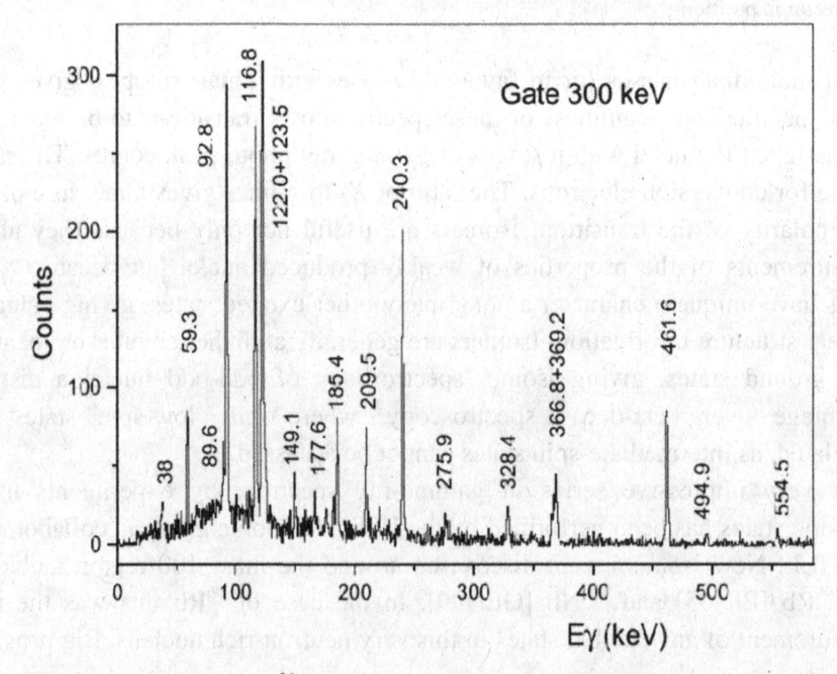

Fig. 3.8.12. Gamma rays from ^{96}Rb measured in 2004 with a modern detector arrangement.

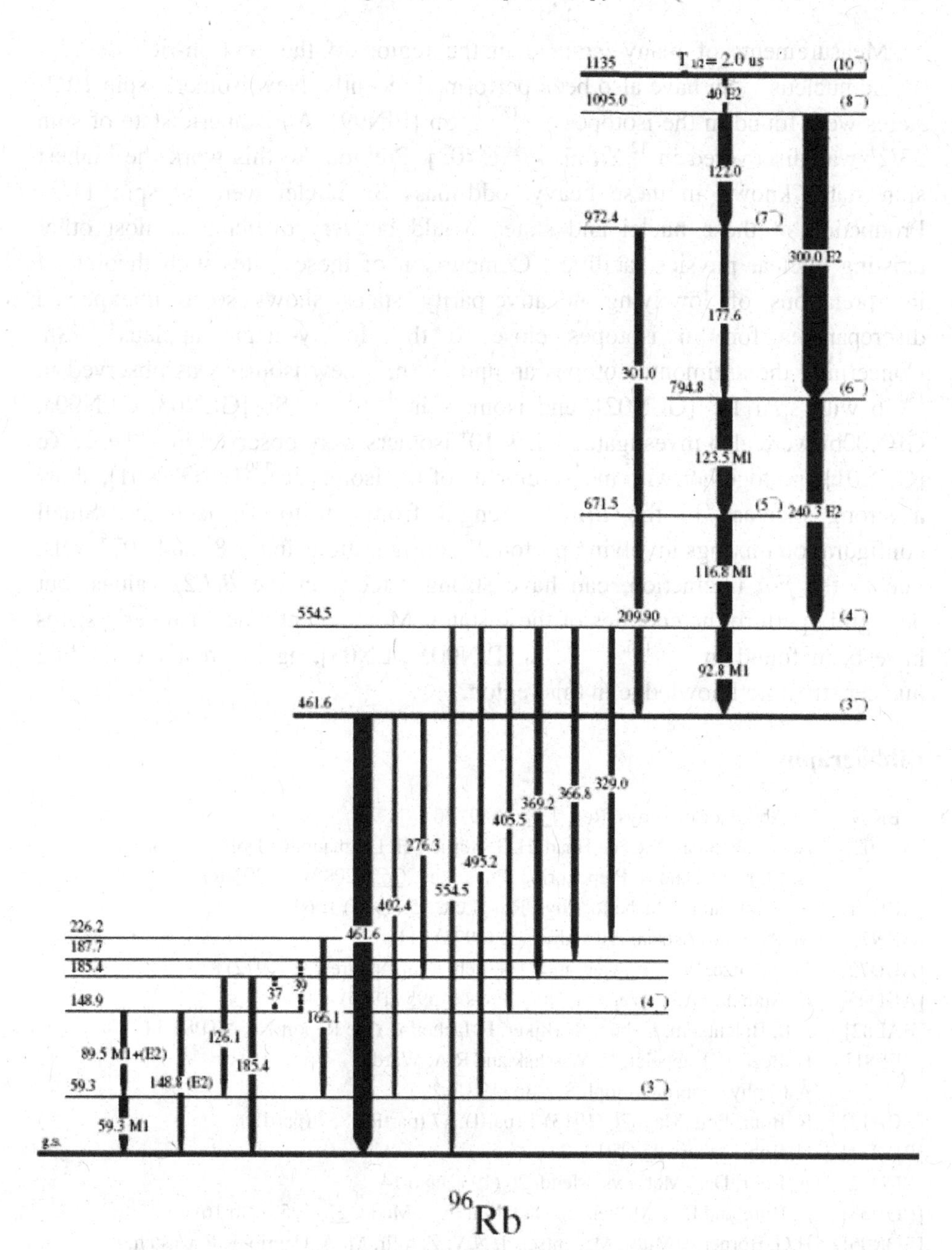

Fig. 3.8.13. Decay scheme of the 2 μs isomer in ^{96}Rb. The low-lying levels and the isomer at 1135 keV have rather spherical configuration but a rotational band develops at 460 keV. The level scheme is based on γ-γ and e-γ coincidences.

Measurements of many isomers in the region of the neutron-rich doubly-magic nucleus ^{132}Sn have also been performed recently. New isomeric spin 19/2$^+$ states were found in the isotopes $^{125, 127, 129}$Sn [PIN99]. An isomeric state of spin 23/2$^+$ was discovered in ^{129}Sn also [GEN02]. Previous to this work the highest spin states known in these heavy, odd-mass Sn nuclei were of spin 11/2$^-$. Production of these nuclei and states would be very difficult at most other existing nuclear physics facilities. Comparison of these states with theoretical interpretations of low-lying negative-parity states shows some unexpected discrepancies for Sn isotopes close to the doubly-magic nucleus ^{132}Sn. Concerning the antimony isotopes around ^{132}Sn, a new isomer was observed in ^{130}Sb with spin 13$^+$ [GEN02], and isomers in $^{129, 131, 133}$Sb [GEN03, GEN00a, GEN00b] were also investigated. New 10$^+$ isomers were observed in ^{132}Te, ^{134}Xe [GEN01] and together with measurements of the isomer in ^{130}Te [GEN01], show a strong increase in the *B(E2)* strength from Sn to Te isotones. Small configuration mixings involving proton 2$^+$ configurations in the 8$^+$ and 10$^+$ levels, due to the *p-n* interaction, can have strong effects on the *B(E2)* values, but negligibly perturb the energies of these states. More recently new isomeric states have been found in $^{123, 125, 127, 129}$In [PIN00, GEN03], again greatly extending nuclear structure knowledge in this region.

Bibliography

[ABR69] A. Abrahamson, Phys. Rev. 178 (1969) 76

[APR02] A. Aprahamian, R.C. de Haan, H.G. Börner, H. Lehmann, C. Doll, M. Jentschel, A. M. Bruce, and R. Piepenbring, Phys. Rev. C65 (2002) 031301(R)

[ARI75] A. Arima and F. Iachello, Phys. Rev. Lett. 35 (1975) 1069

[ARN73] M. Arnould, Astron. Astrophys. 22 (1973) 311

[AUD72] J. Audouze, W.A. Fowler, and D.N. Schramm, Nature 238 (1972) 8

[AUD95] G. Audi and A.H. Wapstra, Nucl. Phys. A595 (1995) 409

[BAL82] A.B. Balentekin, I. Bars, R. Bijker, F. Iachello, Yale Report No. YTP82-11

[BEE81] H. Beer, F. Käppeler, K. Wisshak and R.A. Ward, Astrophys. Journal Suppl. Ser. 46 (1981) 295

[BOH13] N. Bohr, Phil. Mag. 26 (1913) 1 (partI), 47 (partII), 857 (partIII)

[BOH14] N. Bohr, Nature 92 (1914) 231

[BOH52] A. Bohr, Dan. Mat. Fys. Medd. 26 (1952) No 14

[BOH53] A. Bohr and B.R. Mottelson, Dan. Mat. Fys. Medd. 27 (1953) No 16

[BÖR06] H.G. Börner P. Mutti, M. Jentschel, N.V. Zamfir, M. A. Caprio, R.F. Casten, E.A. Mc Cutchan, and R. Kruecken, Phys. Rev. C73 (2006) 034314

[BÖR78] H.G. Börner, H.R. Koch, H. Seyfarth, T. v. Egidy, W. Mampe, J.A. Pinston, K. Schreckenbach, D. Heck, Z. Phys. A286 (1978) 31

[BÖR88] H.G. Börner, J. Jolie, F. Hoyler, S.J. Robinson, M.S. Dewey, G.L. Greene, E.G. Kessler, R.D. Deslattes, Phys. Lett. B215 (1988) 45

[BÖR90] H.G. Börner, J. Jolie, S.J. Robinson, R.F. Casten, J.A. Cizewski, Phys. Rev. C42 (1990) R2271

[BÖR91] H.G. Börner, J.Jolie, S.J. Robinson, B. Krusche, R. Piepenbring, R.F. Casten, A. Abrahamian, J.P. Draayer, Phys. Rev. Lett. 66 (1991) 691 and 2837

[BÖR93] H.G. Börner and J. Jolie, J. Phys. G19 (1993) 217

[BUR02] D.G. Burke, Phys. Rev. C66 (2002) 024312

[CAS78a] R.F. Casten A. I. Namenson, W. F. Davidson, D. D. Warner, H. G. Börner, Phys. Lett. 76B (1978) 280

[CAS78b] R.F. Casten, H.G. Börner, J.A. Pinston, W.F. Davidson, Nucl. Phys. A309 (1978) 206

[CAS78c] R.F. Casten and J. Cizewski, Nucl. Phys. A309 (1978) 477

[CAS80] R.F. Casten, G.J. Smith, M.R. Macphail, D. Breitig, W.R. Kane, M.L. Stelts, S.F. Mughabghab, J.A. Cizewski, H. G. Börner, W.F. Davidson, K. Schreckenbach Phys. Rev. C21 (1980) 65

[CAS98] R.F. Casten et al., Phys. Rev. C57 (1998) R1553

[CAS99] R.F. Casten, D. Kusnezov, and N.V. Zamfir, Phys. Rev. Lett. 82 (1999)5000

[CAU32] Y. Cauchois, J. Phys. Rad. 3 (1932) 320

[CAU34] Y. Cauchois, Ann. Phys. I (1934) 215

[CEJ03] P. Cejnar, S. Heinze, J. Jolie, Phys. Rev. C68 (2003) 034326

[CHA32] J. Chadwick, Nature 129 (1932) 312

[CIZ78] J.A. Cizewski, R.F. Casten, G.J. Smith, M.L. Stelts, W.R. Kane, H.G. Börner, W.F. Davidson, Phys. Rev. Lett. 40 (1978) 167

[CIZ79] J.A. Cizewski, R.F. Casten, G.J. Smith, M.R. Macphail, M.L. Stelts, W.R. Kane, H.G. Börner, W.F. Davidson, , Nucl. Phys. A323 (1979) 349

[COH87] E.R. Cohen and B.N. Taylor. Rev. Mod. Phys. 59 (1987) 1121

[COL85] G. Colvin and K. Schreckenbach, AIP Conf. Proc. No. 125(1985)290

[COR97] F. Corminboeuf, J. Jolie, H. Lehmann, F. Foehl, F. Hoyler, H.G. Börner, C. Doll, P.E. Garrett, Phys. Rev. C56 (1997) R1201

[DAV81] W.F. Davidson, D.D. Warner, R.F. Casten, K. Schreckembach, H.G. Börner, J. Simic, M. Stoyanovic, M. Bogdanovic, S. Koicki, W. Gelletly, G.B. Orr, M.L. Stelts, J. Phys. G7 (1981) 455 and 843

[DAW93] M.S. Daw and M.I. Baskes, Phys. Rev. Lett. 50 (1993) 1285

[DEW06] M. S. Dewey, E. G. Kessler Jr, R.D. Deslattes, H.G. Börner, M. Jentschel, C. Doll, and P. Mutti, Phys. Rev. C73 (2006) 044303

[DIF94] F. DiFilippo, V. Natarajan, K.R. Boyce, D.E. Pritchard, Phys. Rev. Lett. 73 (1994) 1481

[DOL00] Ch. Doll, H.G. Börner, T. von Egidy, H. Fujimoto, M. Jentschel, H. Lehmann, JRNIST 105 (2000) 167

[DOL99] C. Doll, H.G. Börner, S. Jaag, F. Kaeppler, Phys. Rev. C59 (1999) 492

[DUM47] J.W.M. DuMond, Rev. Sci. Instrum. 18 (1947) 626

[EGI79] T. von Egidy, J. Almeida, G. Barreau, H.G. Börner, W.F. Davidson, R.W. Hoff,

P. Jeuch, K. Schreckenbach, D.D. Warner, D.H. White, Phys. Lett. 81B (1979) 281
[EGI81] T. von Egidy, G. Barreau, H.G. Börner, W.F. Davidson, J. Larysz, D.D. Warner, P.H.M. van Assche, K. Nybo, T.F. Thersteinsen, G. Lovhoiden, E.R. Flynn, J.A. Cizewski, R.K. Sheline, D. Decman, D.G. Burke, G. Sletten, N. Kaffrell, W. Kurcewicz, T. Björnstad, G. Nyman, Nucl. Phys. A365 (1981) 26
[EGI84] T. von Egidy, H. Daniel, P. Hungerford, H.H. Schmidt, K.P. Lieb, B. Krusche, S.A. Kerr, G. Barreau, H.G. Börner, R. Brissot, C. Hofmeyr, R. Rascher, J. Phys. G10 (1984) 221
[FAL96] C. Falander et al., Phys. Lett. B388 (1996) 475
[FAR95] D.L. Farnham et al., Phys. Rev. Lett. 75 (1995) 3598
[FER34a] E. Fermi, Z. Phys. 88 (1934) 161
[FER34b] E. Fermi, Nuovo Cim. 11 (1934) 1
[FER46] E. Fermi, Proc. Am. Phys. Soc. 90 (1946) 20
[FOW84] W. A. Fowler, Rev. Mod. Phys. 56 (1984) 149
[GAB07] G. Gabrielse, D. Hanneke, T. Kinoshita, M. Nio, and B. Odon, Phys. Rev. Lett. 99 (2007) 039902
[GAR97] P.E. Garret, M. Kadi, Min Li, C.A. McGrath, V. Sorokin, Minfang Yeh, S.W. Yates, Phys. Rev. Lett. 78 (1997) 4545
[GEN00a] J. Genevey, J.A. Pinston, H. Faust, C. Foin, S. Oberstedt, M. Rejmund, Eur. Phys. J. **A 9** (2000) 191
[GEN00b] J. Genevey J.A. Pinston, H. Faust, C. Foin, S. Oberstedt, B. Weiss, Eur. Phys. J **A 7** (2000) 463-465.
[GEN01] J. Genevey, C. Foin, M. Rejmund, R.F. Casten, H. Faust, S. Oberstedt, Phys. Rev. **C 63** (2001) 054315.
[GEN02] J. Genevey C. Foin, M. Rejmund, H. Faust, B. Weiss, Phys. Rev. **C 65** (2002) 034322.
[GEN03] J. Genevey, J.A. Pinston, H. Faust, R. Orlando, A. Scherillo, G.S. Simpson, I.S. Tsekhanovic, A. Covello, A. Gargano, W. Urban, Phys. Rev. C67 (2003) 054312.
[GEN99] J. Genevey, F. Ibrahim, J.A. Pinston, H. Faust, T. Friedrichs, M. Gross, S. Oberstedt, Phys. Rev. C **59** (1999) 82-89.
[GOL83] R.Golub et al., Z. Phys. B51 (1983) 187
[GÖP49] M. Göppert-Mayer, Phys. Rev. 75 (1949) 1969
[GRE86] G.L. Greene, E.G. Kessler, R.D. Deslattes, and H.G. Börner, Phys. Rev. Lett. 56 (1986) 819
[GRE91] G.L. Greene, M.S. Dewey, E.G. Kessler jr, E. Fischbach, Phys. Rev. D44 (1991) R2216
[HAG68] R.S. Hager and E.C. Seltzer, Nucl. Data Tab. A4 (1968) 1
[HAH39] O. Hahn and F. Strassmann, Naturwissenschaft 27 (1939) 11
[HÄR98] T. Härtlein, M. Heinebrodt, D. Schwalm, C. Fahlander, Eur. Phys. J. A2 (1998) 253
[HAX49] O. Haxel, J.H.D. Jensen, and H.E. Suess, Phys. Rev. 75 (1949) 1766
[HOF82] R.W. Hoff, W.F. Davidson, D.D. Warner, H.G. Börner, T. von Egidy Physical Review C25 (1982) 2232
[HUN83a] P. Hungerford, T. Von Egidy, H.H. Schmidt, S.A. Kerr, H.G. Börner, E. Monnand

Z. Phys. A313 (1983) 339
[HUN83b] P. Hungerford, T. Von Egidy, H.H. Schmidt, S.A. Kerr, H.G. Börner, E. Monnand Z. Phys. A313 (1983) 325
[IAC00] F. Iachello, Phys. Rev. Lett. 85 (2000) 3580
[IAC01] F. Iachello, Phys. Rev. Lett. 87 (2001) 052502
[IAC74] F. Iachello and A. Arima, Phys. Lett. B53 (1974) 309
[IAC81] F. Iachello and S. Kuyucah, Ann. Phys. (N.Y.) 136 (1981) 19
[IAC98] F. Iachello, N.V. Zamfir, and R.F. Casten, Phys. Rev. Lett. 81 (1998) 1191
[JEN96] M. Jentschel, K.H. Heinig, H.G. Börner, J. Jolie, E.G. Kessler, Nucl. Instr. Meth. B115 (1996) 446
[JEU79] P. Jeuch, T. von Egidy, K. Schreckenbach, W. Mampe, H.G. Börner, W.F. Davidson, J.A. Pinston, R. Roussille, R.C. Greenwood, R.E. Chrien, Nucl. Phys. A317 (1979) 363
[JOL97] J. Jolie, N. Stritt, H.G.Börner, Ch. Doll, M.Jentschel, S.J. Robinson, and E.G. Kessler, Z. Physik B102 (1997)1
[KAZ60] A.H. Kazi et al., Rev. Sci. Instrum. 31 (1960) 983
[KEI91] J. Keinonen, A. Kuronen, P. Tikkanen, H.G. Börner, J. Jolie, S. Ulbig, E.G. Kessler, R.M. Nieminen, M.J. Puska, A.P. Seitsonen, Phys. Rev. Lett. 67 (1991) 3692
[KER81] S.A. Kerr and E. Monnand, Annex ann. Report ILL (1981) 42
[KES01] E.G. Kessler et al., Nucl. Instr. Meth. A457 (2001) 187
[KES99] E.G. Kessler Jr., M.S. Dewey, R.D. Deslattes, A. Henins, H.G. Börner, M. Jentschel, C. Doll and H. Lehmann, Phys. Lett. A255 (1999) 221
[KLA81] H. V. Klaptor et al., Z. Phys. A299 (1981) 213
[KLA91a] N. Klay et al., Phys. Rev. C44 (1991) 2801
[KLA91b] N. Klay et al., Phys. Rev. C44 (1991) 2839
[KOC78] H.R. Koch, Jül-Spez-10 (1978) Zentralbibliothek, Kernforschungsanlage Jülich
[KOC80] H.R. Koch, H.G. Börner, J.A. Pinston, W.F. Davidson, R. Roussille, J.C. Faudou, O.W.B. Schult, Nucl. Instr. Meth. 175 (1980) 401
[KRU82] B. Krusche, K.P. Lieb, H. Daniel, T. Von Egidy, G. Barreau, H.G. Börner, R.Brissot, C. Hofmeyr, Nucl. Phys. A386 (1982) 245
[KRU84] B. Krusche, K.P. Lieb, L. Ziegeler, H. Daniel, T. Von Egidy, R. Rascher, G. Barreau, H.G. Börner, D.D. Warner, Nucl. Phys. A417 (1984) 231
[KRU85] B. Krusche, C. Winter, K.P. Lieb, P. Hungerford, H.H. Schmidt, T. v. Egidy, H.J. Scheerer, S.A. Kerr, H.G. Börner, Nucl. Phys. A439 (1985) 219
[KUR95] A. Kuronen, J. Keinonen, H.G. Börner, and J.Jolie, Phys. Rev. B52 (1995)12640
[LEH98] H. Lehmann et al., Phys. Rev. C57 (1998) 569
[MAM78] W. Mampe et al. , Nucl. Instr. 154 (1978) 127
[MEI24] L. Meitner, Z. Phys. 26 (1924) 169
[MEI39] L. Meitner and O.R. Frisch, Nature 143 (1939) 239
[MOH05] P.J. Mohr and B.N. Taylor, Rev. Mod. Phys. 77 (2005) 1
[MOL77] E. Moll *et al.*, Kerntechnik 19 (1977) 374
[NAT93] V. Natarajan, K.R. Boyce, F. DiFilippo, D.E. Pritchard,

Phys. Rev. Lett. 71 (1993) 1998
[NIL55] S.G. Nilsson, Dan. Mat. Fys. Medd. 29 (1955) No 16
[NOL79] P. J. Nolan and J.F. Sharpey-Schaffer, Rep. Prog. Phys. 42 (1979)
[OSH95] M. Oshima et al., Phys. Rev. C53 (1995) 3492
[PDG06] Particle Data Group's Review, J. Phys. G33 (2006) 1
[PET84] D.Petrascheck, H. Rauch, Acta Cryst. A40 (1984) 445
[PIN00] J.A. Pinston *et al.*, Phys. Rev. C61 (2000) 024312
[PIN05] J.A. Pinston, J. Genevey, R. Orlandi, A. Scherillo, G.S. Simpson, I. Tsekhanovich, W. Urban, H. Faust, N. Warr, Phys. Rev. C71 (2005)064327
[POR56] C. E. Porter and G.E. Thomas, Phys. Rev. 104 (1956) 483
[RAI04] S. Rainville, J.K. Thomson, and D.E. Pritchard, Science 303 (2004) 334
[RAI05] S. Rainville J.K. Thompson, E.G. Myers, J.M. Brown, M.S. Dewey, E.G. Kessler Jr., R.D. Deslattes, H.G. Börner, M. Jentschel, P Mutti, D.E. Pritchard, Nature 438 (2005) 1096
[RAI50] J. Rainwater, Phys. Rev. 79 (1950) 432
[ROB92] S.J. Robinson and J.Jolie, ILL internal Report RO15T, 1992
[RÖS78] F. Rösel et al., At. Data Nucl. Data Tab. 21 (1978) 292
[RUT11] E. Rutherford, Phil. Mag. 6 (1911) 669
[SCH82] H.H. Schmidt, P. Hungerford, H. Daniel, T. von Egidy, S.A. Kerr, R. Brissot, G. Barreau, H.G. Börner, C. Hofmeyer, K.P. Lieb, Phys. Rev. C25 (1982) 2888
[STE 75] A. Steyerl, Nucl. Instr. 125 (1975) 461
[STR00] N. Stritt, J. Jolie, M.Jentschel, H.G. Börner and Ch. Doll, J. Res. Natl. Inst. Stand. Technol. 105 (2000)71
[STR97] N. Stritt, J. Jolie, M.Jentschel and H.G. Börner, Phys. Rev. Lett. 78 (1997) 2592
[STR98] N. Stritt, J. Jolie, M.Jentschel, H.G. Börner and Ch. Doll, Phys. Rev. B58 (1998) 2603
[STR99a] N. Stritt, J. Jolie, M. Jentschel, H.G. Börner, C. Doll, Phys. Rev. B59 (1999) 6762
[STR99b] N. Stritt, J. Jolie, M. Jentschel, H.G. Börner and H. Lehmann, Phys. Rev. B60 (1999) 6476
[TAL93] I. Talmi, Simple Models of Complex Nuclei, Harwood Academic Publishers, ISBN 3718605311
[THI79] F. K. Thielemann et al., Astron. Astrophys. 74 (1979) 175
[UDE97] T. Udem, A. Huber, B. Gross, J. Reichert, M. Prevedelli, M. Weitz, T. W. Hänscht, Phys. Rev. Lett. 79 (1997) 2646
[WAR79] D.D. Warner, W.F. Davidson, H.G. Börner, R.F. Casten, A.I. Namenson, Nucl. Phys. A316 (1979) 13
[WAR82] D.D. Warner, R.F. Casten, M. Stelts, H.G. Börner, G. Barreau Physical Review C26 (1982) 1921-1935
[WEI35] C.F. von Weizsäcker, Z. Phys. 96 (1935) 431
[WHI98] D.H. White, R.W. Hoff, H.G. Börner, K. Schreckenbach, F. Hoyler, G. Colvin, I. Ahmad and A.M. Friedman, Phys. Rev. C57 (1998) 1112
[WHI87] D.H. White, H.G. Börner, R.W. Hoff, K. Schreckenbach, W.F. Davidson, T. von Egidy, D.D. Warner, P. Jeuch, G. Barreau, W.R. Kane, M.L. Stelts,

R.E. Chrien, R.F. Casten, R.G. Lanier, R.W. Lougheed, R.T. Kouzes, R.A. Naumann, R. Dewberry, Phys. Rev. C35 (1987) 81

[WIG27] E. Wigner, Z. Phys. 43 (1927) 624

[YSF91] Yearbook of Science and the Future, Encyclopaedia Britannica (1991) 395

[ZAM02] N.V. Zamfir et al H.G. Börner, N. Pietralla, R.F. Casten, Z. Berant, C.J. Barton, C.W. Beausang, D.S. Brenner, M.A. Caprio, J.R. Cooper, A.A. Hecht, M.Krticka, R. Krücken, P. Mutti, J.R. Novak, A. Wolf, Phys. Rev. C65 (2002) 067305

[ZIE85] J.F. Ziegler, J.P. Biersack, U. Littmark, The Stopping and Range of Ions in Solids, Pergammon Press, New York (1985)

Chapter 4

Fission Research at the Institut Laue-Langevin

During 35 years of fission research at the Institut Laue-Langevin (ILL) systematic investigations of fission probabilities and properties of fragments in binary, ternary and quaternary fission were performed. All reactions were induced by slow neutrons ranging from cold to hot neutrons. Many new facets of this complex process were explored exploiting the high neutron fluxes and the clean neutron spectra at curved neutron guides available at the ILL. In the following a survey of results is given.

4.1 Introduction

The discovery in 1939 by O. Hahn and F. Strassmann of barium when uranium was irradiated by neutrons [HAH39] led L. Meitner and O. R. Frisch to identify a new phenomenon called "fission" [MEI39]. Remarkably, only a few months after the discovery, N. Bohr and J. A. Wheeler presented the ground-breaking theory of fission based on the liquid drop model [BOH40]. The model is still a cornerstone for our understanding of the process. A major puzzle remained, however. Asymmetric fission of the actinides became understandable not before the shell model became known. It was Maria Goeppert-Mayer, one of the co-discoverers of the model, who in 1948 pointed to the connection between fragment mass-asymmetry and closed neutron shells [MAY48]. L. Meitner was among the first to conceive symmetric and asymmetric fission as two distinct modes [MEI50]. Further major steps on the experimental side were the discovery of spontaneous fission of uranium by G. N. Flerov and K. A. Petrzak in 1940 [FLE40] and of ternary fission in 1947 by Tsien San-Tsiang [TSI47].

Traditionally nuclear fission was studied by radiochemical and physical methods. The physical techniques employed were photographic emulsions, ionization chambers, solid state and TOF detectors. As a new tool for fission studies, the idea came up to develop a mass separator for fission fragments based on magnetic and electric fields. In the early sixties of last century this idea took shape [EWA63]. With the advent of powerful reactors and with fissile targets placed near the core, the count rates for fission fragments recoiling from the target and being intercepted by the mass separator became competitive. In view of the high resolution for masses and energies of fragments this was an intriguing new option. Credit for having pioneered the research on nuclear fission at the ILL should be given to P. Armbruster who pushed the installation of the unique LOHENGRIN spectrometer (see chapter 2). The instrument became operational in 1975 [MOL75]. This was the start signal for fission studies at the ILL.

Studies in nuclear fission at the Institut Laue-Langevin (ILL) started virtually in coincidence with the reactor start-up and have been pursued up to the present days. As already outlined in the first part of the book, the ILL high flux reactor is a powerful source of neutrons with neutron energies ranging from cold to thermal to hot. Neutron energies available for experiments do not exceed a few 100 meV. This rather narrow energy window has dictated the type of studies to be performed in nuclear fission. On the other hand, the unrivalled high neutron flux at thermal and cold energies has allowed performing detailed analyses of very rare but revealing fission phenomena. Furthermore, the very clean neutron spectrum at the exit of neutron guides, which are in common use at the ILL, has been crucial for several precision experiments like fission cross section measurements.

Following a brief survey on available instruments and methods, topics having been addressed in fission physics at the ILL are presented. Much work has gone into the determination of fragment masses and nuclear charges for target nuclei being fissile with thermal neutrons. The targets ranged form ^{229}Th through ^{251}Cf. Of prime interest were the charge distributions of fragments which – at least for the light fragment group of asymmetric fission – were measured isotope by isotope as a function of the kinetic energy of fragments. Salient features here are the even-odd effects of charge and also neutron number yields, the charge number variance and the discovery of superasymmetric fission induced by thermal neutrons. Likewise energy distributions of fragments were determined with great precision. In particular the discovery of cold fission raised much interest, a process where virtually all of the available energy is exhausted either by the kinetic energies of fragments or by their deformation energies at scission. A further topic of many studies was ternary fission. Systematic investigations were carried out for (i) the ternary yields with alphas and tritons as the light ternary particle accompanying the two main fragments and (ii) the ternary yield distributions for heavier ternary particles up to the Si isotopes. Studying ternary fission with multi-detector arrays led to the discovery of quaternary fission where two light particles (mostly alphas and tritons) are ejected in the fission processes. Finally, in fission induced by polarised neutrons the violation of parity conservation was investigated both for binary and ternary fission. Novel and unexpected asymmetries in the emission pattern of ternary particles were observed when the neutron polarisation was pointing perpendicular to the reaction plane spanned by the two fragments and the ternary particle.

It has to be stressed that all fission investigations at the ILL were performed with cold and thermal neutrons. To put however the data obtained in perspective, results for spontaneous and low energy fission are discussed for comparison.

A shorter version of the present review was reported at conferences [GOE10].

4.2 Instruments and Methods

In fission studies at the ILL the fragment separator LOHENGRIN has from the early days of reactor operation onwards been a versatile work horse [Mol 75]. The instrument has been successfully in operation for more than 3 decades and continues to be in high demand. It serves as a separator for unslowed fission fragments recoiling from a thin target of fissile material placed inside the heavy water moderator close to the reactor core. It is a focusing parabola spectrograph consisting of a magnetic sector field and a cylinder condenser (see Fig. 4.2.1). The magnetic field fixes for fragments the ratio momentum/ionic charge while the electric field establishes the ratio kinetic energy/ionic charge [FAU81]. The fragments have to travel 23 m from the source to the focal plane corresponding to a travel time of about 2 μs. The fission products (after prompt neutron evaporation) are separated along parabolas into spectra of A/q lines with A the mass and q the ionic charge number. The separator has an A/q dispersion perpendicular to the parabolas and a dispersion in energy along the parabolas of fixed A/q but varying energy. The mass resolution can be pushed to $A/\Delta A \approx 1500$ (FWHM) and the energy resolution to $E/\Delta E \approx 1000$ (FWHM). For applications e.g. in spectroscopy of fission products this high energy resolution is not required and a focussing magnet near the focal plane allows to reduce the dispersion [FIO 93] (see Fig. 4.2.1). Due to the exceptionally high thermal neutron flux at the target position of $5.3 \cdot 10^{14}$ n/cm^{2}s^1 also very rare fission phenomena can be studied.

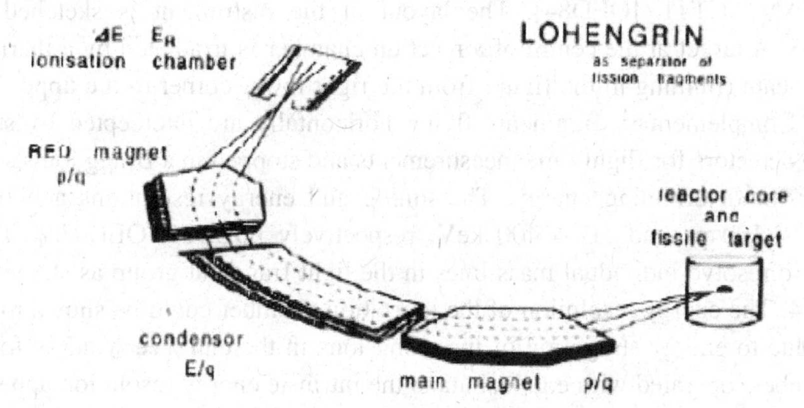

Fig. 4.2.1. The LOHENGRIN mass separator.

In Fig. 4.2.2 a plot of A/Q lines is given where A is the mass number and q the ion charge number of ions having passed the spectrometer. For constant high voltage on the condenser the magnetic field of the dipole magnet was scanned at a constant rate of 0.15 G /s. The identification of A and q for each line is shown in the lower part of the figure. The figure demonstrates the high performance of the instrument. To obtain the yield Y(A) for a given fragment mass A requires to add the contribution to Y(A) for each ionic charge q and to sum over the energy distribution of this mass.

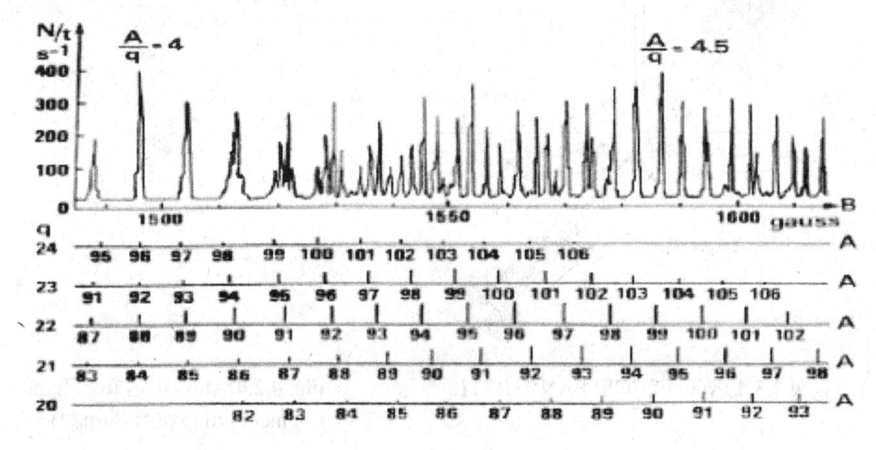

Fig. 4.2.2. Scan of A/q spectrum.

Complementary to LOHENGRIN a spectrometer based on electronic detectors for velocities and energies of fragments was constructed and called

COSI FAN TUTTE [OED84]. The layout of the instrument is sketched in Fig. 4.2.3. A target at the centre of a reaction chamber is irradiated by a thermal neutron beam (running in the figure from the right lower corner to the upper left corner). Complementary fragments flying horizontally are intercepted by start and stop detectors for flight time measurements and stopped in a Bragg ionisation chamber for determining energy. The timing and energy resolutions achieved were $\delta t \approx 100$ ps and $\delta E \approx 400$ keV, respectively [OED81, OED83a]. This enabled to resolve individual mass lines in the light fragment group as shown in Fig. 4.2.4. The energy resolution of the ionisation chamber could be shown to be mainly due to energy straggling of incoming ions in the entrance window foils. For chambers operated with e.g. isobutane the intrinsic energy resolution appears to be better than 100 keV. This outstanding resolution was exploited in the study of true cold fission [MOE75, TRO89, Fig. 4.7.5]. Nevertheless, as to resolution these instruments cannot compete with LOHENGRIN. But the original idea was to study neutron and gamma emission for well defined fragment mass splits. Unfortunately this part of the project could not be completed because at that time no background free neutron beam at the exit of a neutron guide could be made available. Some 25 years later the situation has changed and facilities along the above lines are foreseen to be installed at a neutron guide.

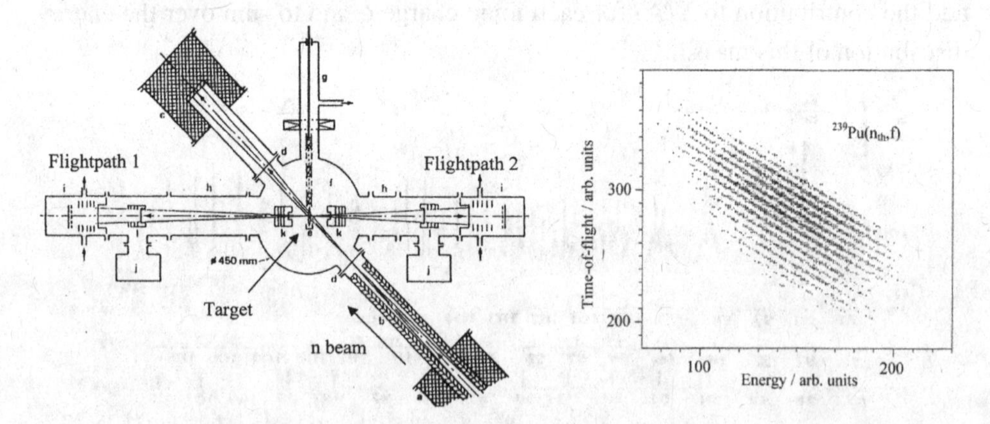

Fig. 4.2.3. Layout of COSI FAN TUITTE.

Fig. 4.2.4. Mass lines from light fragment group of $^{239}Pu(n_{th},f)$.

Based on mass and momentum conservation in the evaluation, usually masses and energies of fission fragments are measured by the double energy (2E) or the double velocity (2V) method, or by a combination of energy and velocity. Many experiments of this type were performed with a variety of energy sensitive detectors like surface barrier diodes, ionisation chambers of different design and multi-wire-proportional chambers.

A detector system specially designed for studies of ternary fission was developed at the ILL by a team from the TU Darmstadt and called DIOGENES [HEE89]. A segment of the circular instrument is on display in Fig. 4.2.5. There are two concentric gridded ionisation chambers centred on a fissile target. The inner chamber is operated at low gas pressure. It is measuring fission fragments while the more energetic ternary particles traverse the chamber. The inner chamber is operated at low pressure (200 mb). Ternary particles are stopped in the outer chamber operated at higher gas pressure (3b). Information on azimuthal angles for the fragments is obtained by segmenting the inner cathode while the angles for the ternary particles are found from the response of a circular position sensitive proportional counter intercalated between the two ionisation chambers. Polar angles for fragments and ternary particles are deduced from a comparison of anode and cathode signals in the chambers. The instrument is hence very versatile and permits to analyse different types of correlations between emission angles and energies.

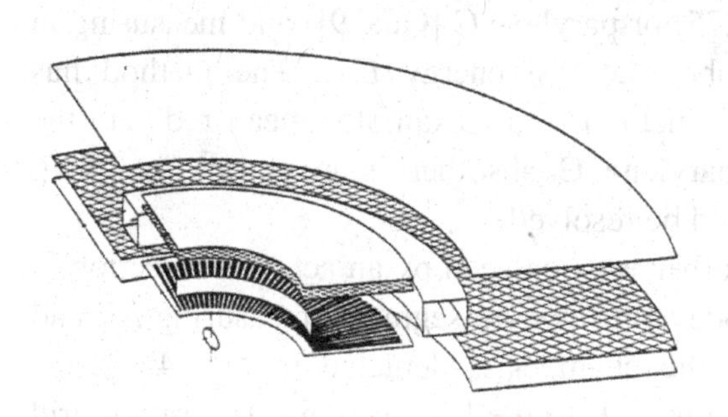

Fig. 4.2.5. Quarter-segment of DIOGENES.

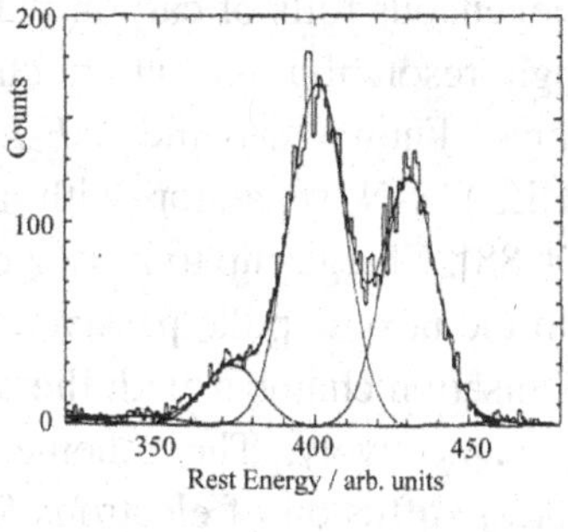

Fig. 4.2.6. Charges Z from ^{235}U(n,f) for A/q = 96/20 at = 103 MeV.

The identification by physical methods of nuclear charges Z of recoiling fragments from low energy fission is difficult and not very satisfactory. Reliable charge distributions have only been obtained for the light mass group from asymmetric fission. It is a drawback that measurements could not be extended at least to symmetric fission of fissile actinides. The difficulty is due to the fact that fragment ions with the above energies are not fully stripped of their electrons (see Fig. 4.2.2) as is required for standard Bragg curve spectroscopy [GRU82]. During slowing down in gases or solids the fragments capture electrons until they become neutral at the end of their range. The Coulomb field carried along with the moving ion is interacting with the electrons of the medium traversed and by exciting or ejecting electrons will loose energy. The specific energy loss $-dE/dx$ is proportional to the square of the ionic charge $q(V,Z)$ which itself depends on velocity V and nuclear charge Z. Besides the electronic there is also a nuclear stopping power when the incoming ions are scattered by the nuclei of the material.

The capture process and atomic encounters of the ions with the constituents of the medium render the theory of specific energy loss very complicated. But extensive tables of energy losses in all elements based on experimental results and guided by theory are available [ZIE08]. The technique of Z measurements relies on the dependence of the Bragg curve of specific ionisation on the velocity and charge of the ions traversing a medium. For the results to be presented below the methods to analyse the shape of the Bragg curve have been rather crude. Good charge resolutions were obtained on LOHENGRIN by slowing down the energy of mass and energy separated fragments in a passive absorber (e.g. thin homogeneous foils of carbon [CLE75] or parylene C [Qua79]) and measuring in a high resolution ionisation chamber the rest energy E_{rest}. The method has become known as the ΔE-E_{rest} method. An example measured at the LOHENGRIN separator with a parylene C absorber is given in Fig. 4.2.6 [BOC88]. Charges up to Z = 42 could be resolved.

In the next step the passive absorber was replaced by an active ΔE section in an ionisation chamber with the anode split into two segments measuring ΔE and E_{rest}, respectively. The scheme of the chamber is depicted in Fig. 4.2.7. To suppress diffusion of electrons from the ΔE to the E_{rest} section a separation grid (labelled 8 in Fig. 4.2.7) was introduced. Nuclear charges Z of fission fragments from the light group were identified up to Z = 40 [BOC87].

A variant of Bragg spectroscopy was developed for use on the COSI FAN TUTTE spectrometer. It exploits the fact that, for given mass and energy of a fragment, the range of the track in an ionisation chamber still depends on the nuclear charge Z. The range can be assessed in a Bragg chamber by measuring the time Δt between the instant the ion enters the chamber and the arrival of the first electrons at the anode [OED83b]. The layout is sketched in Fig. 4.2.8. The start signal is derived from a time pick-off detector in front of the entrance window of the chamber [OED81] and the stop signal is obtained from a threshold trigger whenever the anode pulse reaches a given level. Typically charge resolutions of $Z/\Delta Z \approx 43$ were obtained for charges below $Z = 40$.

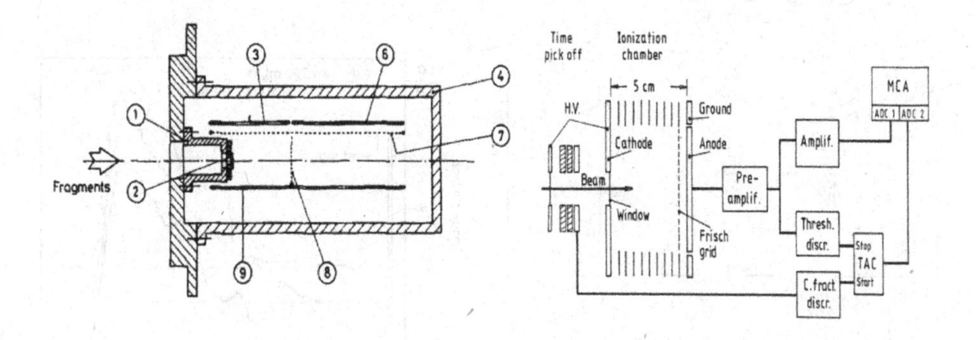

Fig. 4.2.7. Ionisation chamber with segmented anode and separation grid.

Fig. 4.2.8. Bragg chamber combined with a start detector at chamber entrance.

Better charge resolutions should be expected to be achievable by inspecting the full Bragg curve. To this purpose the Bragg curve has to be digitised by flash ADCs and analysed in much finer detail than described so far. First experiments on this line were performed by mounting a detector combination of start detector and ionisation chamber in the focal plane of LOHENGRIN. The Bragg curve of specific ionisation in the chamber is sampled at the anode as a time-dependent electron current. As indicated in Fig. 4.2.9 the Bragg curve and the profile of the current pulse are mirror-shaped. For physical reasons, however, the performance of Bragg curve spectroscopy is limited. Straggling of energy loss of fission fragments passing through thin foils was studied very early [ARM76]. An ultimate limit is the influence of scattering of incoming ions by atoms of the stopping medium in the slowing-down process. As already pointed out above, besides the electronic stopping power, the scattering leads to an additional nuclear stopping power. Examples for these atomic shocks are provided in Fig. 4.2.10 [OED86]. For demonstration purposes an extreme example was chosen. The three Bragg curves shown have been obtained for a heavy fragment with mass $A = 139$ and kinetic energy $E = 68$ MeV. The counting gas in the chamber was the heavy Xe. The bumps in the curves correspond to scattering events on Xe atoms in the stopping process of ions. Evidently, Bragg curve spectroscopy will not be successful for identifying nuclear charges with Xe as counting gas.

Unfortunately these studies from the eighties of last century were not continued because at that time not sufficient computational power was on hand.

As already stated, neutrons and gammas being emitted in the fission process are foreseen to be investigated at the ILL in the near future. Meanwhile, on LOHENGRIN the gamma spectroscopy of fission products following isomeric or

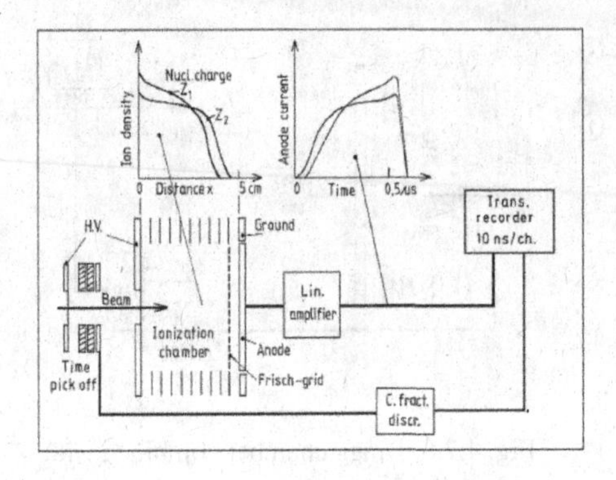

Fig. 4.2.9. Tracing Bragg curves by fast digitizing of current pulse.

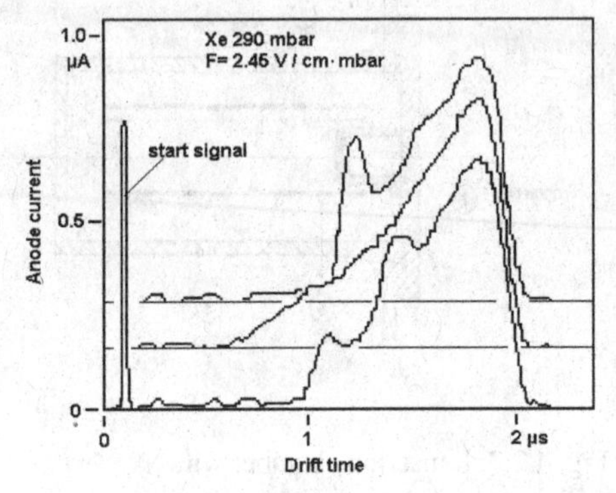

Fig. 4.2.10. Sample of Bragg curves for fragments with A = 139 in Xe.

beta-decay has become an active field of research. Fission products are very neutron rich and, at least in the past, have not been readily accessible for studies of nuclear structure. A survey of results obtained is given in the part of the book on nuclear spectroscopy.

4.3 Mass Distributions of Fragments from Thermal Neutron Induced Fission

Mass distributions of fragments from a decaying nucleus are a characteristic feature of fission. Very early in the history of fission it became evident that the mass distributions of fragments in the lighter actinides are asymmetric with two well separated groups of lighter and heavier fragments. This was in conflict with predictions from the Liquid Drop Model of fission elaborated by N. Bohr and J. A. Wheeler [BOH39]. The puzzle was solved with the advent of the nuclear shell model. Very early it was conjectured that the shell stabilisation of the nascent fragments could drive the nucleus towards asymmetric fission [MAY48, MEI50, FAI62]. A quantitative basis was laid e.g. in the asymmetric two-centre shell model [MAR72]. For the discussion of symmetric against asymmetric fission in physical terms the statistical scission point model introduced by B. Wilkins et al. is very convenient and often advocated [WIL 76]. It is pointed out that besides spherical shells also deformed shells are of importance in fission. In some sense this model has been a precursor to the Brosa-Grossmann-Müller model, for short the Brosa model [BRO90]. The impact of this latter model has been enormous. It has become customary to analyse mass and energy distributions of fission fragments in terms of modes introduced by Brosa. Complementary to the models sketched, purely microscopic theories of fission have been a challenge. Impressive results have been reported more recently e.g. for the fragment mass distribution of a fissioning ^{238}U nucleus [GOU06].

It is well established that in low energy fission of the lighter actinides the mass distributions are asymmetric. But for increasing excitation energy a symmetric fission component rapidly catches up (see e.g. [GON91]. In the heavier actinides like Fm and heavier, symmetric fission becomes dominant even in spontaneous fission. Fm isotopes with neutron numbers larger than N = 156 undergo symmetric fission and mass distributions in the actinides remain symmetric. The switch from asymmetric to symmetric fission appears to be linked to the disappearance of the second barrier in the double-humped barrier profile observed in the lighter actinides. In contrast to the first barrier, the second barrier is known from theory to be asymmetric in mass and hence serves as a gate to asymmetric fission. A special feature for the Fm and some neighbouring isotopes is symmetric bimodal fission where a narrow and a broad mass distribution exhibit an unusually high and a lower total kinetic energy distribution, respectively [HUL90]. For lighter nuclei down to the Businaro–Gallone point mass distributions are known to follow the Liquid Drop Model predictions for symmetric mass splits. Of interest is the transition from asymmetric to symmetric fission. In a remarkable series of experiments it was discovered that symmetric fission already becomes predominant in light, i.e. neutron poor isotopes of the actinides U, Pa and Th [SCH98]. Fission was induced in this work by Coulomb excitation of well identified isotopes with relativistic energies. The excitation was on average 11 MeV. At this excitation energy all lighter elements are fissioning symmetrically.

Right at scission the fission fragments are in general deformed. The deformation rapidly relaxes after scission into intrinsic excitation energy which is added to the one already present. The bulk of this energy is liberated by prompt neutron evaporation in times shorter than 10^{-14} s. When the available energy has dropped below the neutron separation energy, the remaining heat is exhausted by gamma radiation. But even having reached their ground states these secondary fragments or "fission products" are not stable against β-decay. On average it takes three β-decays for the products to arrive to the stability line of the nuclides. It has been the task of patient radio chemists to identify all these isotopes and to find their so-called "independent yields", i.e. the yields with which they emerge after neutron evaporation.

Since theory calculates primary mass distributions which are the ones of physical interest, the task of experimentalists is to find ways how to come close to the primary distributions. At the Institut Laue-Langevin mass distributions for thermal induced fission of actinides ranging from ^{229}Th up to ^{251}Cf as the targets were investigated by physical methods before β-decay could set in. In the following, in most cases secondary masses are reported where it has not been tried to correct the data for neutron evaporation. The reason is that the focus was on the measurement of nuclear charge distributions which are not hampered by the emission of charged particles and therefore faithfully represent the original charge division of the fissioning nucleus. Nevertheless, also the secondary mass distributions carry interesting information on the fission process.

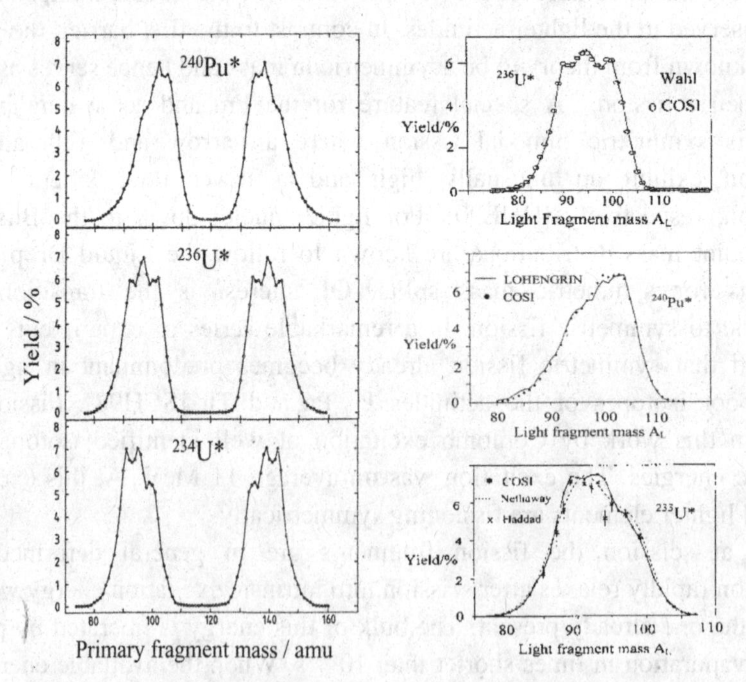

Fig. 4.3.1. Primary mas distributions measured in 2V experiment.

Fig. 4.3.2. Secondary mass distributions measured by different techniques.

There is only one exception where distributions very close to primary mass distributions have been investigated at the ILL. In this work the technique called (2V) method was used based on a measurement of the two velocities of complementary fragments V_1 and V_2. Applying mass and momentum conservation to a binary decay, the primary masses A_i* of the two fragments i=1 and i=2 are readily calculated from the two velocities provided there is no neutron emission. But since the neutrons are to good approximation evaporated isotropically from the fully accelerated fragments, the measured average velocities are identical to the primary velocities. They allow to find the average primary masses $\langle A_i^* \rangle$ with a mass resolution (FWHM) of 1.5 mass units [TER62]. The ILL experiment simulated the (2V) method by applying symmetry considerations to a (1V) measurement [GEL86]. Results for the primary mass distributions of the thermal neutron induces reactions on the targets ^{233}U, ^{235}U and ^{239}Pu are on display in Fig. 4.3.1. The structure in the mass yields *Y(A*)* is due to the even-odd effect in the nuclear charge yields *Y(Z)* favouring even Z charge numbers as to be discussed in detail in chapter 4.6. In the U-isotopes the structure is better pronounced than in plutonium pointing to a stronger even-odd effect in uranium. It is further seen that, compared to ^{236}U*, in the reaction ^{233}U(n_{th},f) the peak near A* = 100 in the light and the mirror-peak near A* = 132 in the heavy mass group has a lower yield. In the Brosa model these peaks correspond to the standard I mode which is driven by the shells Z = 50 and N = 82 in the fragments at scission (see chapter 5). The reason for this rapid change in the yield distribution must be sought in subtle variations of the potential energy surface near the saddle point or on the way to scission. Spectacular variations of yield distributions have been reported for spontaneous fission of five even Plutonium isotopes ranging from ^{236}Pu to ^{244}Pu [DEM97]. It is further of interest to quote the ratio of asymmetric to symmetric fission yield. Long before the more refined Brosa modes came up it was customary to think in terms of two modes in fission, one symmetric and another one asymmetric. For the reaction ^{235}U(n_{th},f) this ratio is usually quoted as the peak-to-valley ratio (530±50); the ratio drops to (420±70) for ^{233}U(n_{th},f) and further to (150±20) for ^{239}Pu(n_{th},f). Moving onwards to thermal neutron induced fission of Cm, Cf and Fm the ratio asymmetric to symmetric yield continues to decrease until symmetric fission takes over in the Fermium isotopes 258 and 259 (see below).

For comparison with the primary mass distributions on display in Fig. 4.3.1, some examples of secondary mass distributions of fragments after neutron emission have been put together in Fig. 4.3.2. Only mass yields for the light group of fragments are presented because these have been at the focus of interest at the ILL. To the left of the figure the distribution of the compound nucleus $^{236}U^*$ prepared by thermal neutron capture in the target ^{235}U is once shown as found with the spectrometer COSI [MOL91], and once as reported from radiochemical studies [WAH88]. The results found agree perfectly well. It has to be noted that the triple humped yield distribution in the pre-neutron data (Fig. 4.3.1) is found back in the post-neutron distribution (Fig. 4.3.2). It means that neutron evaporation does not destroy the basic structure. At very close inspection of the pre- and the post-neutron distributions, however, it is observed that the left wing of the light group stays at a fixed position while the right wing shifts to smaller mass numbers after neutron emission. It indicates in a rather direct way that neutron evaporation from the fragments is not uniform for all fragment masses. Instead, the heavy fragments from the light group emit more neutrons than the lighter ones. In neutron studies of the fission process a similar behaviour was observed for the heavy mass group. This has led to the notion of "saw-tooth" behaviour of neutron evaporation as a function of fragment mass.

For $^{240}Pu^*$ from the $^{239}Pu(n_{th},f)$ reaction two post-neutron mass distributions from studies at the LOHENGRIN and the COSI spectrometers are given in Fig. 4.3.2 [SCH84, KAU91]. Obviously the agreement especially near mass A = 100 is not very good. The structure seen in the pre-neutron distribution in Fig. 4.3.1 is better reproduced in the COSI data of Fig. 4.3.2. Also data from radiochemistry are closer to the COSI results. Finally in Fig. 4 3.2 are plotted post-neutron mass distributions for the rarely studied reaction $^{232}U(n_{th},f)$ leading to the fissioning compound $^{233}U^*$. The distributions of yields obtained by physical methods with COSI [KAU91] are only roughly compatible with those from radiochemistry [HAD88, NET91]. The fine structure in the yields seen in the COSI spectrometer is corroborated by measurements of nuclear fragment charges. As discussed in chapter 4.6 the structure of mass distributions is clearly correlated with the even-odd effect of charge yields. The post-neutron yields of $^{233}U^*$ in Fig. 4.3.2 should not be directly compared to the pre-neutron yields for $^{234}U^*$ in Fig. 4.3.1. It appears nevertheless that the structure in the two neighbouring nuclei is not much different.

As to mass distributions an outstanding discovery has been super-asymmetric fission in reactions induced by thermal neutrons. The existence of superasymmetric fission has since long been postulated by theory [GRE00]. One of the highlights of the LOHENGRIN spectrometer is the possibility to investigate the emission of particles at extremely low yields, either fragments or ternary particles. When extending the measurements of fragment masses to mass numbers below A = 80 it was indeed found that the yields do not plunge smoothly to yield levels too low to be observable. Instead, the slopes of the yield curves have kinks near A = 80 and even exhibit shoulders at masses around A = 70. The lowest yields measured were less than 10^{-6} % per fission near A = 68.

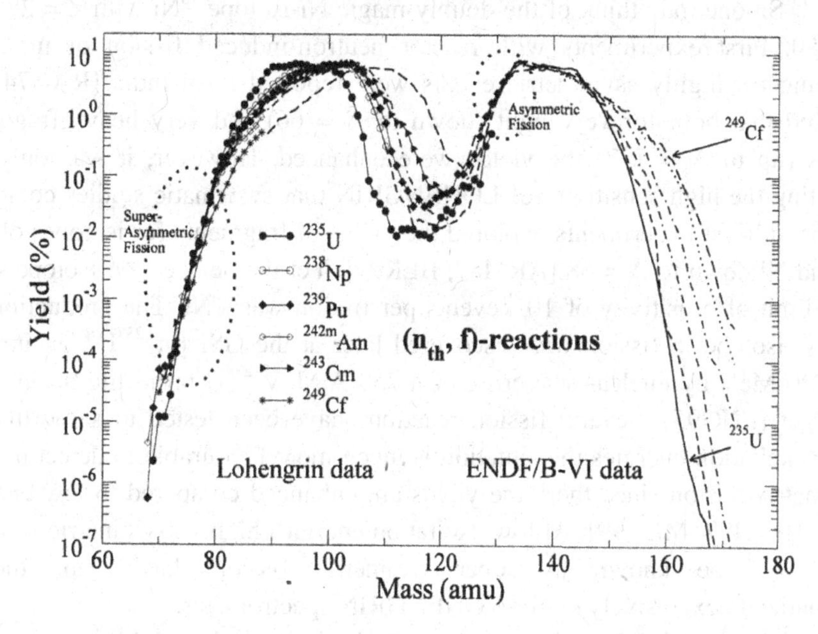

Fig. 4.3.3. Mass distributions in thermal neutron fission. Yields for the light fragment group and for symmetric fission from LOHENGRIN. Yields for heavy group from ENDF/B-VI data tables.

An overview of mass distributions for six reactions having been studied is given in Fig. 4.3.3 [TSE04a]. The reactions and the corresponding references for thermal neutron induced fission are for ^{233}U [LAN80, SID89, TSE04a], for ^{237}Np [MAR90, TSE01], for ^{239}Pu [SCH84, DIT91, TSE04a], for ^{242m}Am [TSE99], 243Cm [TSE04b] and for ^{249}Cf [HEN94]. Evidently, FIG. 4.3.3 summarises a large amount of work by many PHD students. Two phenomena are catching the eye in the figure. They are marked and labelled “Asymmetric Fission” and

"Super-Asymmetric Fission". In the mass regions near the heavy mass $A_H \approx 130$ and the light mass $A_L \approx 75$, respectively, the yield distributions of the six reactions virtually coincide. As to asymmetric fission this is standard knowledge since the very early days of nuclear fission. The abrupt rise of the yield is ascribed to the extra-stability of nuclei near the doubly magic ^{132}Sn with Z = 50 and N = 82. In consequence, the light fragments have to shift to higher masses for increasing compound mass of the fissioning nucleus. This is readily seen in FIG. 4.3.3. But the question has always been, why is it that only in the heavy fragments the magic nuclei play such a crucial role? Why are effects due to magic nuclei in the light mass group not observed? In analogy to the doubly-magic ^{132}Sn one may think of the doubly-magic Ni-isotope ^{78}Ni with Z = 28 and N = 50. First experiments with reactor neutron-induced fission of uranium searching for highly asymmetric events were reported from India [RAO74]. It appeared that both for very light (down to A = 66) and very heavy fragment masses (up to A = 177) the yields were enhanced. However, it was only by exploiting the high sensitivity of LOHENGRIN that systematic studies could be performed. First experiments explored the yields of fragments for isotopes of Fe, Co and Ni down to A = 68 [ARM87, BER91]. Yet the heaviest Ni isotope seen at the limit of sensitivity of 10^{-9} events per fission was ^{76}Ni. The production of the ^{78}Ni isotope in fission was discovered later at the GSI for ^{238}U* excited to about 20 MeV by nuclear scattering of a 750 A·MeV ^{238}U projectile beam on a Be target [ENG95]. Several fission reactions have been tested to show that at higher excitation energies the situation is much more favourable to detect highly asymmetric fission since there the yields are enhanced compared to low energy fission [HUH97, MUL99]. At low excitation energies highly asymmetric fission, having become known as super-asymmetric fission, has been studied systematically exclusively on the LOHENGRIN spectrometer.

The mass region labelled super-asymmetric fission in Fig. 4.3.3 is zoomed in the Figs. 4.3.4 and 4.3.5. In Fig. 4.3.4 [DEN00] two reactions are on display not shown in Fig. 4.3.3 for the sake of clarity: ^{242}Pu* and 246Cm* [FRI98]. Note that in the figure the yields are shifted by a factor 10 from reaction to reaction. Evidently, for all reactions the slope of the yield curve exhibits a kink right at the same mass number A = 80. It may be noticed in Figs. 4.3.6 and 4.3.7 that for mass 80 the dominating charge is Z = 32 (Ge). Since the change in slope is present for all reactions studied this is an indications that the effect is linked to fragment structure. Thinking of shell effects the Ge isotope ^{82}Ge with N = 50 would be a good candidate. Also the N/Z = 1.56 ratio of ^{82}Ge is close to the N/Z ratios 1.55–1.57 of the compound nuclei investigated. Neutron evaporation could then bring the conjectured ^{82}Ge to the observed isotope ^{80}Ge. In passing attention should be called to the fact that for even and odd charge numbers of the fissioning nucleus the kink appears at the same mass number A = 80.

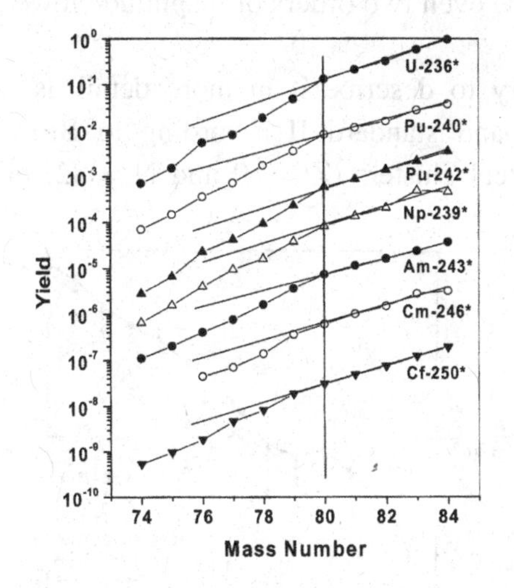

Fig. 4.3.4. Kink in the slope of mass distributions for super-asymmetric fission.

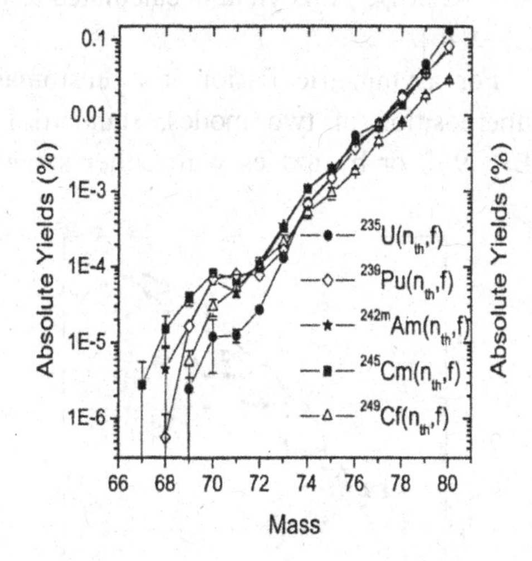

Fig. 4.3.5. Shoulder in the mass distributions For super-asymmetric fission.

More spectacular than the kinks are shoulders in far-asymmetric fission revealed in Fig. 4.3.5 [ROC04]. The peaks of the shoulders are at mass number A = 70 though not equally pronounced for all reactions. Yields in the peak at A = 70 differ by a factor of ten while for mass numbers near A = 75 they are close to within a factor of two. Again looking for a possible structure effect in the fragments, the stabilising effect of Ni-isotopes with charge number Z = 28 comes back to mind. That Ni-isotopes are accountable for the shoulders could directly be proven in experiment. In Fig. 4.3.6 the far asymmetric mass distribution for the reaction ^{235}U(n_{th},f) is decomposed into the contributions of the individual elements [SID89]. Already in this first comprehensive investigation of far asymmetric fission events the preponderance of Ni-isotopes near the peak of the shoulders at A = 70 showed up. As a further example, the same observation of Ni in the peak yield is made for the reaction ^{245}Cm(n_{th},f) in Fig. 4.3.7 [FRI98, ROC04]. The Ni-isotope ^{70}Ni has a N/Z ratio of 1.5 and is hence very close to

N/Z for the compound nuclei at issue while for ^{78}Ni it is 1.786. Of course the N/Z arguments can only serve as a rough guide and help make results plausible. They should not be meant to replace a full-fledged theory. As to the yield for ^{78}Ni from ^{235}U(n_{th},f) it has been estimated by extrapolation form the data in Fig. 4.3.6 that it should be close to $6 \cdot 10^{-11}$ and thus beyond the reach of LOHENGRIN [ENG95]. For ^{245}Cm(n_{th},f) this yield is calculated to be even two orders of magnitude lower [TSE04b].

For asymmetric fission it is customary to describe it in more detail as a superposition of two modes, standard I and standard II according to Brosa [BRO90], or by modes with either spherical clusters (Z = 50 and N = 82) or

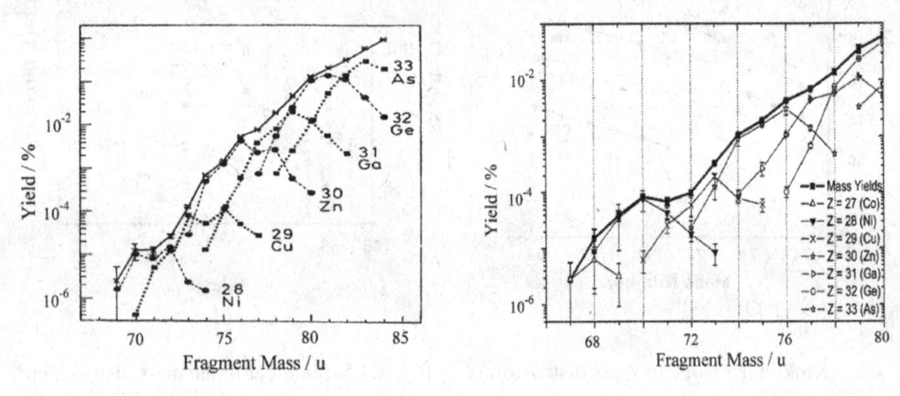

Fig. 4.3.6. Super-asymmetric fission of ^{236}U*. Fig.4.3.7. Super-asymmetric fission of 246Cm*.

deformed shells (N = 88) following Wilkins [WIL76] /see Fig. 4.5.3). For both modes a Gaussian mass distribution is usually assumed though deviations from Gaussians have bee considered [TSE04b, MUL99]. In the evaluation of experiments the task is to determine the centre and the width of the distributions. Taking far-asymmetric called super-asymmetric fission into account new modes have to be introduced. In view of the experimental findings one may think of a single super-asymmetric mode standard III based on ^{78}Ni [RUB01]. Probably a better approach is to foresee two super-asymmetric modes standard III and standard IV based on the shell N = 50 at mass-number A≈ 80, and the shell Z = 28 at mass-number A = 70, respectively [TSE04b]. The modes III and IV are much narrower than the modes I and II.

Finally, the evolution of symmetric fission in the fissile actinides should be addressed once more. For ^{234}U*, ^{236}U* and ^{240}Pu* the relative contributions of asymmetric and symmetric fission were quantified as the peak-to-valley ratio P/V of mass distributions and discussed in connection with Fig. 4.3.1. The decrease of this ratio for heavier systems quoted above is also evident from Fig. 4.3.3. The yields at symmetry increase steadily. The P/V ratio drops in (n_{th},f) reactions from about 500 in the U-isotopes to about 60 in ^{246}Cm*, to about 34 in ^{250}Cf* and to about 3 in ^{245}Es*. For still heavier actinides there are no thermal neutron induced fission data.

In summary, the salient features of mass distributions in the actinides from Th to Cf under study are all evident in Fig. 4.3.3. There is, first, the constancy in the onset of asymmetric fission at the position of Z = 50 and/or N= 82 nuclei near the doubly magic ^{132}Sn nucleus. These nuclei are acting as a cornerstone for the light wing of the heavy mass peak (asymmetric fission). There is, second, in the light peak a similar cornerstone due to the influence of magic Z = 28 and N = 50 nuclei (superasymmetric fission). It fixes the left wing of the light mass peak. Both, in the light and the heavy mass group, only the wings at larger fragment masses are free to move with compound mass. Mass splits have to conform to

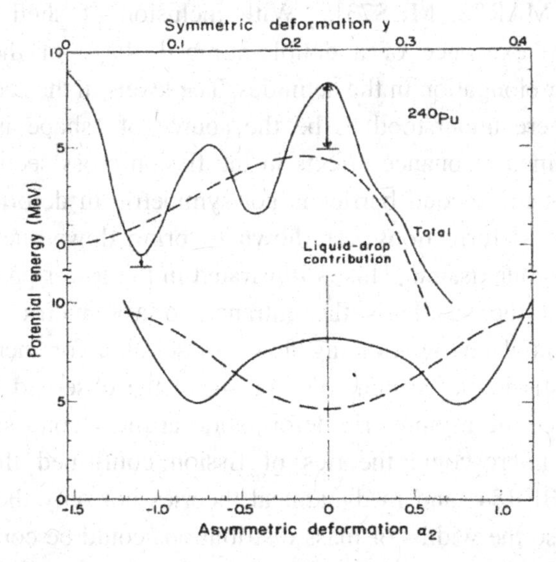

Fig. 4.3.8. Double-humped fission barrier (top) and asymmetry of second barrier (bottom).

mass number conservation and the constraints imposed by the two limiting cornerstones. Therefore mass distributions are becoming wider with characteristic shifts in the wings to the right of the distributions when the mass of the fissioning compound increases. The present simple picture even suggests that mass widths should increase linearly with the mass A_F of the fissioning compound. This is indeed observed (see Fig. 4.6.5).

As a short comment to the theory of mass distributions, the calculated potential energy surface of a deforming nucleus is sketched in Fig. 4.3.8 [BOL72]. The liquid drop model (LDM) applied in 1939 by Bohr-Wheeler [BOH39] to fission remains the backbone of our understanding. However, in particular asymmetric fission prevailing in low energy fission of the lighter actinides is not understood in the framework of the LDM. It was the discovery of the nuclear shell model which sparked the idea to trace asymmetric fission to shell effects in the fragments [MAY48, MEI50]. The suggestions were corroborated by many theoretical groups presenting calculations of potential energy surfaces for the lighter actinides as a function of deformation. The starting point for the calculations is the ground state. For increasing deformation the nucleus moves over the saddle point of potential energy until at the scission point the stability of the mono-nucleus is lost. The nucleus undergoes fission [MÖL70, PAS71, BOL72, MAR72, MUS73]. With inclusion of shell effects a first discovery was the existence of a double-humped shape of the barrier as a function of nuclear elongation in the actinides. The levels in the second minimum when occupied were understood to be the source of "shape isomers". They furthermore explained resonance effects in the fission cross sections. A second discovery was that the second barrier is not symmetric in deformation. At this barrier asymmetric deformations are shown to bring down the height of the barrier and thus favour fission. This is illustrated in the lower part of Fig. 4.3.8. The asymmetry at the saddle is the entrance to asymmetric valleys in the landscape of potential energy running down to scission for increasing overall elongation of the fissioning system. The origin of the observed fragment mass asymmetry is hence the asymmetric deformation at the second saddle point. In later years fully microscopic theories of fission confirmed the early semi-classical results [BER84] and in dynamical theories not only the fact of being asymmetric but also the widths of mass distributions could be correctly assessed [GOU05]. An interesting observation in this context is that e.g. in the neutron-rich Fm isotopes the asymmetric second fission barrier disappears and the fragment mass distributions become symmetric [HUL94].

4.4 Kinetic Energies of Fission Fragments

The energy Q set free in nuclear fission is shared by the total kinetic energy

*TKE** and the total excitation energy *TXE* of the primary fragments of binary fission:

$$Q = TKE^* + TXE \tag{4.4.1}$$

For thermal neutron induced reactions the Q-value is the sum of the neutron mass and the difference in mass between the target nucleus and the mass of the two fragments before prompt neutron emission. For the lighter actinides the energy released is typically $TKE^* \approx 180$ MeV and $TXE \approx 30$ MeV. The excitation energy *TXE* is partly due to the intrinsic excitation already present at scission but mostly due to the deformation energy of fragments at scission which relaxes rather quickly in about $5 \cdot 10^{-21}$ s into internal heat. The energy *TXE* is evacuated by neutron emission in times shorter than about 10^{-15} s followed by the slower γ-emission in times up to 10^{-9} s (90% of all γ´s) and beyond. In experiment the energy *TXE* is found by measuring neutron spectra (multiplicity, kinetic and binding energies of neutrons) and gamma spectra (multiplicity, energies). Since neutrons are evaporated to good approximation isotropically in the moving frame of fragments, the fragment velocity remains on average unchanged (see chapter 4.3). The kinetic energy, however, is getting slightly smaller since the mass of the fragments decreases by neutron loss. For a fragment of primary mass A* = 100 the kinetic energy decreases by ~ 1% per emitted neutron. In most experiments energies and not velocities are measured and, hence, the energy release *TKE* observed has to be corrected for neutron emission to find the initial energy *TKE**.

In nuclear power reactors the main contribution to the heat produced originates from the kinetic energy loss when the fragments are slowed down in the material of the fuel element. One may think that these large kinetic energies make fission easily detectable. It is true that large energies are characteristic of fission but the extremely short ranges of fragments have prevented an early discovery of fission by physical methods. To give an impression of the ionisation density (-dE/dx), a typical heavy fragment HF with mass A_H ~ 140 and initial energy E_H ~ 70 MeV, and a typical light fragment LF with mass A_L ~ 105 and energy E_L ~ 100 MeV have been chosen and their specific energy losses in thin solid foils of thickness 100 μg/cm^2 are given in Table 4.4.1. The losses by the above HF and LF fragments are very similar and therefore only one figure is quoted as a rule of thumb for both. The energy losses are given in MeV. The reason for the large size of the ionisation is the large effective ionic charge which exceeds q = 20e for unslowed light fragments (see Fig. 4.2.2). The effective charge enters as q^2, i.e. quadratically into the ionisation. Evidently the energy losses increase very rapidly the lighter the material traversed is. A corollary of large specific energy losses are short ranges R of fragments in materials. To give two examples: in U metal the range of fission fragments does not exceed 6 μm while in human tissue (skin) the range is inferior to about 25 μm. Precise energy loss data are to be found in [ZIE08]. Despite the large energies, the extremely short ranges impose stringent conditions on entrance windows for detectors of ionising radiations like e.g. Geiger-Müller counters or ionisation chambers.

Energy losses in actinides, which are slightly smaller than for Au in Table 4.4.1 also limit the thickness and thus the amount of material which is acceptable for a target in order to keep losses and corrections in the evaluation under control. For high resolution experiments target thicknesses should not exceed about 100 μg/cm^2. Thin foils are also requested as backings or entrance windows for detectors or reaction chambers filled with gas at low pressure. Very often plastic foils are used. For practical reasons several plastics were tested on LOHENGRIN as to the energy losses for fission fragments [AIT90]. Two examples are given in Fig. 4.4.1. Energy losses are plotted for foil thicknesses of 100 μg/cm^2. At the same kinetic energy the losses are larger for heavy fragments than for light ones. While for heavy fragments the losses appear to be not much dependent on mass, for the light fragments which were of prime interest on the spectrometer LOHENGRIN, the losses vary strongly with mass.

Table 4.4.1. Fragment energy loss in MeV in foils of thickness 100 μg/cm^2.

Target	C	Al	Cu	Au	Formvar
Energy Loss	8.5	7.0	4.0	3.0	8.1

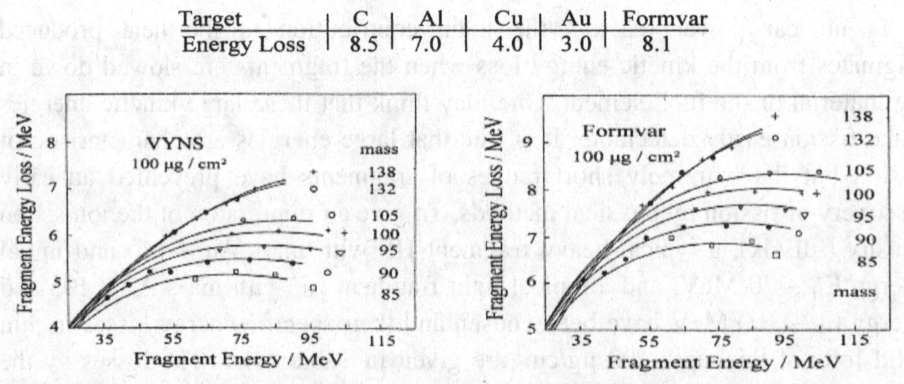

Fig. 4.4.1. Energy losses of light and heavy fragments in thin Formvar and VYNS Foils of 100 μg/cm^2. The curves through the data points are to guide the eye.

The energy calibration of detectors for fission fragments has for a long time been controversial. Just to extrapolate to fission fragments the response of detectors to alfa-particles of known energy has pitfalls since the high ionisation density of fragments in the detector material entails charge recombinations and hence pulse height defects which are not easily assessed. But kinetic energies of fragments are reliably determined by measuring their velocities and masses. Making use of the conservation laws for mass and momentum the basic formula for the primary total kinetic energy TKE^* is

$$TKE^* = E_L^* + E_H^* = (k/2)\, A_C V_L^* V_H^* \tag{4.4.2}$$

with E_L^*, E_H^*, V_L^* and V_H^* the primary energies and velocities of light and heavy fragments before neutron emission, respectively, and A_C the mass of the fissioning compound nucleus. The factor k in eq. (4.4.2) keeps track of the transformation of units, i.e. amu for masses A_C, (cm/s) for velocities and MeV for energies. On the ^{12}C-scale its value is $k = 1.0365$. Velocities are established with high precision by taking the length L of flight paths and the time T it takes the fragment to travel along L. Remarkably total kinetic energies can be calculated from velocities without requiring the knowledge of masses. The average primary energy release is found from eq. (4.4.2) to be

$$<TKE^*> = (k/2) A_C [<V_L^*><V_H^*> + cov(V_L^*, V_H^*)] \tag{4.4.3}$$

where the *cov* term is approximated by linear regression to read

$$cov(V_L^*, V_H^*) = <dV_L^*/dV_H^*>\, \sigma^2(V_H^*) \tag{4.4.4}$$

The cov term is negative and contributes less than 0.5%. The velocity distributions for the three standard fission reactions ^{233}U(n_{th},f), ^{235}U(n_{th},f) and ^{239}Pu(n_{th},f) from (2E,1V) experiments at the ILL are plotted in Fig. 4.4.2 [GEL86]. The two humps for the light and heavy fragment groups at larger and smaller velocities, respectively, are well separated. For the two groups the averages and variances of velocity distributions in eqs. (4.4.3) and (4.4.4) can therefore be determined without ambiguity. The average velocities are indicated in the figure. Typically V_L = 1.4 cm/ns for the light and V_H = 1.0 cm/ns for the heavy fragment. Typical variances are $\sigma^2\,(V) \approx 4{\cdot}10^{-3}$ (cm/ns)^2.

Table 4.4.2 summarises the primary total kinetic energy releases *TKE** as obtained in ILL experiments for (n_{th},f) reactions and as reported form other labs for spontaneous fission of ^{252}Cf (velocity experiments [HEN81, KIE91] evaluated according to eqs. (4.4.3) and (4.4.4)).

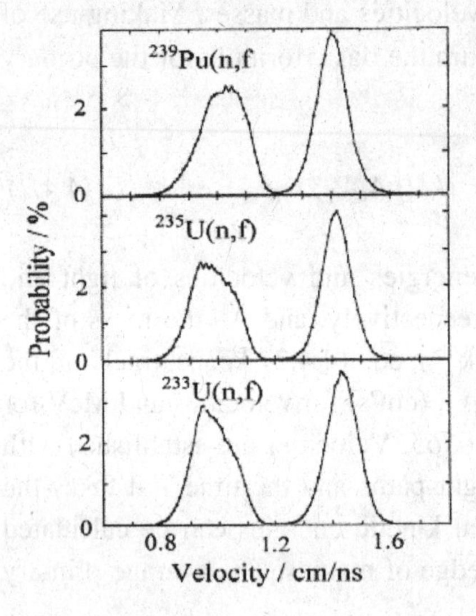

Fig. 4.4.2. Velocity distributions of standard fission reactions.

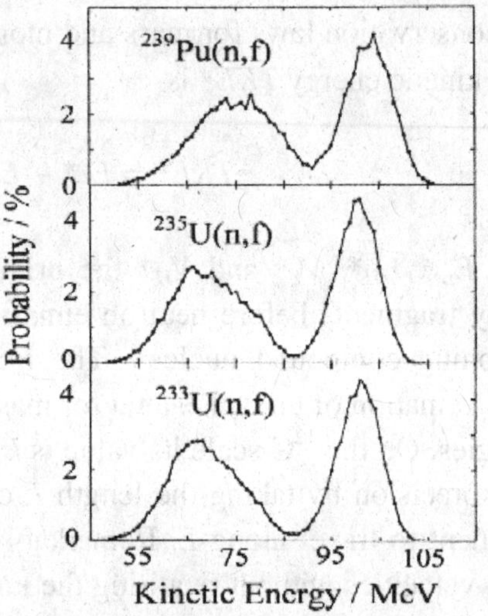

Fig. 4.4.3. Primary kinetic energy distributions.of standard reactions.

Table 4.4.2. Primary and secondary total kinetic energies TKE*and TKE, from ILL experiments. Viola data from survey of fission reaction [VIO85].

Reaction	^{233}U(n_{th},f)	^{235}U(n_{th},f)	^{239}Pu(n_{th},f)	^{252}Cf(sf)
TKE*/MeV	170.1(5)	170.5(5)	177.9(5)	184.0(13)
TKE / MeV	168.1(10)	168.6(10)	175.8(10)	181.0(10)
Viola	170.6(30)	170.1(30)	176.4(30)	188.1(32)

More detailed information involving explicitly fragment masses can be obtained by noting that in (2V) experiments primary fragment masses can be calculated from mass and momentum conservation, viz.

$$A_L^* = A_C\, V_H^* / (V_L^* + V_H^*) \qquad (4.4.5)$$

with an analogue equation for the heavy mass A_H^*. Kinetic energies E_{KL}^* (and analogously for E_{KH}^*) for individual fragments follow from

$$E_{KL}^* = (k/2)\, A_L^*\, V_L^{*2} \qquad (4.4.6)$$

For the (2E,1V) experiments performed at the ILL the evaluation is more complicated because the missing second velocity has to be found from symmetry considerations. We will not go into these details here. The results for the primary fragment kinetic energies are on display in Fig. 4.4.3. By comparison with Fig. 4.4.2 it is seen that the separation between the light and the heavy group is not as clear-cut as for velocities. It is therefore worthwhile to apply a procedure outlined above for the total average kinetic energy $\langle TKE^* \rangle$ to assign reliable averages for mass and energy for each group. We quote here only results from velocity measurements for the $^{235}U(n_{th},f)$ reaction. Averages for the groups L and H before neutron evaporation are given in Table 4.4.3[HER78, GEL84, MUL84].

Table 4.4.3. Average velocities, masses and energies for the L and H group from $^{235}U(n_{th},f)$.

$\langle V_L^* \rangle$/cm/ns	$\langle V_H^* \rangle$/cm/ns	$\langle A_L^* \rangle$/amu	$\langle A_H^* \rangle$/amu	$\langle E_{KL}^* \rangle$/MeV	$\langle E_{KH}^* \rangle$/MeV
1.420(5)	0.983(5)	96.4(2)	139.6(2)	100.6(5)	69.8(5)

In by far most studies of nuclear fission the energies of fragments are measured by solid state detectors or ionisation chambers. The fragments intercepted have evaporated all prompt neutrons. They are secondary fragments and the question arises how the response of these detectors can be calibrated. Also for this task the LOHENGIRN spectrometer has proven to be very useful [WEI86]. To this purpose a time-of-flight system comprising a fast time pick-off for the start and a surface barrier detector for the stop was installed in the focal plane of the instrument. The stop detector was movable so the distance start-stop could be varied between a short and a long flight path. The timing difference between the two detector positions was measured for a fixed setting of fields on LOHENGRIN. Velocity distributions were investigated for a wide range of fragment mass numbers A. Note that LOHENGRIN is a velocity filter and that mass numbers are identified by mere inspection of the *A/q* pattern (see Fig. 4.2.2). Deriving energies form velocities and mass numbers one should bear in mind that the fragment velocities are roughly at 3% of the speed of light c and hence relativistic effects cannot be neglected. With the rest mass m_o the relativistic kinetic energy is for V << c

$$E = \tfrac{1}{2} m_o V^2 \left[1 + \tfrac{3}{4}(V/c)^2\right] \qquad (4.4.7)$$

It is a 0.1% effect. A simple way to take this effect into account is to balance the relativistic mass increase with the mass defect of fission fragments and to calculate kinetic energies of unslowed fragments with

$$E = (k/2)\,AV^2 \qquad (4.4.8)$$

where A is the mass number in amu. With velocities in units of (cm/ns) the energy is in units of MeV. Energies calculated with eq. (4.4.8) are underestimated by less than 100 keV which is smaller than the uncertainties introduced by the finite accuracy of the velocity measurements. With E and A being known, the pulse height X in the surface barrier detector was registered. A total of 6 detectors all from the same brand (ORTEC F-series) were investigated.

As proposed already in the early sixties by H.W. Schmitt et al. [SCH63] the detector response is parameterised as

$$E = (a + a'A)X + (b + b'A) \qquad (4.4.9)$$

with a, a′, b and b′ being calibration constants. Remarkably the formula tells that for given A the response X is a linear function of the incoming energy E, and that for given pulseheight X the energy depends linearly on mass A. Within the precision of $\delta E = 150$ keV achieved for the determination of kinetic energies in the present experiment the parameterisation was confirmed to be excellent. However, it would not be practical to suggest that any solid or gaseous ionisation detector should be calibrated on LOHENGRIN or any other similar facility. The way out was again proposed by H.W. Schmitt [SCH65]. The idea is to use the pulseheight spectrum of spontaneous fission of ^{252}Cf as reference for calibration. At LOHENGRIN, following calibration of each detector, the beam was stopped and a Cf source was inserted. Due to asymmetric fission of ^{252}Cf the spectrum is double-humped and the average pulse-heights X for the light and heavy peak P_L and P_H, respectively, are readily determined. But the task remains how to determine the four constants in eq. (9) with only two parameters P_L and P_H. The solution was found by observing that, for given pulse heights P_L and P_H, the dependence of the energy response on fragment mass is virtually identical for all detectors of a given brand. This simply shows that the mass dependence of the pulse height defect is universal. It means that starting from the two well established triples ($\langle E_L \rangle$, P_L, $\langle A_L \rangle$) and ($\langle E_H \rangle$, P_H, $\langle A_H \rangle$) two further triples ($\langle E_L' \rangle$, P_H, $\langle A_L \rangle$) and ($\langle E_H' \rangle$, P_L, $\langle A_H \rangle$) with pulse heights P_L and P_H interchanged should be identical for all detectors. In experiment this was demonstrated to be true within 250 keV but not better. The four triples thus obtained are then used to find the four calibration constants in Eq. (4.4.9). Numerical results for the brand of surface barrier detectors ORTEC F-series are given in [WEI86]. It should be stressed that for other types of Si detectors or operating conditions the calibration constants are at most only approximately valid. The present results are thought to settle longstanding issues on precise average kinetic energy data for low energy fission of the lighter actinides [GON91]. As a complementary example to Table 4.4.3 we put together in Table 4.4.4 post-neutron data for the light and heavy group of fragments from spontaneous fission of ^{252}Cf from the references [HEN81] and [WEI86]. With the reliably calibrated surface barrier detectors the post-neutron kinetic energies of fragments were investigated. The total *TKE* release after neutron emission has already been given in Table 4.4.2 for a convenient direct comparison with the pre-neutron data. As expected the neutron emission reduces the kinetic energies, the reduction being largest for ^{252}Cf(sf) having the largest neutron multiplicity among the examples shown.Though not further discussed here, it should be mentioned that many investigations combining velocity and energy measurements have been conducted with the aim to extract the prompt neutron multiplicity of fragments from a comparison of primary and secondary mass

Table 4.4.4. Average post-neutron velocities, masses and energies for the L and H group from ^{252}Cf(sf).

$\langle V_L \rangle$/cm/ns	$\langle V_H \rangle$/cm/ns	$\langle A_L \rangle$/amu	$\langle A_H \rangle$/amu	$\langle E_{KL} \rangle$/MeV	$\langle E_{KH} \rangle$/MeV
1.367(6)	1.034(4)	106.2(2)	142.1(2)	102.6(6)	78.4(6)

distributions. This indirect method is complementary to direct counting of neutrons. In (n,f) fission studies with energetic neutrons the indirect method avoids problems due to background when neutrons evaporated from fragments have to be discriminated against the neutrons inducing fission [MUL84].

Fragment velocity measurements on LOHENGRIN were technically of importance to check whether the calibration constants of the instrument relating magnetic and electric field settings to masses, energies and velocities of fragments are consistent with the theoretical design parameters. A very good agreement was found. It could also be made sure that the usual calibration of the spectrometer with a neutron reaction like ^{6}Li(n,α)τ yields reliable results compatible with the more direct calibration with fragment velocities.

Finally, in the above Table 4.4.2 experimental results are compared to an empirical relation in the bottom line reviewing a large body of fission data ranging from Fe to the heaviest elements [VIO85]. The relation is often used as reference and reads

$$TKE^* = 0.1189(11)\, Z^2/A^{1/3} + 7.3(15) \qquad (4.4.10)$$

According to eq. (4.4.10) the kinetic energy set free in fission increases proportional to the Coulomb parameter $Z^2/A^{1/3}$ of the fissioning nucleus. It indicates that the main contribution to the kinetic energy has to be attributed to the Coulomb repulsion between fragments following scission. But also the pre-scission kinetic energy of the system on the way from the saddle point to scission will increase with the Coulomb parameters $Z^2/A^{1/3}$ since the potential energy drop from saddle to scission is getting larger for larger fissilities Z^2/A. However, as shown below in Fig. 4.4.6 the contribution by the pre-scission energy is only about 5 % of the total energy *TKE**. The formula reproduces the trend of experimental findings rather well though the deviations showing up are a warning that eq. (4.4.10) should not be adopted for making predictions of kinetic energies from isotope to isotope.

The search for reliable figures of the average *TKE* in fission has a long history, as retraced in the above. By contrast, from the outset it was established that the distribution of *TKE* is well reproduced by a Gaussian. A result obtained in the fifties of last century is on display in Fig. 4.4.4 [MIL58]. The *TKE** before neutron emission found at that time for spontaneous fission of ^{252}Cf was $TKE^* = 181.9\ MeV$, thus only $\approx 2MeV$ smaller than the figure recommended nowadays after decades of research (s. Table 4.4.2). The shapes of the distributions remain Gaussian through the actinides until bimodal fission obtains in Fm, Md and No isotopes. The variance σ^2_{EK} of the distribution, however, increases with fissility Z^2/A by more than a factor 2 when going from Th to Es isotopes. This is shown in Fig. 4.4.5 [GON91]. The main contribution to the kinetic energy of fragments is the Coulomb repulsion between fragments after scission once the influence of the nuclear interaction has dwindled away. This energy is proportional to the product of nuclear fragment charges Z_1 and Z_2 divided by the distance R of the fragments centres of masses Z_1Z_2/R. A first difficulty for the calculation of *TKE* immediately appears. At scission the fragments are in general strongly deformed to a degree which is not a priori known and for deformed bodies facing each other higher multipoles of the Coulomb interaction have to be taken into account.There is still another component which has to be considered. When nuclear matter slides down form the saddle point at the fission barrier to the scission point a certain amount ΔV of potential energy is set free. This energy will partly go into collective and partly into intrinsic degrees of freedom (dof). The main collective motion is the stretching mode of the nucleus describing the descent from saddle to scission. There are still other collective modes like rotations around the fission axis or around an axis perpendicular to the fission axis to be studied in fission induced by polarised neutrons. Further collective modes like bending, wriggling and twisting are discussed in connection with the angular momentum carried by fragments. In the following only the stretching mode is considered. The kinetic energy of this mode at scission is labelled as E_K^{sci}. Usually E_K^{sci} is called pre-scission kinetic energy. The remaining part of ΔV is absorbed by excitation of intrinsic dofs. It is customary to invoke friction during the stretching process for this transfer of collective into intrinsic energy and therefore the intrinsic excitation at scission E_X^{sci} is often called the dissipated energy E_{DIS}. However, by contrast there are also reasons to believe that the motion from saddle to scission could be superfluid. In this scenario another mechanism for the transfer should come into play, viz. excitations due to the sudden snapping-off of the neck right at scission. This will be further discussed below in the chapter on charge distributions.

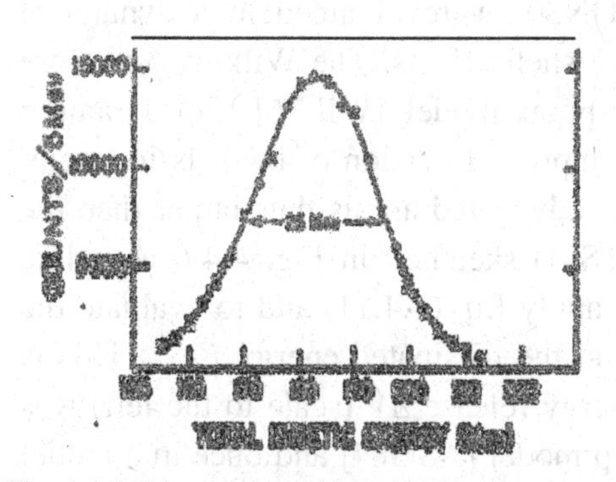

Fig. 4.4.4. Distribution of the total kinetic energy for fragments from ^{252}Cf(sf).

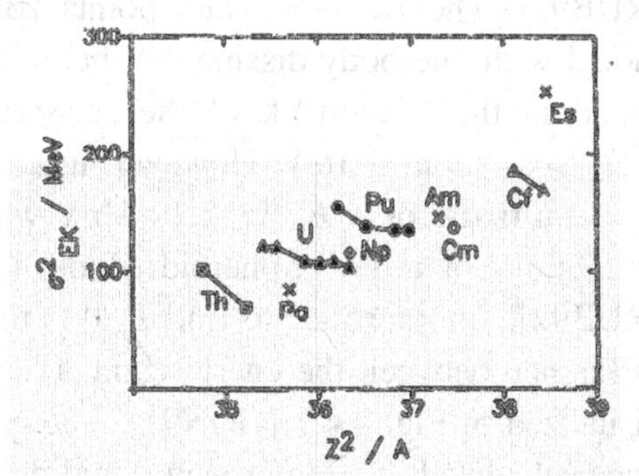

Fig. 4.4.5. Variance σ^2_{EK} of TKE distributions vs fissility Z^2/A in near-barrier fission.

Calling the energy showing up in the intrinsic dofs E_X^{sci}, whatever the mechanism may be, one has

$$\Delta V = E_K^{sci} + E_X^{sci} \tag{4.4.11}$$

In eq. (4.4.11) the contribution of the energy at the saddle $(Q-V_{SAD})$ with V_{SAD} the potential energy at the saddle is neglected. This approach is justified for thermal neutron fission of fissile nuclei where $(Q-V_{SAD})$ is small and on average only about 1 MeV. Unfortunately E_K^{sci} in eq. (4.4.11) is not directly accessible to experiment. Instead, several model calculations have been performed. Results of four different approaches to the average pre-scission energies E_K^{sci} as a function

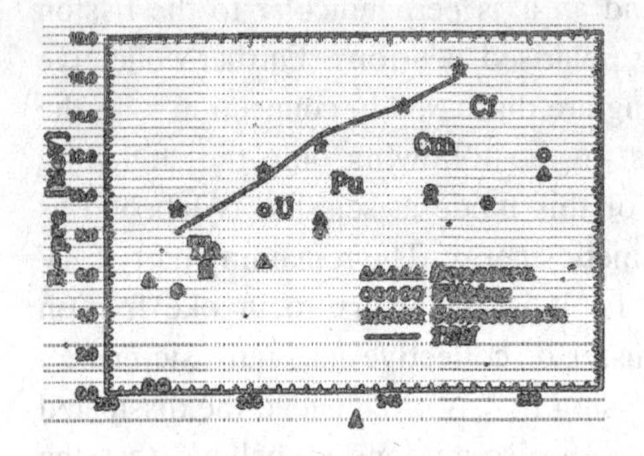

Fig. 4.4.6. Pre-scission kinetic energy $<E_K^{sci}>$ vs mass number of the fissioning nucleus.

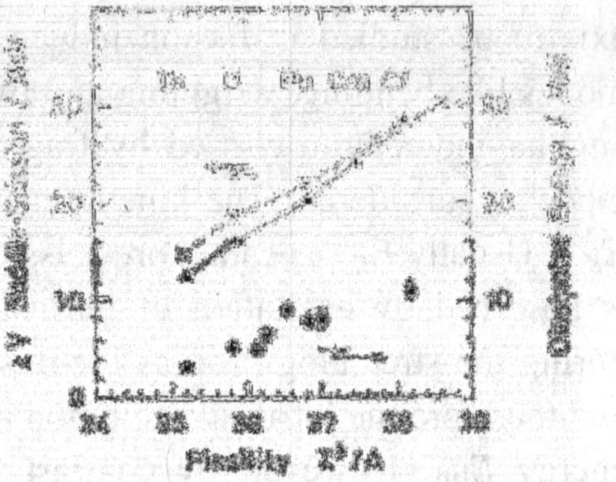

Fig. 4.4.7. Potential energy gain ΔV between saddle and scission and energy dissipated E_{DIS} vs Z^2/A.

of the mass number A of the fissioning nucleus are reviewed in Fig.. 4.4.6 [RUB92]. The Bonasera data points [BON86] were obtained in a dynamical model with one-body dissipation including shell effects. The Wilkins data were found in the framework of the scission-point model [WIL76]. For Uranium $<E_K^{sci}>$ is about *9 MeV*. However, in an improved version of the scission point model a figure of $< E_K^{sci}> \approx 14$ MeV was advocated and is thus larger than the predictions of a Two-Spheroid Model (TSM) sketched in Fig. 4.4.6 as a line [RUB92]. Another idea to find E_K^{sci} is to apply Eq. (4.4.11) and to evaluate the difference between the energy gain ΔV and the dissipated energy E_X^{sci}. This is visualized in Fig. 4.4.7 [GON91]. The energy release ΔV (scale to the left) was once calculated in an improved liquid drop model [ASG84] and once in a model of fluctuation-dissipation-dynamics [ADE85]. The two very different approaches yield concordant results. For the average dissipated energy $<E_X^{sci}>$ (scale to the right) the even-odd effect of fragment charge distributions was analysed. The evaluation of e-o effects to assess excitation energies depends on models with parameters not well known (see chapter 4.6).

The energies plotted in Fig. 4.4.7 are thought to be an upper limit. The difference $(\Delta V-E_X^{sci})$ is given in Fig. 4.4.6 as stars and labelled Gonnenwein.

Comparing the four approximations for E_K^{sci} there is evidently a large scatter and one should try to find indications which approximation could be given more credit. Some evidence comes from ternary fission, more specifically from trajectory calculations for the three outgoing particles. In these calculations the starting conditions at scission have to be adapted to best reproduce the angular and energy distributions of ternary particles. The starting energies of the fission fragments are amongst the parameters to be adjusted. For the reaction ^{235}U(n,f) in one calculation $E_K^{sci} \approx 10$ MeV [GUE91] and in another calculation $E_K^{sci} \approx 12$ *MeV* [GUS07] gave the best fits. It appears therefore that in Fig. 4.4.6 the stars and the continuous line from TSM come closest to the truth.

A further lesson may be drawn from the measurement of total kinetic energies *TKE* by applying energy conservation to deduce the total excitation energy in fission *TXE*. The total energy set free in (n,f) fission is the Q-value which is readily found from mass tables as the difference in masses of the compound nucleus and the two fission fragments taking into account the excitation energy of the compound nucleus due to the binding (and possibly kinetic) energy of the neutron. Energy conservation then yields the relation already quoted as eq. (4.4.1) $Q = TKE^* + TXE$ and *TXE* is obtained as the difference $(Q - TKE^*)$. The average $\langle TXE \rangle$ for thermal neutron induced reactions of fissile isotopes is displayed in Fig. 4.4.8 as a function of fissility Z^2/A. From ^{229}Th(n,f) to ^{249}Cf(n,f) *TXE* increases on average from *19 MeV* to *36 MeV*. The rise of $\langle TXE \rangle$ with fissility is found back in the increase of the number of neutrons evaporated from the fragments. The study of neutron and gamma emission from the fragments allows evaluating *TXE* independently from eq. (4.11) and hence checking the energy balance in fission. Energy conservation requires

$$TXE = \nu\,(S_n + \eta) + E_\gamma \qquad (4.4.12)$$

where ν is the neutron multiplicity, S_n the neutron binding (separation) energy, η the kinetic energy of neutrons in the cm frame of the fragments and E_γ the total energy release by gammas. The average *TXE* is approximately determined by taking the averages on the right side of eq. (4.4.12). To give an example: for thermal neutron induced fission of $^{236}U^*$ one has $\langle \nu \rangle = 2.42$, $\langle S_n \rangle \approx 5$ MeV, $\eta = 1.5$ MeV and $\langle E_\gamma \rangle = 6.43$ MeV [KNI91, NIS98, PLE72]. The $\langle TXE \rangle$ turns out to be 22.3 MeV and thus within error bars in good agreement with the results on display in Fig. 4.4.8. It proves that the experimental energy balance is satisfied. As a side remark it is worth mentioning that the gamma energy release E_γ is correlated with neutron evaporation. It may be parameterized as

$$<E_\gamma> = p \cdot <\nu> + q \text{ MeV} \quad (4.4.13)$$

where p and q further depend on fissility Z^2/A [NIF73, HAM03]. The first term on the right of eq. (4.4.13) takes into account the competition between neutron

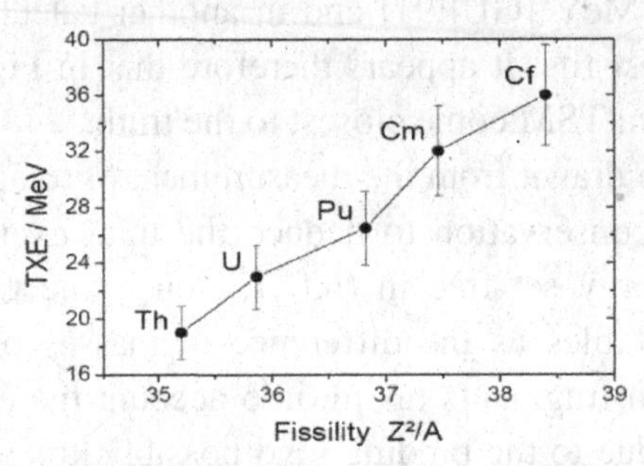

Fig. 4.4.8. Average total excitation energy TXE vs. fissility for thermal neutron reactions.

and gamma emission near the end of the neutron evaporation cascade, while the second term may be viewed as the average gamma energy in neutron-less fission with $<\nu> = 0$. For $^{236}U^*$ excited by capture of thermal neutrons the parameter p is p = 1.1 and the energy q is given to be 3.5 MeV [NIF73].

Finally, the energies *TKE** and *TXE* contributing to the total energy *Q* released in eq. (4.4.1) may be written explicitly as

$$TKE^* = E_K^{sci} + V_{coul} \quad (4.4.14)$$

and

$$TXE = E_X^{sci} + V_{def} \quad (4.4.15)$$

V_{coul} and V_{def} are the energies which at scission are still bound as potential Coulomb (plus nuclear) energy and as deformation energy, respectively. From the data reported it is then worth evaluating the fractions E_K^{sci}/TKE^* and E_X^{sci}/TXE. From Fig. 4.4.6 and Table 4.4.2 one finds that E_K^{sci} contributes from 6.5% for $^{236}U^*$ to 8.5% for $^{250}Cf^*$ to the total final *TKE*. Similarly, comparing E_X^{sci} labelled as dissipated energy in Fig. 4.4.7 with *TXE* from Fig. 4.4.8 it is inferred that the contribution of E_X^{sci} to the total excitation energy *TXE* ranges from 16% for $^{230}Th^*$ to 30% for $^{250}Cf^*$.

4.5 Mass-Energy Correlations

Well before the start of the ILL reactor important work on mass and energy distributions of fission fragments and their correlations had already been performed all around the world. This knowledge served as basis for more detailed and refined studies at the ILL. Outstanding pioneering work was performed at Oak Ridge, USA. Two examples of mass-energy distributions for ^{235}U thermal neutron fission are reproduced in Figs. 4.5.1 and 4.5.2 [SCH66]. These experiments were among the first to make use of surface barrier detectors in (2E) fission studies. In panel (a) of Fig. 4.5.1 solid points are the pre-neutron mass distribution in thermal-neutron fission of $^{235}U(n,f)$. It has been obtained by applying a correction to the double-energy data for neutron evaporation from the fragments. The distribution compares favourably with the primary mass distribution as evaluated from 2V measurements (see Fig. 4.3.1). The open circles in the figure are from radiochemical work and hence depict post-neutron mass yields. Particularly the high yield of the secondary mass 134 is standing out. It is brought about by the saw-tooth structure of neutron emission in the graph multiplicity versus fragment mass. For mass numbers decreasing in the heavy mass group from 155 to 130, the multiplicity is getting continuously smaller and approaches zero multiplicity for masses in the neighbourhood of the magic mass number 132 with 50 protons and 82 neutrons. Fragments in this mass region therefore accumulate yield from heavier neighbours. In panel (b) of Fig. 4.5.1 it is noteworthy that the average kinetic energy of the light fragment stays virtually constant while for the heavy fragment the kinetic energy decreases with increasing asymmetry of mass partition. For increasing asymmetry it should be expected that the light fragment energy increases because its share of the available energy Q increases. However, for larger asymmetries the Q-value of the fission reaction is getting smaller and this explains why the energy is bent down and becomes almost constant. This phenomenon is observed for all fissile actinides.

A further observation is the dip in total kinetic energy when comparing symmetric to asymmetric fission. For thermal neutron fission of $^{236}U^*$ the dip of *TKE* in Figs. 4.5.1 and 4.5.2 is near 25 MeV. The dip is most pronounced in the lighter actinides and tapers off for the heavier actinides Cf and Es [GON91]. For still heavier elements Fm and beyond mass-energy distributions with completely different characteristics show up, in particular symmetric bimodal fission [ITK09].

Complementary to *TKE* the total excitation energy *TXE* is plotted in Fig. 4.5.2 as a function of fragment mass. The dip in kinetic energy when approaching symmetry is compensated by a hike in the excitation energy. Upon comparing symmetric to asymmetric fission the hike of *TXE* is about 17 MeV and hence smaller than the dip in *TKE*. This mismatch is due to the decrease of the available energy Q by about 9 MeV when checking Q for the mass ratio 132/104 against Q for the ratio 118/118 in $^{236}U^*$ thermal neutron fission. For very asymmetric fission a rise of *TXE* is likewise found. According to Eq. (4.4.15) *TXE* receives contributions from both, the excitation energy E_X^{sci} and the potential energy of deformation V_{def} at scission. As already indicated in the foregoing chapter, E_X^{sci} is small compared to *TXE* (see Figs. 4.4.7 and 4.4.8) and the deformation energy V_{def} is carrying the lion's share of *TXE*.

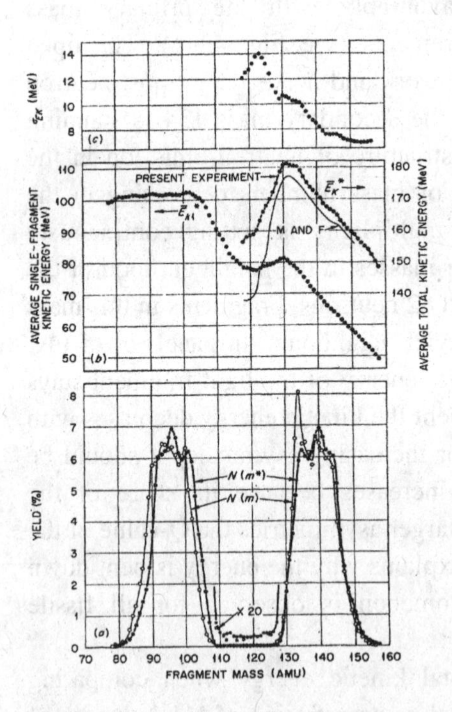

Fig. 4.5.1. Post-neutron and pre-neutron mass Distributions N(m) a nd N(m*), respectively, for $^{236}U^*$ thermal neutron fission (panel a). Single fragment E_{K1}^* and total pre-neutron emission kinetic energy E_K^* as a function of mass (panel b). Root-mean-square width σ_{EK} of total kinetic energy (panel c).

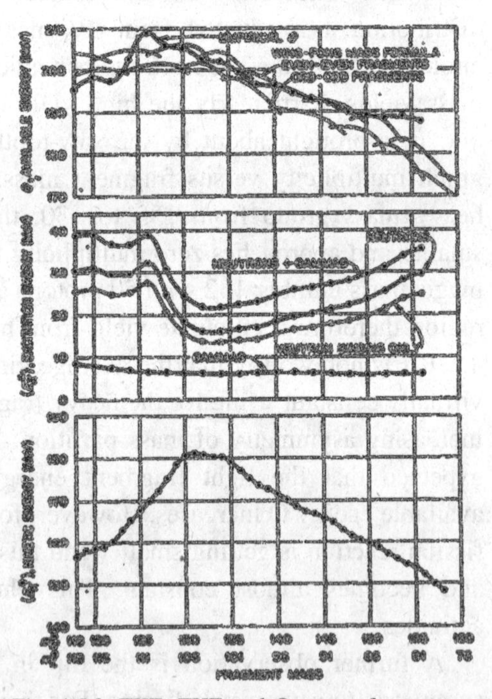

Fig. 4.5.2. TKE, TXE and Q versus fragment mass for thermal neutron fission of $^{236}U^*$(bottom, middle and top panel, respectively). The neutron and gamma contributions to TXE are shown separately.

The characteristics of both, the total kinetic and the total excitation energy *TKE* and *TXE*, respectively, are well understood in scission–point models [WIL76]. In these models the deformability and deformation energies of fragments are calculated by introducing into the liquid drop model shell effects in the fragments. For large deformability the quasi-equilibrium scission configuration will be stretched to two elongated spheroids and, hence, the kinetic energy *TKE* will be small and the excitation energy *TXE* large. By contrast, fragments with a pronounced shell structure will tend to be spherical corresponding to more compact scission configurations. It ensues that the Coulomb repulsion between fragments and therefore the total kinetic energy *TKE* will be large at the expense of excitation energy *TXE*. In Fig. 4.5.2 *TXE* is further decomposed into the contributions by neutrons and gammas. It is evident that above all neutron evaporation capitalizes on any energy not exhausted by kinetic energy while the gamma energy changes little with fragment mass ratio.

Coming back to Fig. 4.5.1, the variance in panel (c) has a characteristic peak at mass A* = 123, i.e. near the transition region from symmetric to asymmetric fission and further a shoulder near mass A = 135. It is a first indication of different fission modes contributing to fission already attracting attention in 1966 [SCH66]. The issue of fission modes has been revived by the theory of Brosa-Grossmann-Müller [BRO90]. It has become a standard exercise to analyse mass-energy distributions in terms of Brosa modes. An example is provided in Fig. 4.5.3, once more for thermal-neutron fission of $^{235}U(n,f)$ [KNI87]. Three modes called superlong, standard I and standard II are assumed to contribute to the mass yield (top panel), the total average kinetic energy <*TKE*> (middle panel) and the standard deviation (bottom panel).

For each mode the primary mass distributions are taken to be Gaussians with characteristic average values and widths. For the superlong mode no shell effects stabilizing the fragment shape at scission are present (hence called "liquid drop mode") and the ensuing stretched scission configuration explains the low *TKE* (see the middle panel). The standard I mode is traced to the influence of the near-spherical magic ^{132}Sn fragment in the heavy mass group where the compact scission configuration is responsible for both, the bump in the *TKE* and the maximum *Q*-value (see the middle panel). The broad standard II mode is interpreted in [WIL76] to be due to shell stabilized but deformed fragments with neutron number A ≈ 88 in the heavy fragment. The energy release for the standard II mode is in between the one for the superlong and the standard I mode.

The experimental data (solid points) are seen to be reasonably well reproduced by the superposition of the three modes (continuous line). The bottom panel in Fig. 4.5.3 finally shows the standard deviation σ_{EK}. Again the data points are well understood as a superposition of modes pictured as a continuous curve. Comparing the top with the bottom panel it is observed that the peak in σ_{EK} appears right in the transition from symmetric to asymmetric fission while the broad shoulder of σ_{EK} is observed in the overlap of the two standard modes.

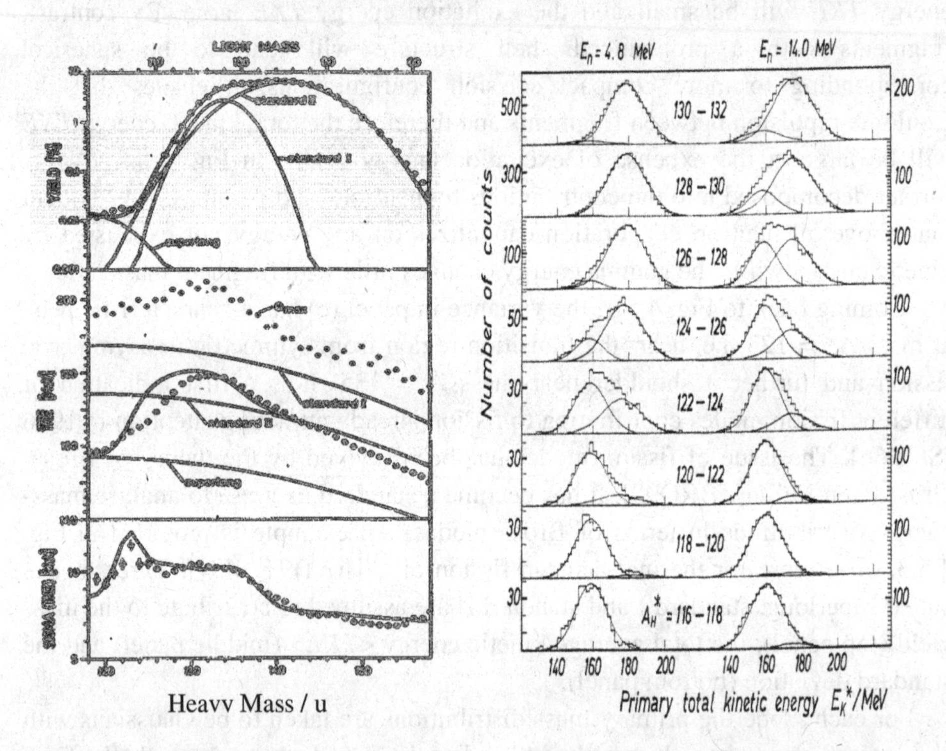

Fig. 4.5.3. Fit of three modes (superlong, Standard I and standard II) to the mass yield, the average kinetic energy <TKE> and its standard deviation σ_{EK} for ^{235}U (n,f) at thermal neutron energies.

Fig. 4.5.4. Deconvolution of total kinetic energy for windows near mass symmetry in ^{232}Th(n,f) at neutron energies of 4.8 and 14.0 MeV.

An even more detailed insight at what happens near the crossover from symmetric to V_{def} asymmetric fission is provided by inspecting not simply variances but, instead, the full kinetic energy distributions. Fig. 4.5.4 gives an example for the reaction ^{232}Th(n,f) induced by neutrons of energy 4.8 MeV (left panel) and 14.0 MeV (right panel) [PFE70]. For mass windows from symmetry at A* = 116 up to mass A* = 120 the distributions are single Gaussians for both neutron energies. At the neutron energy 4.8 MeV (first chance fission), however, the distributions for mass numbers A* = 120 to A* = 128 are clearly a superposition of two Gaussians centred at a lower <*TKE*> near symmetry and at a higher <*TKE*> when moving towards asymmetry. For still more asymmetric masses A* = 128 to A* = 132 the distributions become single Gaussians again. Yet, the most interesting feature in the figure is the observation that the two Gaussians in the intermediate mass range are best described as two modes, symmetric and asymmetric, retaining their characteristic <*TKE*> when moving through the mass windows. What changes with mass is the relative yield of the two modes (often quoted as the peak-to-valley ratio of mass distributions). This leads to skewed distributions in the crossover and a maximum width of the TKE distribution near A* ≈ 124 when the two modes have equal weights. It has to be stressed that the two modes behave independently without any mixing or crosstalk between them. This is precisely in the spirit of the "two-mode hypothesis" postulated as early as 1951 [TUR51]. From the viewpoint of Brosa modes the historical asymmetric mode is an amalgam of standards I and II.

Since with increasing excitation energy of the fissioning nucleus the yield of the symmetric mode rises faster than the asymmetric mode it should be expected that the equi-yield mass is moving to heavier masses. This behaviour is in fact found at the higher incoming neutron energy of 14.0 MeV in the ^{232}Th(n,f) reaction on display on the right-hand side of Fig. 4.5.4. The mass number with equal yield from both modes is now shifted from A * ≈ 124 to A* ≈ 128. At lighter masses the symmetric mode prevails and the *TKE* distribution is skewed to the right due to the surge of asymmetric events with larger *TKE*, while for heavier masses the asymmetric mode takes over and the distributions are skewed to the left due to the symmetric mode with lower *TKE* fading away. It means that even at the higher excitation of the compound nucleus studied the separation and independence of the two modes is still clear-cut. The two-mode phenomenon has been studied for many reactions, e.g. ^{234}U and ^{238}U(n,f) [HOL77], ^{232}Th(p,f) [NAG96] always with results as regards the independence of modes similar to those described.

These experimental findings are corroborated by many calculations of the potential energy surface (PES) near the saddle point. A comprehensive analysis of PES is to be found in [MOL09 with references therein]. It is well established that the dominating asymmetric mass splits in low energy fission of actinides are a direct consequence of asymmetric barriers at the saddle point being lower than for symmetric mass splits. In the present context, however, it is most interesting that a high ridge surges between the symmetric and asymmetric valleys which develop from the barrier down to scission. The ridge is preventing nuclear matter to flow from one valley to the neighbouring one. In [MOL09] this is visualized for the fissioning ^{232}Th nucleus in Fig. 4.6.17 of chapter 4.6. Of course, raising the excitation energy of the fissioning nucleus further and further, a point should be reached where the ridge separating the modes will be inundated by the nuclear flow towards scission. At the energies studied in Fig. 4.5.4 this has not yet happened.

Changing the topic, two examples of mass distributions for the rarely studied reactions ^{232}U(n,f) and ^{237}Np(n,f) are given in Figs. 4.5.5 and 4.5.6. The distributions were taken in the high thermal neutron flux available at the ILL reactor by the (2E) double energy method. Surface barrier diodes served as the energy detectors for fragments. The distributions for five energy windows E_L of the light fragment are plotted in Fig. 4.5.5 for thermal neutron fission of ^{232}U as a function of the provisional heavy mass μ_H [ASG81]. It has to be pointed out that the windows are chosen to lie in the high energy tail of the energy distribution. The so-called provisional masses shown in the figure are evaluated in (2E)-experiments in those cases where, like in the present reaction, the neutron multiplicity as a function of fragment mass is unknown. Provisional masses are calculated under the assumption that no neutrons at all are evaporated from the fragments. For the windows at large energies being inspected in the present example this is a valid assumption. In Fig. 4.5.5 the fine structure in the distributions is seen to vary smoothly as the energy window of constraint is shifted. Throughout the energy region analysed, peaks of the fine structure protrude at intervals of five masses. The phenomenon is traced to the even-odd effect of nuclear fragment charges favouring for even-Z compound nuclei the yields for even-Z fragments. This proton pairing effect is discussed in more detail in chapters 4.6 and 4.7. The five-mass interval is just a consequence of the fact that fragments try to maintain rather closely the neutron-to-proton ratio N/Z of the fissioning compound. For the stable actinides at issue this ratio is $N/Z \approx 1.5$. It means that for a shift in charge ΔZ from one even charge to the next even one with $\Delta Z = 2$ the neutron shift ΔN should be very nearly $\Delta N = 3$. Hence the shift in mass ΔA is $\Delta A \approx 5$. Besides pairing also shell effects for both, protons and neutrons, are modulating the mass yields.

The next example of a (2E) measurement at the ILL is for the reaction ^{237}Np(n,f) [WAG81]. The target is an odd-Z and even-N nucleus. With thermal neutrons fission proceeds by sub-barrier fission and, hence, the fission cross section is small. The cross section was measured to be σ_{fi} = 26 ± 5 mb. The high available thermal neutron flux of Φ_{th} = of 5 · 10^9 n/cm²s and the large ratio of thermal to fast neutrons Φ_{th}/Φfast = 5·10^4 nevertheless allowed to take reliable data with good statistics. Provisional fragment mass distributions $N(\mu_H)$ for rather high kinetic energy windows E_L of the light fragment are plotted in Fig. 4.5.6. Compared to the reaction ^{232}U(n,f) the fine structure is much less pronounced and the five mass interval from Fig. 4.5.5 is reduced to a 2.5 mass interval. The even-odd effect in the nuclear charges invoked for ^{233}U* is apparently absent for ^{238}Np*. For the odd-Z nucleus Np this appears to be not too surprising. The 2.5 mass structures correspond to charge shifts ΔZ of ΔZ = 1. The fine structure is then understood by bearing in mind that for any given charge there is virtually always a fragment mass being the most probable. It should, however, be mentioned that later LOHENGRIN experiments have disclosed strong even-odd effects for very asymmetric fission [TSI01] (see chapter 4.6).

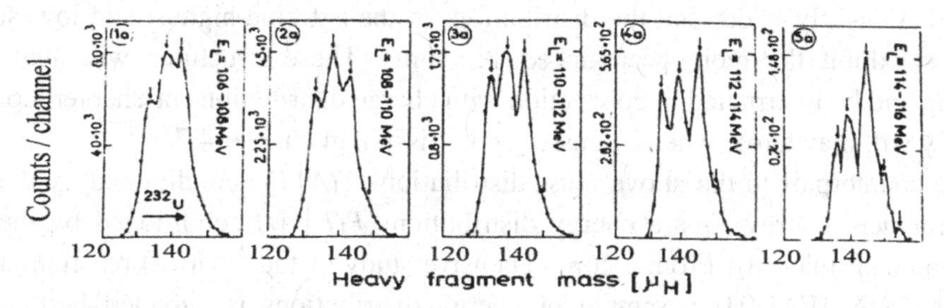

Fig.4.5.5. Provisional mass distributions conditioned by energy windows for the light mass from the reaction ^{232}U(n,f) induced by thermal neutrons.

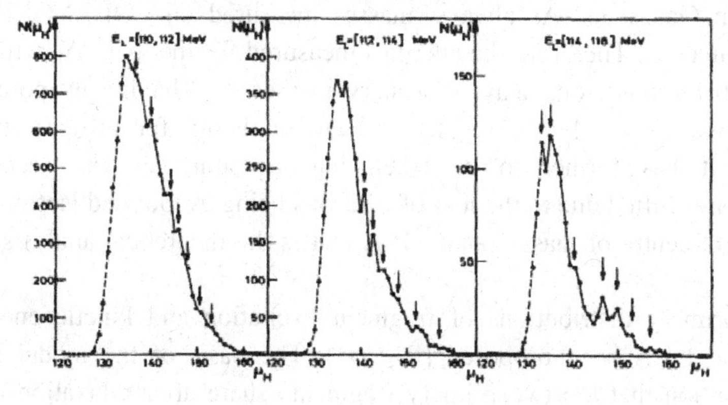

Fig. 4.5.6. Provisional fragment mass distributions N(μ_H) for the reaction ^{237}Np(n,f) constrained by windows in the kinetic energy E_L of the light fragment.

On LOHENGRIN and partly also on COSI mass-energy correlations were investigated for a series of fissile isotopes. The analysis was hereby focused on the light fragment group because the main interest was the study of fragment charge distributions which for the above spectrometers is only feasible in the light group. Typical examples for post-neutron mass distributions constrained by the kinetic energy E_L of the light fragment are on display in Fig. 4.5.7 for the reactions ^{233}U(n,f) and ^{235}U(n,f) induced by thermal neutrons. To the left are results from LOHENGRIN [QUA88] and to the right from COSI [MOL92]. It is notable that at large kinetic energies (bottom corners to the right) the mass distributions exhibit fine structure roughly every five masses apart, as was already evident in Fig. 4.5.5. With energy decreasing below the range studied in Fig. 4.5.5 the structure gets lost and at energies near E_L = 90 MeV has virtually disappeared. At still lower energies, however, some structure reappears. For comparable energies E_L the two sets of data are very similar. At COSI the energy range was slightly wider and the distributions at the extreme highest and lowest energies exhibit the more pronounced structures. These structures will again show up and be interpreted in connection with charge distributions of chapter 4.6, and they will play a role when discussing cold fission in chapter 4.7.

The counterpart to the above mass distributions $Y(A|E_L)$ conditioned by the light fragment energy E_L are energy distributions $P(E|A_L)$ conditioned by the light fragment mass A_L. From a comprehensive study of the ^{233}U(n,f) reaction at LOHENGRIN [FAU04] a sample of energy distributions is depicted in Fig. 4.5.8. The experimental energy distributions (data points) are seen to be to good approximation Gaussians. As always, masses measured on LOHENGRIN are post-neutron masses. Therefore the energies measured for the mass A_L will have received contributions from heavier primary masses A_L* having evaporated ν neutrons so that $A_L = A_L^* - \nu$. Starting with a theory for primary energy distributions it has further to be taken into account that the secondary distributions are shifted due to the loss of neutrons being evaporated isotropically in the fragment centre of mass system. This makes the theoretical analysis quite involved.

For the primary distributions of fragment excitation and kinetic energy a statistical model has been proposed [FAU04]. The basis of the model is the daring assumption that the two primary fragments share after relaxation of the deformation the same temperature T and that this temperature follows the simple law to be proportional to the Q-value of the reaction leading to a specific mass and charge split of the fragments, i.e. $T = f \cdot Q$ with f a constant. The constant is fitted to be f = *0.0045*. The average excitation energies and their distributions depend as usual on temperatures and level densities. It is further assumed that the excitation energies E_L*and E_H* of two complementary fragments are independent

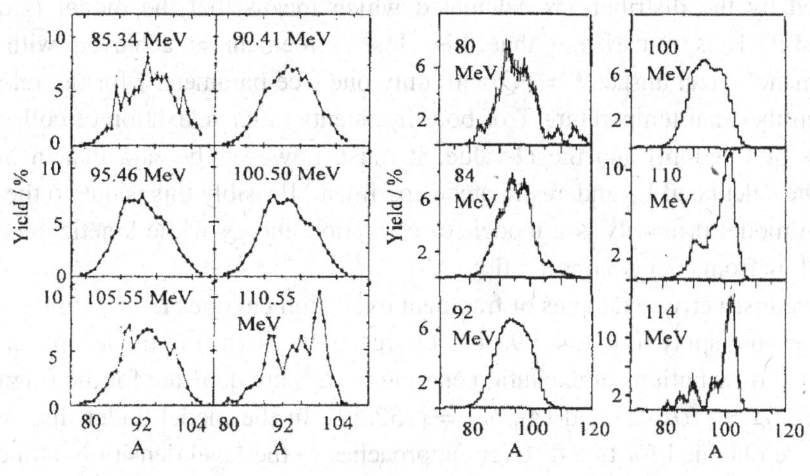

Fig. 4.5.7. Conditional mass distributions for fixed energy E_L of the light fragment in ^{233}U(n,f) to the left (LOHENGRIN) and ^{235}U(n,f) to the right (COSI).

within the constraints of energy conservation. In a Monte Carlo calculation two excitation energies E_L* and E_H* are selected and their contributions properly weighted. Their sum yields the total primary excitation energy *TXE*. The total primary kinetic energy *TKE** then simply follows from *TKE** = *Q* – *TXE*. It remains to partition *TKE** among the two fragments to deduce primary fragment kinetic energies E_L* and E_H*. Finally, neutron evaporation will eventually transform the primary into secondary fragments which are accessible to experiment at LOHENGRIN. As already stated this will shift the distributions to slightly smaller energies E_L and E_H.

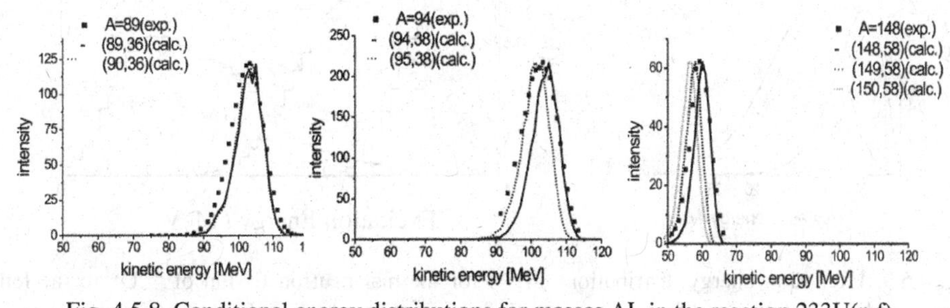

Fig. 4.5.8. Conditional energy distributions for masses AL in the reaction 233U(n,f).

In Fig. 4.5.8 besides the measured data points continuous and broken lines describe the calculated secondary energy distributions pending on neutron multiplicity. In the examples given the experimental distributions are accurately bounded by the distributions calculated which means that the model is quite successful. It is remarkable that this close agreement is achieved with the phenomenological ansatz $T = f{\cdot}Q$ with only one free parameter f for the relation between the final temperature T of both fragments (after relaxation of collective degrees of freedom) and the Q-value. It must, however, be said that in many cases the calculated E_L underestimates experiment. Possibly this is due to the fact that the model primarily is a model for excitation energy while kinetic energies just follow from energy conservation.

Some instructive examples of fragment excitation energies E^* as a function of mass are on display in Fig. 4.5.9. For the reaction ^{233}U(n,f) in the left panel, the calculated distributions of excitation energies $P(E^*)$ are depicted for the fragment pair $(A_L,Z_L) = (102,42)$ and $(A_H,Z_H) = (132,50)$. In the model under discussion they were obtained for two different approaches to the level density (continuous or pointed lines). In the graph the curves to the right belong to the light and those to the left to the heavy fragment. Obviously, the prediction of excitation energies for the heavy fragment depends strongly on the proper choice of level densities [FAU04]. For comparison, in the right panel of the figure experimental excitation energy distributions $P(E^*)$ for spontaneous fission of ^{252}Cf are shown.They were obtained by evaluating experimental neutron and gamma emission data [MAE84]. Labels on the graphs indicate fragment masses. The wide variety of spectrum shapes is astonishing and the more it appears to be difficult to predict them from theory.

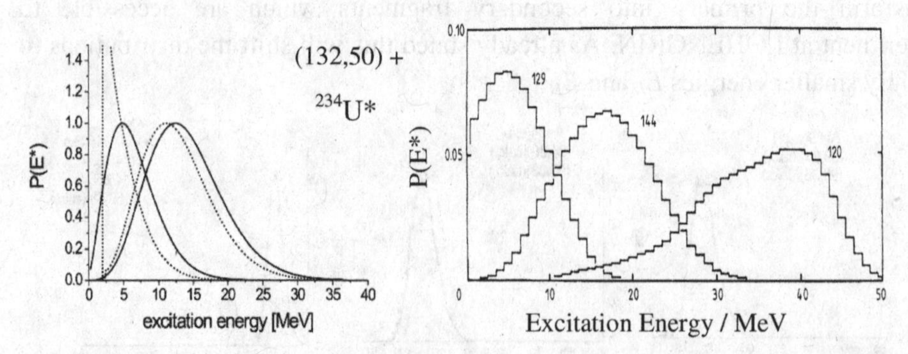

Fig. 4.5.9. Excitation energy distributions P(E*) for thermal neutron fission of ^{234}U* to the left from a model calculation and for spontaneous fission of ^{252}Cf to the right from experiment.

A particularity was disclosed on LOHENGRIN when studying the variances of kinetic energies for the reactions ^{233}U(n,f) and ^{235}U(n,f) [SID89]. It was found that the widths of the distributions exhibit prominent spikes for specific mass numbers. In Fig. 4.5.10 a first spike stands out at the light mass $A_L \approx 109$ for both isotopes and a second one at the heavy mass $A_H \approx 123$ for ^{234}U* and at $A_H \approx 124$ for ^{236}U*. The peak in the light group is readily understood by inspection of the fragment energy versus mass on the right of the figure. Near mass 109 the slope of the kinetic energy as a function of mass is very steep. At the same time, right at this mass the neutron multiplicity in fission of U-isotopes reaches its maximum with average value $<\nu> \approx 2$. Therefore several primary masses with very different energies come together in the given post-neutron mass and, hence, the width of the energy distribution for the secondary fragments measured becomes wider. The peaks in the heavy group can not be explained in the same way since for masses around 123 to 124 the neutron multiplicity has a minimum with average value $<\nu> \approx 0.5$. The peaks are therefore present in the primary energy distribution. Going back to Fig. 4.5.3, in thermal neutron fission of ^{235}U a spike in the variance of the energy distribution was also observed there at the primary mass $A_H^* = 123$. It was argued that the spike comes about when the symmetric and the asymmetric modes have equal weight since the modes do not merge but,

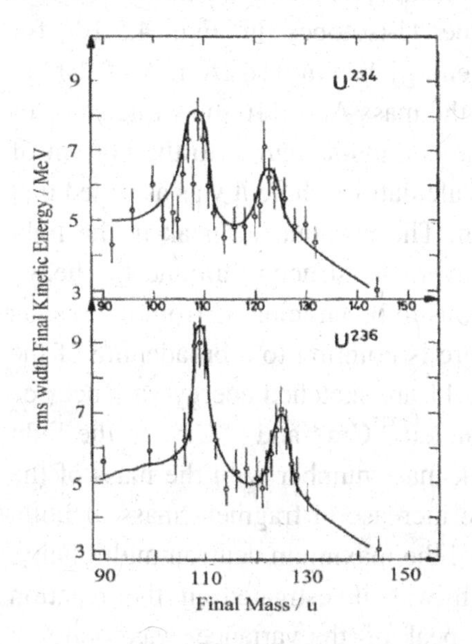

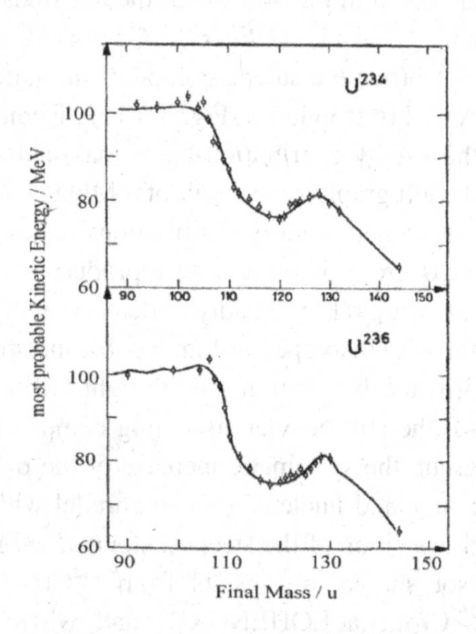

Fig. 4.5.10. Variances of the kinetic energies of fragments near symmetry for thermal neutron fission of ^{234}U*and ^{236}U* (to the left) and fragment kinetic energies near symmetry(to the right).

instead, retain their energy characteristics (see Fig. 4.5.4). Relative to each other the energy distributions of the two modes are shifted. Let us recall that the distinct energy distributions reflect the elongated scission configurations for symmetric liquid drop fission in contrast to the more compact configurations for the asymmetric cluster modes. The peaks in the heavy mass group in Fig. 4.5.10 are hence in line with the findings in Fig. 4.5.3. They just reflect the independent evolution of symmetric and asymmetric fission modes.

But also for the peak in the light group the same argument may be valid. Since the complement to the heavy masses $A_H^* \approx 123$ and 124 are the primary masses $A_L^* \approx 111$ and 112 for the U-isotopes $^{234}U^*$ and $^{236}U^*$, respectively, which after neutron emission will become secondary masses $A_L \approx 109$, the overlap of two energy distributions with different mean energies will contribute to the broadening of the energy width also in the light group. This broadening effect should further reinforce the effect linked to neutron emission from fragments with very different kinetic energies as argued above.

It is to be expected that the spikes in the variance of the energy distributions are also present in the heavier actinides investigated at LOHENGRIN. Examples are given in Fig. 4.5.11 for the reactions $^{242}Am(n,f)$ [GUT91], $^{245}Cm(n,f)$ [FRI98] and $^{249}Cf(n,f)$ [HEN92]. Compared to the U-isotopes in Fig. 4.5.10, for $^{242}Am(n,f)$ the steepest slope of the kinetic energy has moved from $A_L \approx 109$ to $A_L \approx 116$ (top left in Fig. 4.5.11) . Right for the mass $A_L \approx 116$ the variance σ_E of the energy distribution has a maximum. The continuous curve to the bottom of the left graph is the result of a Monte Carlo calculation where it was assumed that the primary energy distributions are smooth. The pronounced peak in the light mass group is quite well reproduced. However, the structure around the heavy mass $A_H \approx 123$ is badly underestimated and, as to be anticipated from the results for the U-isotopes in Fig. 4.5.10, the structure is pointing to a broadening of the primary distribution. To the right of Fig. 4.5.11 are sketched energy variances σ_E for the still heavier fissioning compound nuclei $^{245}Cm^*$ and $^{250}Cf^*$. In the light group the systematic increase of the σ_E peak mass number with the mass of the compound nucleus goes in parallel with the increase in fragment mass of both, the position of the steepest slope dE_K/dA_L and the maximum neutron multiplicity. Not shown are results from $^{252}Cf^*$ which was investigated in the reaction $^{251}Cf(n,f)$ at LOHENGRIN and where the peak of the variance was found at mass number $A_L \approx 122$ [BIR07]. This result should be compared to spontaneous fission of ^{252}Cf where the maximum of the neutron multiplicity is also found for the primary mass $A_L^* = 122$ in the light fragment group [BUD88].

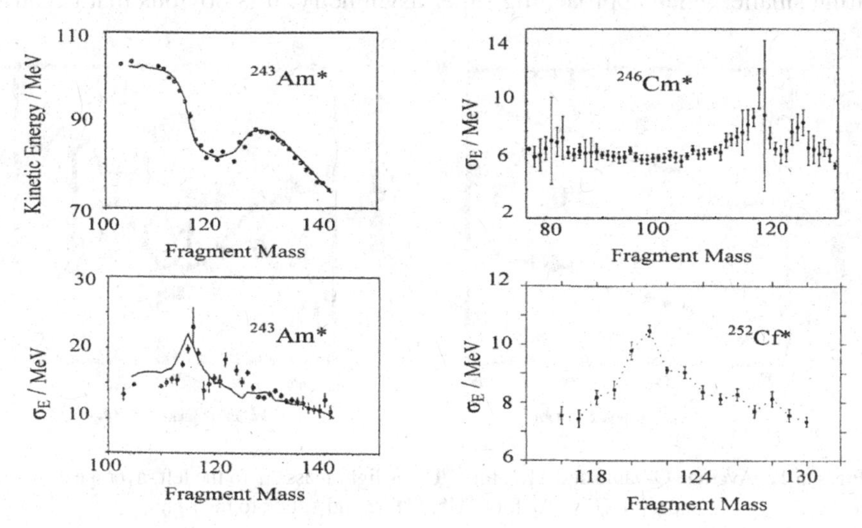

Fig. 4.5.11. Variances of the kinetic energy distributions for fragments from ^{242}Am (n,f) (bottom left), ^{245}Cm(n,f) (top right) and ^{249}Cf(n,f) (bottom right) vs fragment mass. Top left: fragment kinetic energies versus mass for ^{242}Cm(n,f).

By contrast, in the heavy mass group the position of the peak (or structure) stays virtually constant from the U-isotopes up to ^{246}Cm* at masses between $A_H = 123$ to 125. It is still clearly emerging at mass 125 for ^{246}Cm* while for ^{250}Cf* it is barely visible. This is not surprising since for ^{250}Cf the mass A = 125 is at exact symmetry while, as visualized in Fig. 4.3.3, the hump of the light mass distribution has come very close to the hump of the heavy fragments making the identification of a symmetric mode difficult. What could be expected is therefore only a broadening of the energy distribution around mass symmetry. The near–constancy of the mass number for the peak in the energy variance in the heavy mass group supports the interpretation outlined for the U-isotopes where it was argued that in the overlap region of the symmetric and asymmetric mode the energy distributions become broader.

A short glance on mass-energy correlations in super-asymmetric fission is given in Fig. 4.5.12. To the left is plotted the average <*TKE*> and the average <*Q*> down to light masses A = 70 in ^{235}U(n,f) [SID89]. Both energies decrease towards super-asymmetry. In the figure to the right the mean excitation energies

<*TXE*> labelled E^* = <*Q*>–<*TKE*> for three (n,f) reactions ^{236}U*, ^{240}Pu* and ^{250}Cf* are compared [HEN94]. In all three cases the excitation energy is getting smaller when approaching super-asymmetry. It is obvious that because of

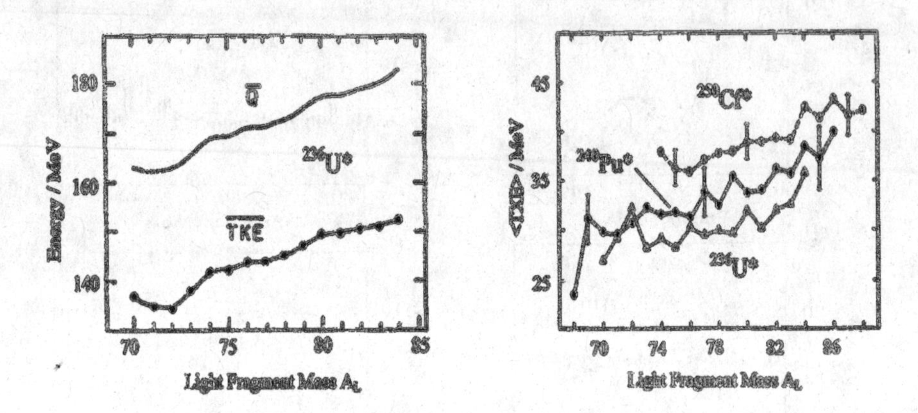

Fig. 4.5.12. Average Q-value and TKE for ^{236}U* vs light mass A_L to the left, average excitation energy TXE vs A_L for ^{236}U*, ^{240}Pu* and ^{250}Cf* to the right.

the smaller available energy *Q* at super-asymmetry there is less and less energy to be distributed to kinetic or excitation energy <*TKE*> or <*TXE*>, respectively, and hence both energies decrease. Of interest are the differences in excitation energies which increase by about 5 MeV from ^{236}U* to ^{240}Pu* and again by about 5 MeV from ^{240}Pu*to ^{250}Cf*.

4.6 Charge Distributions

4.6.1 *General Properties*

The investigation of fragment charge distributions has been for many years at the focus of research on the instruments LOHENGRIN and COSI of the ILL. Throughout physical methods based on the inspection of the ionization loss curve were employed to identify nuclear fragment charges (see chapter 2). Unfortunately these reliable and comprehensive methods are limited to measurements in the light fragment group. Since the time required for identification is much shorter than any intervening β-decay the charges measured are those of primary fragments. The charge numbers in the heavy mass group follow therefore unequivocally from the conservation of charge, i.e. $Z_{comp} = Z_L + Z_H$. The Z-balance is valid to very good approximation and only spoiled by ternary fission with emission of light charged particles. But this is a very rare process. Nevertheless a regrettable drawback is the fact that any charge resolution is lost for charges coming close to Z = 45 which means that charges in the symmetry breaking of even the lightest actinide Th studied with thermal neutrons are not accessible to measurement. For the comparative study of charge distributions in symmetric and asymmetric fission the Th isotopes are particularly well suited because the modes are well separated as brought to evidence by radiochemical studies (see chapter 4.6.4).

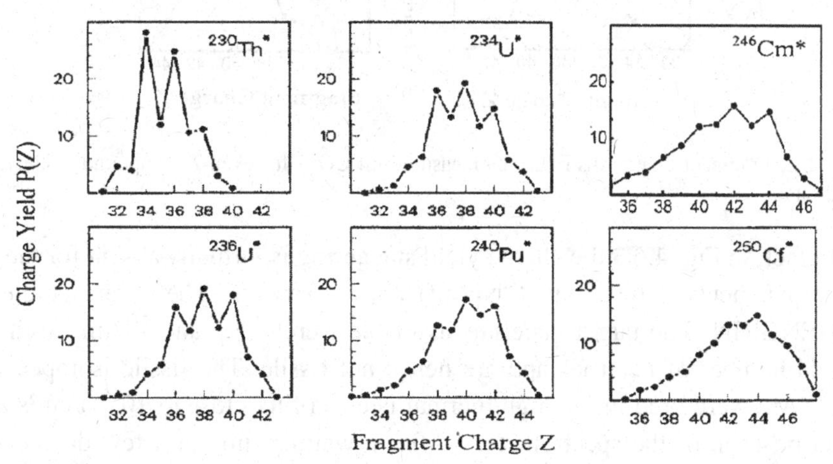

Fig. 4.6.1. Fragment charge distributions measured on LOHENGRIN for even-Z compound nuclei ^{230}Th*, ^{234}U*, ^{246}Cm*, ^{236}U*, ^{240}P'u* and ^{250}Cf*.

Charge distributions were measured for even-Z compound nuclei on LOHENGRIN for the reactions ^{229}Th(n,f) [BOC90], ^{233}U(n,f) [QUA88], ^{235}U(n,f) [CLE75,SIE76, LAN80, SID89], ^{239}Pu(n,f) [SCH84, DIT91], ^{241}Pu(n,f) [FRI98], ^{243}Cm(n,f) [TSE04], ^{245}Cm(n,f) [FRI98, ROC02, ROC04] and ^{249}Cf(n,f) [DJE89, HEN92, HEN94]. They were collected for review papers [BOC89, FIO93] and are on display in Fig. 4.6.1. There is a pronounced even-odd staggering in charge yields, most prominent in the lightest reaction ^{230}Th* investigated, gradually tapering away for the heavier nuclei and almost disappearing for ^{250}Cf*. It has to be stressed that even charge splits always exhibit higher yields.

Charge distributions for complementary nuclides were studied on the COSI spectrometer. The even-Z isotopes analyzed in thermal neutron fission were ^{229}Th(n,f) [Bou89], ^{232}U(n,f) [KAU92], ^{235}U(n,f) [MOL91, SCH94b], ^{239}Pu(n,f) [Bou91], ^{241}Pu(n,f) [SCH94b]. Two examples for ^{233}U* and ^{242}Pu* are given in

Fig. 4.6.2. Comparing Figs 4.6.1 and 4.6.2 the size of the yield staggering is very similar, both for the three U-isotopes and for the two Pu-isotopes. As discussed in more detail below, the evident cause for the staggering favouring even charge splits is proton pairing in fragments from fissioning even-Z compounds.

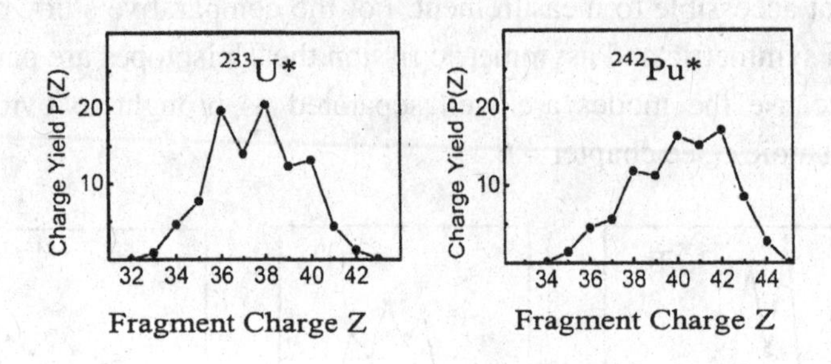

Fig. 4.6.2. Fragment charge distributions measured on COSI for even-Z compound nuclei 233U* and 242Pu*.

In fact, in Fig. 4.6.3 the charge yield staggering is virtually absent for the odd-Z thermal neutron reactions ^{238}Np(n,f) and ^{242}Am(n,f). They were studied on LOHENGRIN. The target materials available were ^{237}Np and ^{241}Am. Both have an odd number of neutrons and are hence not fissile. The fissile isotopes ^{238}Np and ^{242}Am were bred in the high thermal neutron flux close to 10^{15} n/cm²s at the target position of the spectrometer. After a waiting time of a few days enough fissile material was produced and the experiment could start. This is a rather unique feature of LOHENGRIN. Details of these non-standard experiments are to be found for Np in [MAR90, DAV98, TSE01] and for Am in [SIE91, STU95, TSE99]. The unpaired proton of the fissioning nucleus has apparently no preference to go to the light or heavy fragment and there is no even-odd effect in

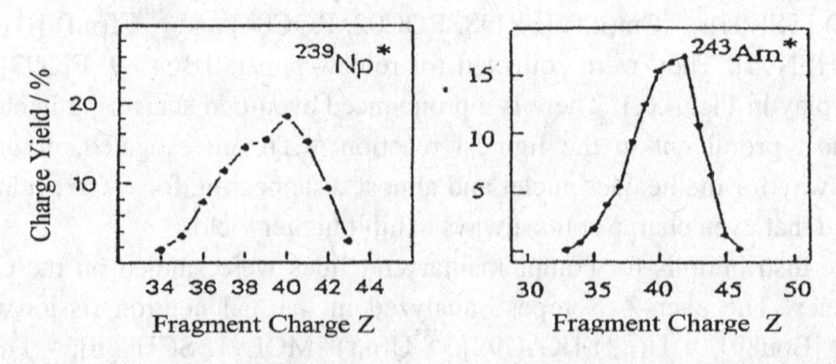

Fig. 4.6.3. Fragment charge distributions measured on LOHENGRIN for odd-Z compound nuclei.

the charge yields. However, in very asymmetric fission it was surprisingly observed that an even-odd staggering shows up (see chapter 4.6.3).

For even-Z nuclei there is a link between the structures of charge yields (Figs. 4.6.1, 4.6.2) and mass yields (Fig. 4.3.1). This is visualized in the 3-dimensional plot of independent yields *Y(A,Z)* for the light fragment group from thermal fission of ^{232}U in Fig. 4.6.4 (left panel) [GOE92]. Obviously, the distribution of yields *Y(A,Z)* is not perfectly smooth. The fine structure is brought to clear view by summing independent yields *Y(A,Z)* either for Z = const or A = const to yield the global mass or charge distributions *Y(A)* or *Y(Z)*, respectively.

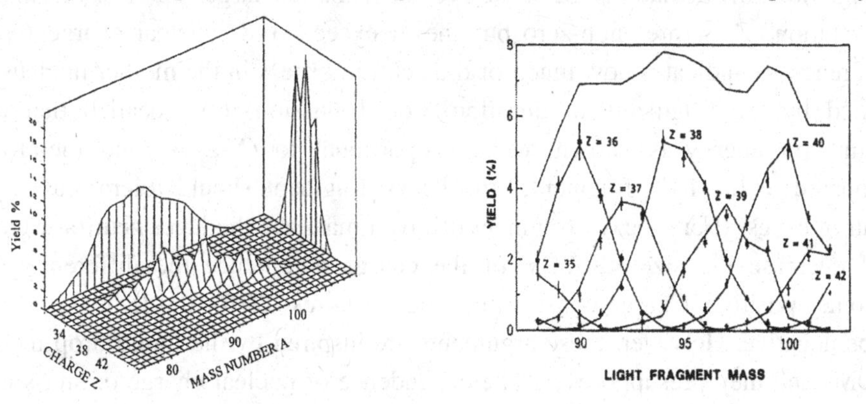

Fig. 4.6.4. Independent yields Y(A,Z) for ^{233}U* (left panel); decomposition of global into isotopic mass distributions for ^{236}U* (right panel).

In the projections of Fig. 4.6.4 the structure is most pronounced for the charge distribution *Y(Z)* with even charges being favoured compared to odd charges. But as seen in the projection, the mass distribution *Y(A)* is affected by the charge structure with the charge even-odd staggering being responsible for the 5-amu fine structure in mass yields. This becomes even more evident in the decomposition of the global mass distribution into the contributions from isotopic mass distributions $Y(A|Z)$ in Fig. 4.6.4 (right panel) [SCH94b]. The boost in the yields of even charges imparts the mass distributions a fine structure.

For increasing masses A_F of the fissioning compound the average masses of the light fragment group move towards higher mass numbers since the heavy group is stabilized in position by shell effects (see Fig. 4.3.3). Average charges in particular of the light group in the global charge distributions *Y(Z)* depend likewise on A_F. The average charges follow the same trend since the mass-to-charge ratios A^*/Z of the fragments tends to stay close to the ratio $A_F/Z_F \approx 2.5$ of the compound. This is demonstrated in Fig. 4.6.5 where for given primary masses A* the differences of isobaric charges $<Z(A^*)>$ from an unchanged charge distribution (UCD) are plotted as a function of $<Z(A^*)>$ (left panel]. The difference $\Delta Z = Z_{UCD} - <Z(A^*)>$ is shown for fission of $^{236}U^*$ and $^{237}Np^*$ [BOC89]. Thereby Z_{UCD} is the charge $Z_{UCD} = A^* \cdot (Z_F/A_F)$ to be expected in case the neutron excess of the fissioning compound is equally shared by the fragments. The data points correspond to given masses A*. Yet, on the abscissa instead of A* the average charges $<Z(A^*)>$ are indicated. As to be seen to the left in Fig. 4.6.5, the deviations ΔZ of the average isobaric charges $<Z(A^*)>$ from the expectation Z_{UCD} are non-zero but never exceed 0.8 nuclear charges. The difference ΔZ indicates how much of the neutron excess in the mother nucleus is carried by the fragments in question. For discussion it is recalled that the asymmetry energy is known to be proportional to $(N-Z)^2/A$ and therefore, compared to the light fragment, the heavy fragment should tolerate a larger neutron excess. For a heavy fragment of given mass A* the lager neutron excess will decrease the size $<Z(A^*)>$ of the charge and hence the difference ΔZ becomes positive. Charge conservation then requires ΔZ for the light fragments to be negative. However, these arguments are inspired by the liquid drop model (LDM) and they presuppose that the dependence of nuclear charge on mass is a smooth function - which is not the case. Instead, the optimum balance in the partitioning of neutron excess has to be considered case by case. For $^{236}U^*$ in Fig. 4.6.5 the charge ΔZ turns positive for $Z \approx 42$ in the light group violating the LDM expectation. For sure, for strictly symmetric fission $\Delta Z = 0$ will be valid.

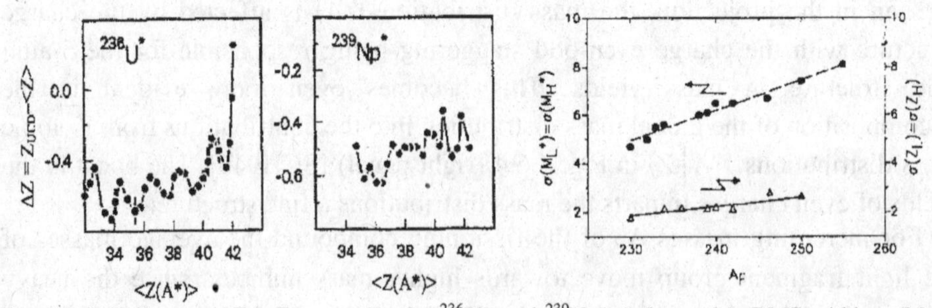

Fig. 4.6.5. Difference $\Delta Z = Z_{UCD} - <Z>$ for $^{236}U^*$ and $^{239}Np^*$ (left panels); RMS widths $\sigma(M_L^*)$ and $\sigma(Z_L)$ of primary mass and charge distributions versus mass A_F of fissioning nucleus(right panel).

The calculation of the average charge $<Z(A^*)>$ for a given primary fragment mass A* has been a challenge for theory since the early days of fission. One of the first approaches was the Minimum Potential Energy (MPE) model which was exploited in variants of increasing sophistication over the years [see GOE91]. The basic idea is to visualize the scission configuration as two almost touching spheres and to compute in its simplest version the corresponding potential energy $V(Z,A^*)$ from

$$V(Z_L,A_L^*) = E_{LDM}(Z_L,A_L^*) + E_{LDM}(Z_H,A_H^*) + Z_L Z_H e^2/R \qquad (4.6.1)$$

with Z and A* the charges and primary masses for the light (L) and heavy (H) fragment, E_{LDM} the eigenenergies of the fragments in a LDM approach, and the last right-hand term the Coulomb interaction between to charged spheres with their centres at a distance R. The evident constraints are $Z_F = Z_L+Z_H$ and $A_F^* = A_L^*+A_H^*$ with F as the label for the fissioning nucleus. The most probable charge for Z_L is then found by minimizing the potential energy $V(Z_L,A_L^*)$ with respect to Z_L from

$$\frac{\partial V\left(Z_L, A_L^*\right)}{\partial Z_L} = 0 \qquad (4.6.2)$$

The idea behind is that for given mass A* the charge Z is settled supposedly at the last moment at scission. The results of MPE models but also other global approaches like statistical models or scission point models have never been fully satisfactory in explaining e.g. the unanticipated change of sign for ΔZ (see Fig. 4.6.5 for $^{236}U^*$). The more it is remarkable that in microscopic calculations of charge distributions a complex dependence of charge ΔZ on mass has been found reproducing the features observed in experiment [DUB08].

For even-Z fissioning nuclei like U in Fig. 4.6.5 fluctuations of the charge polarization ΔZ are a well pronounced feature. It is observed that the fluctuations are linked to the even-odd character of the nuclear charge $<Z(A^*>$. Even fragment charges induce stronger deviations from Z_{UCD} than odd charges. For odd-Z fissioning nuclei like Np in Fig. 4.6.5 only some weak structure remains. Surveying a large body of thermal neutron induced and spontaneous fission reactions from Th up to Cf it was shown in radiochemical work that the slope $d(\Delta Z)/dA^*$ of the polarization $<\Delta Z(A^*)>$ increases with fissility Z_F^2/A_F of the fissioning nucleus [NAI07].

Coming back to the global distributions $Y(Z)$, the RMS widths of primary mass and charge distributions *Y(A*)* and *Y(Z)* are plotted in Fig. 4.6.5 (right panel) for thermal fission of the actinides studied at the ILL. In connection with the survey of mass distributions in Fig. 4.3.3 it was suggested that the widths of the distributions should increase linearly with the mass A_F of the compound. The suggestion is fully confirmed in the figure. Since the charge closely follows the mass, the widths of the charge distributions should observe the same trends. In fact, the charge widths are seen to increase linearly for increasing mass A_F of the fissioning nucleus, the ratio for mass-to charge widths staying close to 2.5, i.e. the mass-to-charge ratios A_F/Z_F of the fissioning compounds under study.

Having shown in Fig. 4.6.5 the average charges $<Z(A^*)>$ of isobaric charge distributions $Y(Z|A^*)$ for the reaction ^{235}U(n,f), the second moment of the distribution, the RMS width, is on display for ^{233}U(n,f) in Fig. 4.6.6 [QUA88]. A pronounced even-odd fluctuation is found which is related to the basic charge even-odd effect in Fig. 4.6.1. This can be understood by referring to Fig. 4.6.4 (right panel) where the independent yields $Y(A^*,Z)$ for ^{236}U* are presented. For example, inspecting for an even charge Z a mass chain A* near the maximum contribution to the isotopic distribution $Y(A^*|Z)$, the yield of this mass will be dominated by one single charge and thereby bring down the width $\sigma(Z|A^*)$. Even-odd fluctuations taken apart, the isobaric RMS width increases gently with the average charge $<Z(A^*)>$ in Fig. 4.6.6.

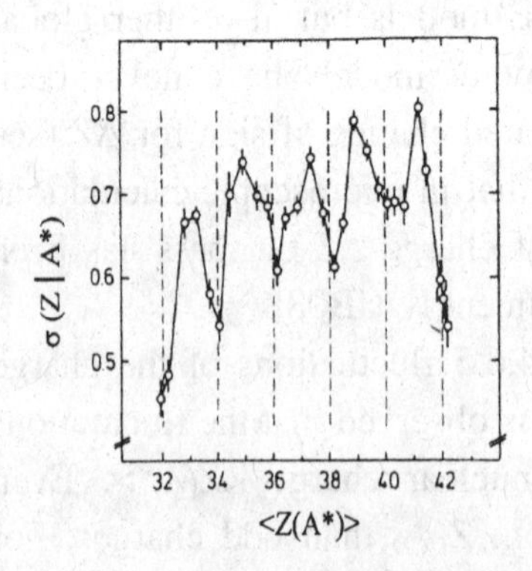

Fig. 4.6.6. RMS width of isobaric charge distribution $\sigma(Z|A^*)$ vs. average charge $<Z(A^*)$ for ^{233}U(n,f).

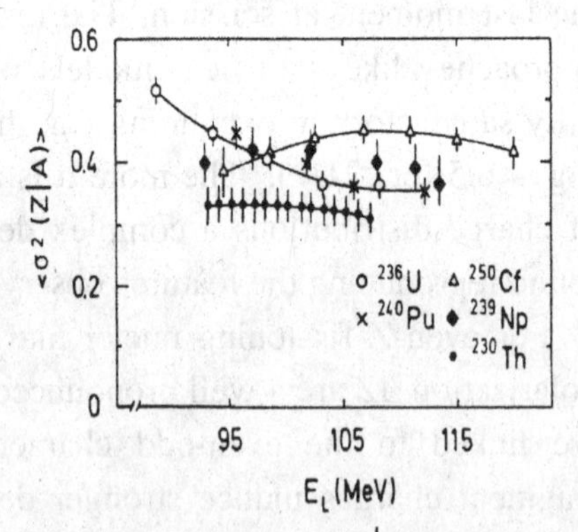

Fig. 4.6.7. Variance $<\sigma^2(Z|A>$ averaged over mass A vs. kinetic energy EL of fragment for (n,f) reactions.

Having been widely studied in heavy-ion collisions the variance of isobaric charge distributions $\sigma^2(Z|A^*)$ at fixed fragment mass is an interesting quantity also in fission. Several theoretical models for assessing its size have been proposed. In order to get rid of the even-odd staggering it is common use to consider the variance $<\sigma^2(Z|A^*)>$ averaged over mass A*. This quantity is plotted in Fig. 4.6.7 as a function of the kinetic energy E_L of the light fragment [BOU89]. Because the reactions shown behave differently, it is not clear from the graph whether there is a dependence on energy at all. Yet there is a definite dependence on the mass of the fissioning nucleus when comparing the Th- with the Cf-result, though ^{236}U*,^{239}Np* and ^{240}Pu* (and likewise ^{233}U* and ^{235}U* not shown) have very similar variances.

The dependence on compound mass is borne out in Table 4.6.1 where isobaric charge variances $<<\sigma^2(Z|A)>>$ averaged over masses and kinetic energies are given for even-Z fissioning compounds:

Table 4.6.1. Isobaric charge variances averaged over fragment mass and energy.

Nucleus	^{230}Th*	^{236}U*	^{240}Pu*	^{250}Cf*
$<<\sigma^2(Z\|A)>>$	0.32(2)	0.385(20)	0.40(2)	0.46(3)

Averaged over fragment mass the variances $<\sigma^2(Z|A)>$ appear to vary smoothly as a function of kinetic energy E_L in Fig. 4.6.7. However, for individual masses the energy dependences can be widely different from one mass to the next. Three examples are provided in Fig. 4.6.8 for the reaction ^{233}U(n,f) [QUA88]. For mass numbers 81 and 82 the variance measuring the spread for the individual charge yields remains non-zero up to the highest energies studied while for mass 89 the variance goes to zero. In most but not all cases the charge maximising the Q-value for that mass becomes preponderant at high energies. There is evidently no simple general for individual fragment masses.

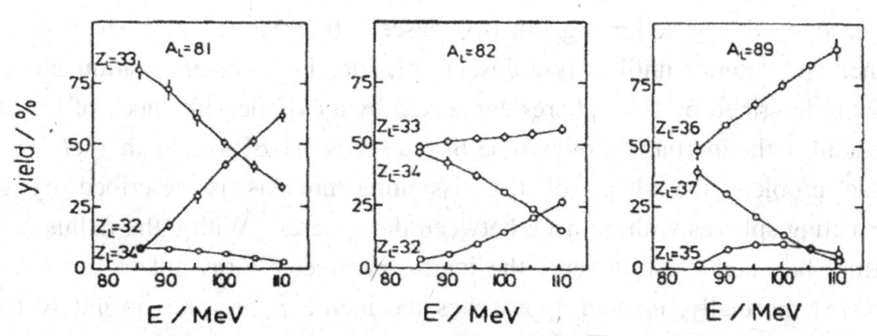

Fig. 4.6.8. Charge yield Y(Z) as a function of the kinetic energy of the light fragment for three light fragment masses.

The theoretical interpretation of isobaric charge variances follows two lines of reasoning. Either they are based on the study of collective charge equilibration or on the independent exchange of individual particles between fragments receding from each other. In the first approach it is argued that while the fragments are still together, taking shape and establishing a well defined mass a giant dipole resonance governs charge equilibration. The mode is driven thermodynamically. Since there are fewer protons than neutrons, the protons have to move faster to keep up with the change in mass. This mode relaxes to equilibrium in times of the order 10^{-22}s [BRO78] and is thus much faster than the time 10^{-21}s it takes the nucleus to move from the saddle to the scission point. Since in low energy fission temperatures involved are very low, the dipole mode exhibits the quantal behaviour of zero-point oscillations of an harmonic oscillator and the average isobaric variance, written σ^2_Z for short, is given by

$$\sigma^2_Z = \hbar\omega / 2C \tag{4.6.3}$$

where $\hbar\omega/2$ is the zero-point energy and C is the stiffness constant of the oscillator. The variance is independent of temperature or excitation energy as suggested by the constancy of the variance as a function of kinetic energy in Fig. 4.6.7 [BER79]. With the inertia parameter B for the charge oscillation the frequency squared of the oscillator is

$$\omega^2 = C / B \tag{4.6.4}$$

For the inertia B it has to be taken into account that the flow of nuclear matter is squeezed by the neck joining the two nascent fragments. The neck is getting thinner and thinner until it ruptures at scission. For a configuration close to scission described by two spheres connected by a cylinder–like neck of length d and radius c the inertia is shown to be $B \sim (d+2c)/c^2$ [HER81]. In another version to the problem the shape of the fissioning nucleus is described by two intersecting spheres with no neck between the spheres. With c the radius of the aperture between the fragments the inertia parameter turns out to be $B \sim 1/c$ [BRO78]. Formally, in both approaches the inertia B becomes infinitely large when the neck or aperture radius c goes to zero at scission. Hence the oscillator frequency ω and the charge variance σ^2_Z are both predicted to become nil, at variance with experiment. The way out is the realization that the charge exchange cannot remain adiabatic down to the very instant of scission. A rather strict condition for adiabaticity to hold is the requirement that the change ΔT of the period of oscillation T within a period has to be smaller than the period itself. With the time rate of change of the period dT/dt the change in one period is $\Delta T = (dT/dt) \cdot T$ and the adiabaticity condition becomes

$$\Delta T \leq T \qquad \text{or} \qquad dT/dt \leq 1 \tag{4.6.5}$$

This condition is violated near scission since $dT/dt = -(2\pi/\omega^2) \cdot d\omega/dt$ and hence $dT/dt = -(\pi/\omega^3) \cdot d\omega^2/dt$; upon assuming – as done below – that the neck closure velocity dc/dt is constant near scission, for $B \sim 1/c$ in eq. (4.6.4) also the decrease $d\omega^2/dt$ of the oscillator frequency squared is constant so that in conclusion $dT/dt \sim 1/\omega^3$. Hence, for oscillation frequencies approaching zero the adiabaticity condition is eventually violated with the consequence that the charge equilibration can no longer follow the change in shape and the charge variance $<\sigma^2(Z|A^*)>$ is frozen in [NIF79].

Since the inertia B in eq. (4.6.4) depends on the size c of the neck which is shrinking, the inertia is time-dependent and hence also the oscillator frequency ω. For a quantitative analysis, therefore, the properties of a time-dependent harmonic oscillator have to be investigated. The task was tackled by several groups. The isobaric width is calculated for an assumed time dependence of ω^2 with the speed of decrease $d\omega^2/dt$ taken to be constant [NIF80, MYE81]. Upon approaching neck closure adiabaticity is found to be lost and the charge variance is frozen in at a finite value. As already argued above, in the inertia model leading to $B \sim 1/c$ [BRO78] the constant speed for decrease of ω^2 entails a constant speed $dc/dt < 0$ of neck closure. The faster the neck is closing, the larger is the speed of decrease $|d\omega^2/dt|$ and the adiabaticity condition in eq. (4.6.5) is violated for larger angular velocities ω. According to eq. (4.6.3) the frozen-in variance σ^2_Z will hence become larger. From the measured increase of the isobaric variance $<\sigma^2(Z|A)>$ from Th to Cf it is thus inferred that the speed of neck closure dc/dt is getting faster from Th to Cf [NIF80, BOC89]. This result is not unreasonable since – as shown in Fig. 4.4.6 – the pre-scission kinetic energy and hence the velocity of descent from saddle to scission increases with the mass of the fissioning compound and therefore also the speed of deformation and in particular of neck closure is expected to follow the same trend. The experimental variances $\sigma^2_Z \approx 0.4$ for the U-Np-Pu cluster in Fig. 4.6.7 correspond to a speed of neck closure *dc/dt* of about 6 fm/10^{-21}s. Adiabaticity is lost in the very last moment, only between $2{\cdot}10^{-22}$s and $4{\cdot}10^{-22}$s before neck closure [MYE81]. Thereby it has to be kept in mind that the results quoted were obtained for an assumed constant speed dc/dt of neck closure. This is a valid first approach but the question is whether it is realistic. A better approach should be to assume that the nascent fragments are drifting apart at constant or even increasing velocity. Modelling with this assumption the shape of the fissioning nucleus by two intersecting spheres [BRO78] the speed of neck closure dc/dt would dramatically increase near scission. The loss of adiabaticity at pinch-off of the neck becomes even more compelling in this picture.

A strong argument put forward in favour of frozen quantal collective fluctuations is the constancy of the variance σ^2_Z when analyzed as a function of excitation energy. As to be seen in Fig. 4.6.7 the variance barely depends on the kinetic energy of the light fragment which by energy conservation also tells that at least averaged over fragment mass the variance is independent of the total excitation energy *TXE*.

As already stated when starting the present discussion of charge variances there are alternative interpretations where the reasoning is not based on collective oscillations. It is suggested that the stochastic exchange of single protons could be responsible for the charge variance. The exchange of protons and neutrons is not completely independent because a correlation develops due to the asymmetry energy preventing large deviations of the N/Z ratio from the initial one in the fissioning nucleus. For heavy-ion reactions it is shown that the independence of the charge variance on the excitation energies of the reaction partners, which is known from experiment, is reasonably well reproduced [SCH81, MER81]. Unfortunately, for low energy fission so far only attempts have been made to describe mass distributions but not charge distributions by nucleon exchange at scission [PRA79].

4.6.2 *Charge Even-Odd Effect*

Following the inspection of general properties of charge distributions we now address the most salient features of charge distributions, viz. the prominent even-odd (e-o) staggering of yields. Charge yields are displayed in the Figs. 4.6.1, 4.6.2 and 4.6.3. The staggering is clearly observed for even-Z fissioning compounds while for odd-Z nuclei there is no evidence for even-odd fluctuation in the mass range shown. It is further seen that the amplitude of the e-o staggering decreases when moving from the lighter ^{230}Th* to the heavier ^{250}Cf* nucleus. Commonly the size of the e-o effect is quantified by calculating from the charge yields [CLE75]

$$\delta_Z = \frac{Y_e - Y_o}{Y_e + Y_o} \tag{4.6.6}$$

with Y_e and Y_o the sum of yields for even and odd charges, respectively. It is common use to normalize the sum $(Y_e + Y_o)$ to $(Y_e + Y_o) = 100$ and to quote the charge even-odd effect in %. In Table 4.6.2 have been collected the charge e-o effects δ_Z as measured on LOHENGRIN, except for ^{233}U* from COSI [KAU91] and ^{239}Pu* from radiochemistry [HAD88, NAI04]. All reactions are thermal neutron fission. The obvious decrease from Th to Cf of the e-o staggering in Figs. 4.6.1 and 4.6.2 is well reproduced by the drop of the charge e-o effect δ_Z. In the table it is eye-catching that within error bars the three U-isotopes on one hand and the three Pu-isotopes on the other hand have very similar values of δ_Z. This is also borne out in Fig. 4.6.9 where δ_Z is plotted as a function of mass number A of the fissioning nucleus. For thermal neutron fission on display in Fig. 4.6.9 there is a jump of the e-o effect from element to element. For a given element the e-o effect δ_Z does apparently not depend on neutron number. The main parameter steering the e-o effect δ_Z thus turns out to be the charge number Z of the fissioning nucleus. Therefore, in Fig. 4.6.10 the e-o effect is plotted on a logarithmic scale as a function of the charge number Z of the fissioning nuclei. From Th to Cf the e-o effect decreases exponentially.

Table 4.6.2. Even-odd effect δ_Z for thermal neutron induced fission.

Nucleus	^{230}Th*	^{233}U*	^{234}U*	^{236}U*	^{239}Pu*	^{240}Pu*	^{242}Pu*	^{246}Cm*	^{250}Cf*
δ_Z/ %	41.2(10)	20.3(15)	22.1(21)	23.4(10)	13.3(40)	11.7(5)	10.0(15)	9.3(12)	4.6(7)

Commonly, in the literature the e-o effect is presented either as a function of fissility Z^2/A [GON91] or the Coulomb parameter $Z^2/A^{1/3}$ [BOC89]. Also there the decrease of δ_Z for the heavier nuclei is exponential. None of the three parameters, Z or Z^2/A or $Z^2/A^{1/3}$, is singled out by a compelling a priori justification. Anyhow, the linear correlation of $ln\ \delta_Z$ vs. Z shown in Fig. 4.6.10 appears to be the smoothest one.

Trying to understand the characteristic features of the charge e-o effect it will be expedient to look for correlations of the effect with other quantities in fission. A first correlation is found between the charge e-o effect δ_Z in Fig. 4.6.10 and the total average excitation energy *TXE* from Fig. 4.4.8 for thermal neutron fission of

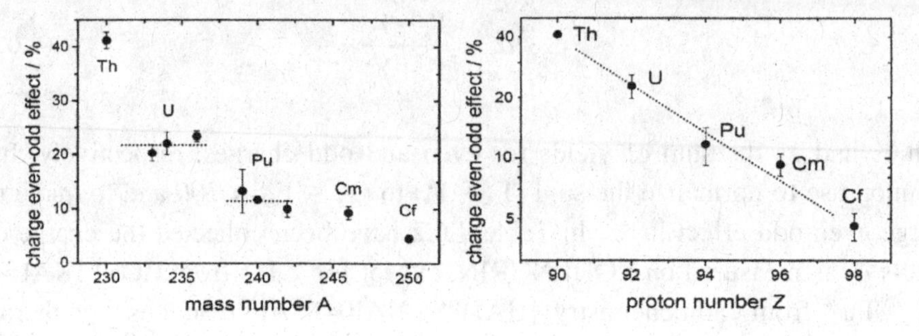

Fig. 4.6.9. Charge even-odd effect δ_Z vs. mass number of fissioning nucleus.

Fig. 4.6.10. Charge even-odd effect δ_Z vs. proton number of fissioning nucleus.

the elements from Th to Cf. For *TXE* the grand average is taken for both, fragment masses and kinetic energies. In Fig. 4.6.11 (left panel) the charge e-o effect decreases in a logarithmic plot linearly with the rise of the average excitation energy. Another charge–energy correlation is the one between the charge e-o effect and the kinetic energy of fragments. For three thermal neutron reactions the e-o effect is plotted as a function of the kinetic energy of the light fission fragment in Fig. 4.6.11 (middle panel) [DJE89]. It should be noted that due to momentum conservation, from the kinetic energies of the light fragments the energies of the heavy fragments are found which enables to evaluate the total *TKE*. A rise or fall of the light fragment energy is mirrored in the total *TKE*.

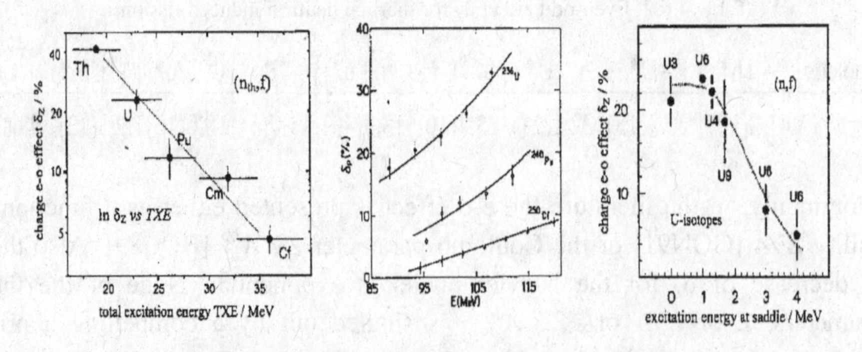

Fig. 4.6.11. Charge e-o effect δ_Z vs. average TXE (left panel), δ_Z vs. kinetic energy E of the light fragment (middle panel) and δ_Z vs. excitation energy E^*_{saddle} at the saddle point (right panel).

Since from eq. (4.4.1) the total excitation energy is $TXE = Q - TKE$, a large light fragment energy in Fig. 4.6.11 corresponds to a small excitation energy. The two figures to the left and in the middle of Fig. 4.6.11 hence convey the same noteworthy information: the charge e-o effect decreases for increasing *TXE*. The graph to the right in Fig. 4.6.11 is probably the most telling [BOC89]. For U-isotopes with e-o effects lumped together in Fig. 4.6.9 the size of the effect is displayed as a function of the intrinsic excitation energy E^*_{saddle} at the saddle point (the energy is given in MeV). The label U3 stands for $^{232}U(n_{th},f)$, U4 for $^{233}U(n_{th},f)$, U6 for $^{235}U(n_{th},f)$, $^{235}U(n_{2MeV},f)$ and $^{235}U(n_{3MeV},f)$, and finally U9 for $^{238}U(n_{3MeV},f)$. The excitation energy at the saddle E_{SAD}^* is

$$E_{SAD}^* = B_n + E_n - E_{bar} \tag{4.6.7}$$

with B_n the neutron binding energy, E_{bar} the height of the barrier and E_n the energy of the incoming neutron. For thermal neutron fission $E_n \approx 0$ and for fissile isotopes in the actinides the excitation at the saddle

$$E_{SAD}^* = (B_n - E_{bar}) < 2\Delta \tag{4.6.8}$$

is below the pairing gap at the saddle $2\Delta \approx 1.7\ MeV$. Hence only collective transition channels are excited. Remarkably, in Fig. 4.6.11 up to an excitation of about 2Δ above the barrier the e-o effect stays constant. For fission induced by neutrons with energies larger than 1 MeV the energy at saddle E_{SAD}^* will become $E_{SAD}^* > 2\Delta$. As soon as the gap energy is overcome intrinsic qp-excitations are populated already in the saddle. In Fig. 4.6.11 the e-o effect is observed to drop down sharply at these energies. This means that by way of nucleon pair breaking the e-o effect is sensitive to the intrinsic single particle degrees of freedom. The e-o effect may hence be considered as a sensor for the energy E_X^{sci} dissipated in the course of fission into intrinsic degrees of freedom. This is a valuable feature allowing to disentangle the contributions of E_X^{sci} and the deformation energy V_{def} at scission to the total excitation energy $TXE = E_X^{sci} + V_{def}$ (see eq. (4.4.15). We recall that *TXE* is readily found as difference between the total available energy Q and the total kinetic energy TKE^* (see eq. (4.4.1)).

The noteworthy result that for excitations at the saddle point which do not exceed the pairing gap the e-o effect stays constant has also been reported from photofission experiments with bremsstrahlung photons. For all three targets studied, ^{238}U [POM93], ^{232}Th [PER97a] and ^{235}U [PER97b], it was found that only for excitation energies in excess of the pairing energy 2Δ the charge e-o effects drop away. A very interesting observation was that the odd neutron in the ^{235}U(γ,f) reaction did not significantly alter the virtually constant charge e-o effect known for the U-isotopes. This is further evidence that the charge e-o effect only depends on the charge number of the fissioning nucleus.

Evidently the question is, how can the energy dissipated E_X^{sci} be assessed quantitatively from a measurement of the charge e-o effect δ_Z. The discussion of this question is controversial. There are, first, statistical models where it is assumed that the intrinsic degrees of freedom reach equilibrium before scission is reached. An example of such a model will be considered in the next subsection 4.6.3. In a second class of models thermal equilibrium is not required. It is pointed out that the coupling between collective fission degrees of freedom and intrinsic degrees of freedom is weak and that therefore fissioning even-even Z nuclei could remain fully paired up to the loss of adiabaticity right at scission discussed in the foregoing section 6.1. Only at this moment proton pairs could be broken with the constituent protons ending up in different fragments, thus giving rise to a charge even-odd effect smaller than 100%.

One of the first models proposed is an original combinatorial analysis of the probability distribution of pairbreaking where the mechanism of breaking is left open [NIF82]. Starting from a superfluid model for the nucleus, pairbreaking is brought about by quasi-particle excitations. With the definitions for

N = maximum number of 2 q-p excitations available with N depending on excitation energy
q = probability to break a pair when the energy is available,
ε = probability for broken pair to be a proton pair, and
p = probability for nucleons from a broken pair to go into complementary fragments

the charge e-o effect δ_Z from eq. (4.6.6). is calculated to be

$$\delta_Z = (1 - 2p\varepsilon q)^N \qquad (4.6.9)$$

The energy consumed, on the other hand, is given by the average number <N> of broken pairs and the energy 2Δ required to break a pair. Since $\langle N \rangle = qN$ it therefore follows

$$E_X^{sci} = 2\Delta \cdot \langle N \rangle = 2\Delta \cdot qN \qquad (4.6.10)$$

The only parameter known with some certainty in these equations is the gap parameter which at the saddle point is $2\Delta \approx 1.7\ MeV$. It is further plausible that the probability ε for proton pairs to be broken is proportional to the number of respective nucleons whence one expects $\varepsilon = Z/A$ with Z and A the charge and mass number of the fissioning nucleus. Therefore $\varepsilon \approx 0.4$ should be a good approximation. For a proton pair broken in the descent from saddle to scission the two protons may be conjectured to stay either together in one fragment or to go into complementary fragments. This would give $p = 0.5$. However, for pairbreaking near neck rupture p is probably more close to $p \approx 1$. The size of the remaining parameter q remains unknown. Assuming $q = 0.5$ as a first guess and with $2\Delta \approx 1.7$ MeV and $p = 0.5$ one finds

$$E_X^{sci} \approx -3.8\ ln\ \delta_Z\ MeV \qquad (4.6.11a)$$

The formula tells that $E_X^{sci} = 0$ for no pair-breaking ($N = 0,\ \delta_Z = 1$), while E_X^{sci} increases the more pairs are broken ($\delta_Z \rightarrow 0$). The energy dissipated is plotted as a function of the charge number Z of the fissioning nucleus in Fig. 4.6.12 for the even-odd effects δ_Z collected in Fig. 4.6.10. Since in this latter figure ln δ_Z decreases linearly with the charge Z, the dissipated energy E_X^{sci} increases linearly with Z in Fig. 4.6.12.

Within conceivable limits the choice for q from $q = 0.2$ to $q = 1$ does not change much the energies predicted for E_{DIS}. For $q = 0.2$ one finds $E_{DIS} = -4.1\ ln\ \delta_Z$ MeV while for $q = 1$ the energy is $E_{DIS} = -3.4\ ln\ \delta_Z$ MeV. There are, however, more uncertainties. For pairbreaking near neck rupture the increase of the gap

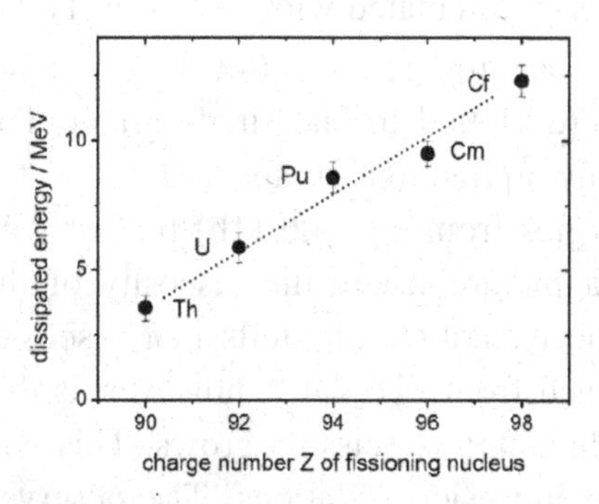

Fig. 4.6.12. Dissipated energy E_X^{sci} vs charge number Z of fissioning nucleus.

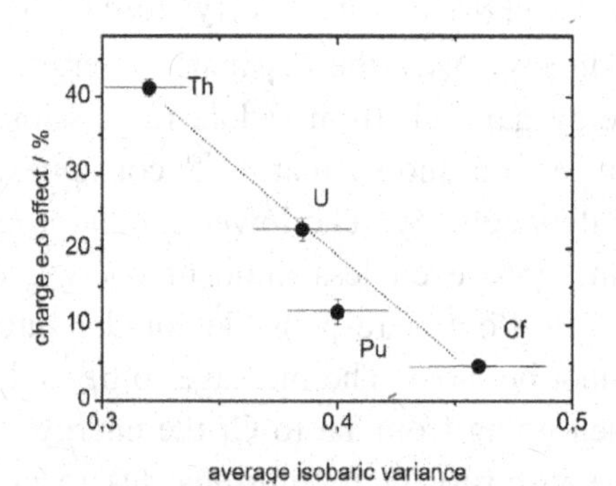

Fig. 4.6.13. Correlation between charge e-o effect and average isobaric charge variance.

parameter Δ with deformation has to be accounted for. Near scission the parameter Δ is reported to be larger by a factor of $2^{1/2}$ than at the saddle point [REY00]. With $\varepsilon \approx 0.4$ and $q = 0.5$ but, as suggested above for this case, $p \approx 1$ the dissipated energy is calculated to be

$$E_X^{sci} \approx -2.2 \ln \delta_Z \text{MeV} \tag{4.6.11b}$$

Taken at face value eq. (4.6.11b) implies that at scission the energy required to overcome pairing is smaller than in the descent from saddle to scission. By the way, smaller dissipation energies also reproduce better the drop of the e-o effect in Fig. 4.6.11 for fission induced by MeV neutrons. This can be understood from the fact that for MeV neutrons a sizable fraction of pair breaking already occurs at the compound stage of the nucleus where the gap energy 2Δ is smaller compared to the situation once fission is progressing.

Anyhow, in spite of large uncertainties the main message from the above analysis is that energies dissipated in fission are small. Likewise small dissipation energies have been reported earlier from the analysis of symmetric/asymmetric mass yield ratios as a function of compound nucleus excitation [GIN82]. With MeV neutrons of increasing energy the (n,f) reactions with ^{232}Th, ^{238}U and ^{235}U were studied. The results reported were E_{DIS} = (1.5 ± 0.5) MeV for ^{232}Th, (4.5 ± 1.5) MeV for ^{238}U and (5 ± 2) MeV for ^{235}U. It is further worth mentioning that from an empirical analysis of energy dissipation it was found that for ^{252}Cf(sf) only 7 MeV of the potential energy gain are transformed into intrinsic excitations and collective vibrational and rotational modes [SCH78].With δ_Z = 0.108(22) for ^{252}Cf(sf) [NAI04] this is very close to the energy calculated with eqs. (4.6.11).

In Fig. 4.4.7 the dissipated energy E_X^{sci} according to eq. (4.6.11a) and the energy gain ΔV from saddle to scission were plotted in one single graph. The comparison shows that E_X^{sci} consumes only a fraction of about 35% of the available energy. The lower excitation energies from eq. (4.6.11b) predict with some 25% even less drain of energy. In a picture where the viscosity of the nuclear flow is responsible for the intrinsic excitation this tells that viscosity cannot be large. The message to be underlined from Fig. 4.4.7, however, is that when going from Th to Cf the energy ΔV liberated at scission grows. This will give viscosity increasing chances to create intrinsic excitations. The observed growth of the energy E_X^{sci} dissipated as a function of compound charge number Z in Fig. 4.6.12 hence finds an obvious explanation in this approach.

In the picture discussed so far pairbreaking occurs between saddle and scission due to viscosity heating up the nucleus. At variance with this picture is the hypothesis that the downward motion is adiabatic and superfluidity is preserved up to the very last moment of scission. As discussed in connection with the interpretation of isobaric charge variances $\sigma^2(Z|A)$ the exchange of protons between the nascent fragments is stopped when close to neck rupture adiabaticity is lost because the speed of neck closure becomes too fast for the charge equilibration to follow. The question is whether also in this view of the charge e-o effects can be understood. As brought to evidence in Fig. 4.6.13 the well pronounced correlation between the charge e-o effect δ_Z and the averaged isobaric charge variance $\langle\langle\sigma^2(Z|A)\rangle\rangle$ supports the idea that nucleon pairs from a perfectly paired superfluid could be broken by a non-adiabatic process in the very last moment. The combinatorial model of the charge e-o effect is flexible enough to accommodate both, the viscosity model or the superfluidity model. In particular the formulas (4.6.9) to (4.6.11) for the size of the o-e effect and its relation to the energy transferred to intrinsic excitations remain valid. The rise of the energy dissipated for nuclei with increasing charge number Z (as in Fig. 4.6.12) or increasing fissility (as in Fig. 4.4.7) is qualitatively also understood in the superfluid model. From Th to Cf the gain in potential energy ΔV is getting larger and in a superfluid motion the velocity of descent and the pre-scission kinetic energy E_K^{sci} will therefore likewise increase as indeed observed in experiment (see Fig. 4.4.6). As discussed by Landau-Zener, higher velocities at scission induce higher non-adiabatic jump probabilities. This entails the creation of a larger numbers N of broken proton and neutron pairs [BOUZ98]. For increased numbers of broken pairs N also the energy E_X^{sci} dissipated will according to eq. (4.6.10) become larger in proportion. But for increasing excitation energy E_X^{sci} the charge e-o effect in eq. (4.6.11) is expected to decrease. The smaller e-o effects for elements when Z is getting larger as observed experimentally in Fig. 4.6.10, thus find a quite convincing interpretation. On the other hand, it is evident that for increasing speeds of deformation or separation of nascent fragments also the neck closure speeds will become larger. The connection between speed of neck closure dc/dt and isobaric charge variance $\sigma^2(Z|A)$ was already discussed in section 4.6.1. The correlation seen in Fig. 4.6.13 between charge e-o effects decreasing and isobaric charge variances $\sigma^2(Z|A)$ increasing when moving from Th to Cf is hence not surprising.

4.6.3 *Charge Even-Odd Effects in Superasymetric Fission*

So far the even-odd effects of global charge distributions were considered. With a look to Figs. 4.6.1 and 4.6.2, however, one may suspect that the size of the even-odd fluctuations is not the same over the full range of fragment charges. To assess possible variations the notion of "local even-odd effects" was introduced. In the early sixties and seventies of last century it was first noticed that isobaric charge distributions $Y(Z|A)$ from thermal neutron fission may to first approximation be represented by Gaussians [WAH62]. In the next step the pronounced even-odd effects were described as deviations $\Delta(Z|A)$ of charge yields from a perfect Gaussian [AMI73]. At about the same time it was proposed to quantify local charge even-odd effects $\delta_Z(Z)$ at the average charge $<Z> = (Z+1,5)$ of 4 consecutive charges Z, Z+1, Z+2 and Z+3 by the expression

$$\delta_Z(Z) = \tfrac{1}{8}\,(-1)^{Z+1}[\{lnY(Z+3)-lnY(Z)\}-3\{lnY(Z+2)-lnY(Z+1)\}] \qquad (4.6.12)$$

with *ln* the natural logarithm of charge yields for $(Z+\nu)$ [TRA72]. The peculiar definition is inspired by noting that for a perfect Gaussian with no e-o staggering at all the even-odd effect in eq. (4.6.12) vanishes. It may further be shown that when averaged over Z the local $\delta_Z(Z)$ coincides with the global even-odd effect considered in the preceding section 4.6.2. The definition in eq. (4.6.12) for the local effect is thus not at variance with the definition in eq. (4.6.6) for the global effect. But it must be stressed that the local even-odd effect following eq. (4.6.12) makes only sense in case the global charge distribution is indeed a Gaussian with e-o fluctuations superimposed. The approach appears to be reasonable for most distributions shown in Figs 4.6.1 and 4.6.2 but not for all. It has been proposed to get rid of this limitation by simply applying eq. (4.6.6), viz. $\delta_Z = (Y_e-Y_o)/(Y_e+Y_o)$, also for isobaric charge distributions since there no conditions on the shape of the distributions are presupposed [KNI92]. It can be proved that indeed after averaging over 5 consecutive masses the isobaric e-o effect following the definition of eq. (4.6.6) coincides with the definition of the local e-o effect as given in eq. (4.6.12) for basic Gaussian distributions [GON92]. However, this model independent definition was not generally adopted for the evaluation of charge distributions. Justified or not, all experimental data on local charge e-o effects continue to be given in terms of the definition in eq. (4.6.12).

Examples of local charge e-o effects have been collected in Fig. 4.6.14. For ^{230}Th* on top left the effect grows continuously towards asymmetric fission

[DJE84, BOC90]. The results for $^{236}U^*$ are more spectacular and came as a surprise [SID89]. While for charge numbers carrying the main yield the e-o effect fluctuates around $\delta_Z = 23\%$ (see Table 4.6.2), below $\langle Z \rangle = 32.5$ the effect suddenly surges and for high kinetic fragment energies reaches 80%. Fortuitous or not, it is striking that the rise of the e-o effect coincides with the mass and charge region characteristic for super-asymmetric fission. Recalling Figs. 4.3.4 to 4.3.7, right at the fragment (mass, charge) = (80,32) there is a kink in the slope of the mass yield curve which appears to be linked to the neighbouring magic neutron number N = 82. At the magic charge number Z = 28 of Ni even a bump in the yield curve shows up. Large e-o effects in super-asymmetric fission were not predicted by the then available combinatorial model of charge distributions. It was therefore felt important to investigate whether super-asymmetric fission of $^{236}U^*$ is a special case or whether large e-o effects are typical for this mass region. An experiment for $^{240}Pu^*$ gave the same clear signature. The size of the e-o effect increases significantly in super-asymmetric fission for charges below Z = 32.5 [DIT91]. For the heavier actinides Cm and Cf the same observation holds with a net increase of the e-o effect in super- asymmetric fission. The similar behaviour in thermal neutron fission of $^{236}U^*$, $^{246}Cm^*$ and $^{250}Cf^*$ to be seen in the left lower panel of Fig. 4.6.14 is obvious [ROC04]. In view of these new findings also the charge distributions of nuclei with odd Z were revisited. The charge e-o effects in the reactions $^{238}Np(n,f)$ and $^{242}Am(n,f)$ were thought to be virtually compatible with $\delta_Z = 0$. But again, when experiments were pushed to super-asymmetric fission, sizable e-o effects appeared as demonstrated in the lower middle and right panels of Fig. 4.6.14 [SIE91, TSE99, TSE01].

Large e-o effects in super-asymmetry were not anticipated in the combinatorial model outlined in section 4.6.2. Yet, if super-asymmetric fission is induced by shell effects in the light fragments, shell effects could also come into play for steering charge e-o fluctuations. One can imagine that magic fragments will not pick up protons from broken pairs. In eq. (4.6.9), viz.

$$\delta_Z = (1 - 2p\varepsilon q)^N$$

this could formally be taken into account by choosing the parameter p governing the probability for nucleons from a broken pair for going into complementary fragments. In the limiting case one could postulate $p = 0$ which means that both nucleons from broken pairs would go into one single fragment, in the present case the heavy fragment. With $p = 0$ the charge e-o effect is boosted to δ_Z = *100%*. The parameter setting $p = 0$ is compatible with both, pairs being broken in the descent from saddle to scission and pairs being broken at scission. Let us recall that for the latter model it was speculated to the contrary in the discussion above that a reasonable choice could be $p = 1$. This may still be valid for standard fission not being steered by shell effects like super-asymmetry.

With the interpretation of local e-o effects based on light fragment shells one should expect that also shell effects in the heavy fragment around Z=50 and N=82 will induce pronounced effects. There are indeed indications for the local

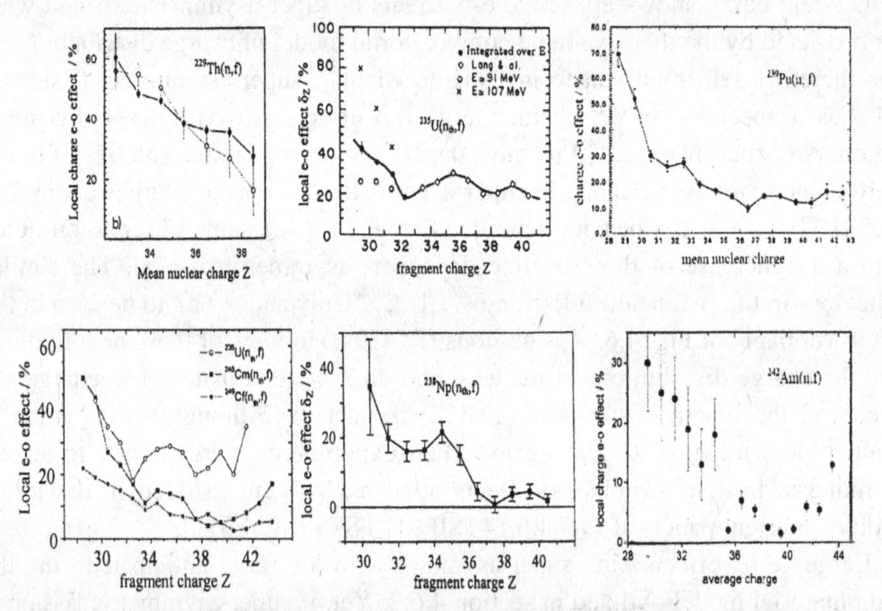

Fig. 4.6.14. Local e-o effect δZ(Z) in fission of 230Th*, 236U*, 240Pu*, 246Cm*, 250Cf*,239Np*, 243Am*.

$\delta_Z(Z)$ getting large when fragments with the above magic numbers are involved. LOHENGRIN experiments for $^{236}U^*$ and $^{240}Pu^*$ point in this direction [SCH84]. However, the measurements in this charge region are not complete and the results achieved so far should not be construed to be a firm proof.

With the interpretation of local e-o effects based on light fragment shells one should expect that also shell effects in the heavy fragment around Z=50 and N=82 will induce pronounced effects. There are indeed indications for the local $\delta_Z(Z)$ getting large when fragments with the above magic numbers are involved. LOHENGRIN experiments for $^{236}U^*$ and $^{240}Pu^*$ point in this direction [SCH84].

However, the measurements in this charge region are not complete and the results achieved so far should not be construed to be a firm proof.

A quite different interpretation of charge e-o effects was developed for the evaluation of experiments conducted in inverse kinematics at the GSI/Darmstadt [SCH00]. Fragmentation of a 1 AGeV ^{238}U beam yielded secondary beams which were analysed as to mass and charge and excited by electromagnetic interactions in a secondary target. Fission at about 11 MeV of excitation energy was analysed by measuring nuclear fragment charges. An impressive series of charge distributions was obtained for neutron-poor isotopes of U down to At. The model proposed appears to be well adapted to the GSI studies but was also used to discuss the charge e-o effects in thermal neutron fission [REJ00]. It is a rigorous statistical approach applied to the nuclear model of a superfluid. Statistical equilibrium of intrinsic degrees of freedom is assumed being reached somewhere before scission. The local charge e-o effect $\delta_Z(Z)$ at fragment charge Z for fissioning nuclei with even charge number Z_{CN} is given as

$$\delta_Z(Z) = P_0^Z + P_2^Z[1 - 2p(Z)]^2 + \ldots \qquad (4.6.13)$$

The P_0^Z and P_2^Z are the probabilities that zero or two quasi-particle proton states are excited, respectively, and $p(Z)$ is the sticking probability of unpaired protons to the fragment with charge Z. Note that p(Z) in eq. (4.6.13) is related but not identical to the parameter p of the combinatorial model in eq. (4.6.9). The sticking probability is set proportional to the proton single particle level density in the fragment Z and upon neglecting possible effects due to shell structure the result is

$$p(Z) = Z/Z_{CN} \qquad (4.6.14)$$

telling that in asymmetric fission the sticking probability to the heavier fragment is larger than to the light one. At symmetry with $p(Z) = 1/2$ only the first term P_0^Z in eq. (6.13) contributes. The probability P_0^Z for the fissioning system to remain fully paired plays a prominent role in the model. The reason is that the charge e-o effect $\delta_Z(Z)$ at symmetry is according to eq. (4.6.13) identical to the survival probability P_0^Z of the fully paired proton subsystem. P_0^Z is a function of the intrinsic excitation energy $U = E_X^{sci}$ near scission. The survival probability $P_0^Z(U)$ or charge e-o effect at symmetry $\delta_Z(Z)$ is calculated as a function of the dissipated energy U. For comparison with the model the experimental δ_Z values have first to be corrected for an asymmetry effect (see below) in order to find the true e-o effect at symmetry. These can then be inserted in the model calculation yielding the corresponding dissipated energy. It turns out that the energies found in the statistical model are surprisingly close to the energies deduced in the combinatorial model.

Moving away from symmetric fission towards asymmetric fission the charge e-o effect is predicted according to eq. (4.6.13) to rise quadratically with Z^2. For thermal fission of Th up to Cf this law reproduces quite well the experimental findings on display in Fig. 4.6.14.

It remains to look into the predictions of the model for fissioning compound nuclei with an odd charge Z_{CN} like the nuclei Np and Am investigated at the ILL. In these nuclei there is at least one unpaired proton. For odd compound charges Z_{CN} the charge e-o effect $\delta_Z(Z)$ is in analogy to eq. (4.6.13)

$$\delta_Z(Z) = P_1^Z[1 - 2p(Z)] + P_3^Z[1 - 2p(Z)]^3 + \ldots \qquad (4.6.15)$$

with P_1^Z and P_3^Z the probabilities for one and three quasi-particles, respectively [STE98]. With $p(Z) = Z/Z_{CN}$ the dependence is cubic in the fragments charge Z. Again the comparison with experiments for Np and Am in Fig. 4.6.14 is very good. It has to be pointed out that in addition to the unpaired proton from the odd Z_{CN} there appears to be at least one additional proton pair being broken providing 2 more unpaired protons. The shape of the local effect as a function of fragment charge is only reproduced when taking these additional protons into account.

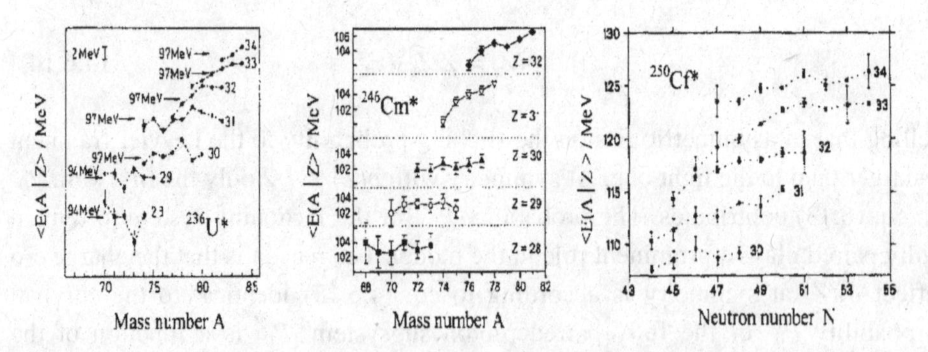

Fig. 4.6.15. Kinetic energy of fragments with fixed charge Z as a function of fragment mass A or neutron number N. Left panel 236U*. Middle panel 246Cm*. Right panel 250Cf*.

There is an additional interesting feature emerging from the study of charge distributions in super-asymmetric fission. It concerns the question of neutron emission. In mass regions with sufficiently large yields the neutron multiplicity can either be measured directly with neutron detectors or determined indirectly by comparing pre- and post-neutron mass distributions. The measurement of primary and secondary mass distributions was outlined in chapter 4.3. These experiments led to the famous sawtooth of neutron multiplicity as a function of fragment mass. However, due to low yields these methods fail in super-asymmetric fission. Some information on neutron emission in super-asymmetric fission is gained indirectly by studying the kinetic energy versus mass or neutron number for given charges of the fragments. In Fig. 4.6.15 three examples are provided for the fissioning nuclei $^{236}U^*$ [SID89], $^{246}Cm^*$ [HEN94] and $^{250}Cf^*$ [ROC04]. In all three cases the average kinetic energies for fragments with charge numbers below $Z \approx 30$ stay constant for all masses shown while for $Z > 30$ the kinetic energy increases strongly and regularly with mass or neutron number. The latter behaviour is understood to be due to neutron evaporation. For every neutron evaporated the fragments loose about 1.0 MeV of kinetic energy. Hence, the more neutrons are emitted the lower the fragment kinetic energy will be. The constancy of the kinetic energy for the lowest charges $Z = 28$ and $Z = 29$ is therefore an indication that there is no neutron emission at all. Recalling Figs. 4.3.6 and 4.3.7 from chapter 4.3, in particular the magic $Z = 28$ (Ni) is precisely the charge responsible for the bump in the mass yields of super-asymmetric fission. The vanishing neutron multiplicity $\nu = 0$ for the magic fragments with $Z = 28$ has to be seen in analogy to the nil multiplicity $\nu \approx 0$ for heavy fragments near the doubly-magic ^{132}Sn with $Z = 50$ and $N = 82$. Neutron emission missing for magic fragments is one of the main characteristics of the neutron sawtooth.

4.6.4 *Charge Even-Odd Effects in Symmetric Fission*

In the study of fission at low excitation energies of standard actinides symmetric fission is singled out in the mode analysis of fragment mass and energy distributions. A specific so-called "superlong" mode is attributed to it [BRO99]. This denomination expresses the experimental fact that compared to asymmetric fission there is a dip in kinetic energy for symmetric fission [GON91]. For example, in the Th- and U-isotopes the dip is close to 20 MeV for near-barrier fission. This was already shown in Figs. 4.5.2 and 4.5.3 for $^{236}U^*$ and in Fig. 4.5.4 for $^{233}Th^*$. The comparatively low kinetic energies are attributed to very long descent paths form saddle to scission leading to more deformed nuclear shapes at scission. In consequence the separation distances between the charge centres will be larger and the kinetic energy of fragments fed by Coulomb repulsion will be lower. It is important to stress that in Figs. 4.5.3 and 4.5.4 it is evident that the symmetric modes do not mix with the neighbouring asymmetric modes. It follows that charge e-o effects for symmetric and asymmetric modes are independent of each other. It should therefore be interesting to compare in experiment the charge e-o effects for the two modes. Unfortunately, for the time being, there is no physical method known to measure fragment charges near symmetry for thermal neutron fission. Of course there are plenty of radiochemical data critically evaluated and published.

Though fragment charges near symmetric fission are not measurable on LOHENGRIN, mass distributions can be readily determined for the full range from light to heavy masses. When discussing mass distributions in chapter 4.3 it was already foreshadowed that the structures 5 mass units apart in the mass distributions of Figs .4.3.1 and 4.3.2 have to be attributed to a charge e-o effect. Recall that the 5 amu staggering is understood by noting that the fissioning compounds under consideration have a neutron to charge ratio of $N_{CN}/Z_{CN} \approx 1.5$. Hence, from even charge Z to even charge (Z+2) the neutron number N increases by $\Delta N \approx 3$ and the mass A of the fragment by $\Delta A \approx 5$. The charge e-o effect with increased yields for even charges therefore induces a staggering in the mass distribution 5 amu apart. This phenomenon may be exploited to get at least some indication on charge e-o effects by inspecting the mass distributions near symmetry. On LOHENGRIN this was done for the nuclei ^{236}U*, ^{240}Pu*, ^{236}Cm* and ^{250}Cf*. For thermal neutron fission of ^{229}Th this method was not employed because the fission cross section σ_{fi} = 30 b is rather small and the yield of ^{230}Th* at symmetry is by almost 3 orders of magnitude lower than for asymmetric fission. For this reaction one has to inspect radiochemical data.

In Fig. 4.6.16 mass distributions covering symmetric fission of isotopes from Th to Cf are on display. In the mass distribution for ^{230}Th* obtained by radiochemical methods [WAH02] it is first the large peak-to-valley ratio P/V for asymmetric-to-symmetric yields which is catching the eye. This ratio is about P/V ≈ 500. Second, there are clearly two separated modes, a symmetric and an asymmetric one. It is remarkable that the mass yield distribution for the symmetric mode, as indicated in the figure, follows a Gaussian and it may safely be assumed that also the charge distribution is to be described by a Gaussian.

Provided the data were detailed enough this would allow evaluating a local charge e-o effect as proposed in eq. (4.6.12). At least a global charge e-o effect encompassing the charges 42 to 47 may be found based on evaluated data [ENDF349]. A very large value of δ_Z = *57(20) %* is calculated for ^{230}Th*. It should be recalled that the identification of two modes in the mass distributions of ^{230}Th* is corroborated by the distributions of kinetic energies *TKE* at fixed mass windows for ^{232}Th(n,f) induced by MeV neutrons shown in Fig. 4.5.4..

In the right and lower panels of Fig. 4.6.16 mass distributions as measured with the LOHENGRIN spectrometer for ^{236}U* [TSE04], ^{240}Pu* [TSE04], ^{246}Cm* [FRI98] and ^{250}Cf* [DJE89, HEN92] are shown. For ^{236}U* and ^{240}Pu* physical and radiochemical data may be compared. There is a non-negligible scatter of results but it is thought that the yields from LOHENGRIN are consistent because

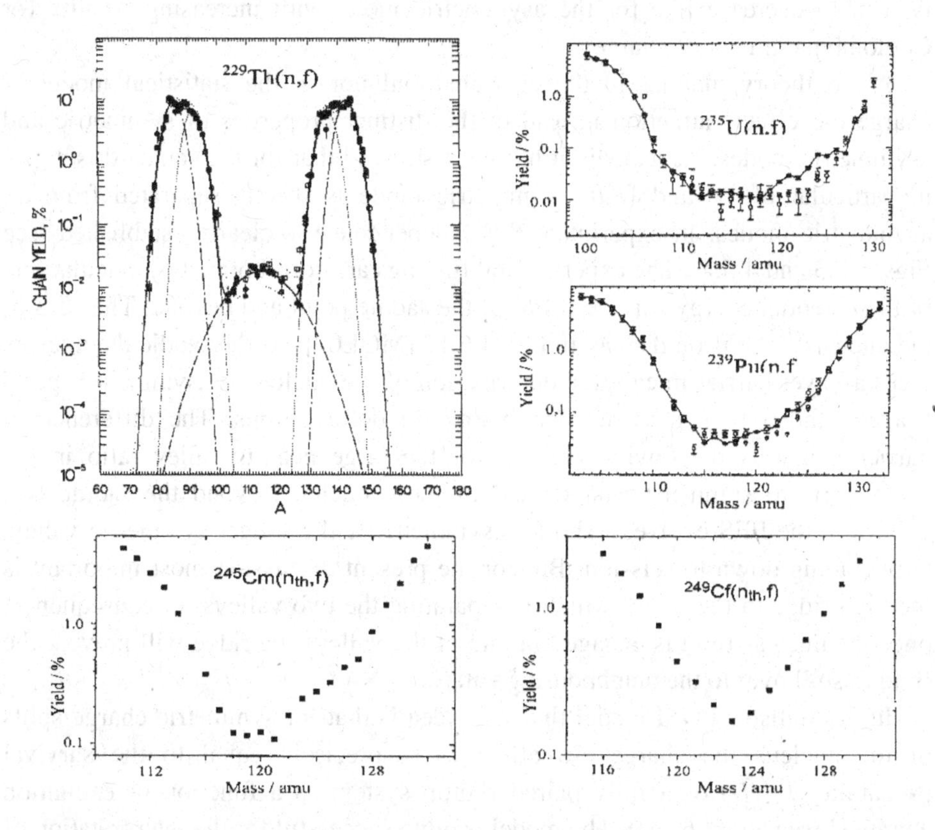

Fig. 4.6.16. Mass distributions across symmetry for (n_{th},f) of ^{229}Th, ^{235}U, ^{239}Pu, ^{245}Cm, ^{249}Cf. For ^{236}U and ^{240}Pu full and open points are from LOHENGRIN and radiochemistry, respectively.

they were measured without changing the technique. For $^{236}U^*$ there is a clear 5 amu oscillation with increased yields for masses 125, 120, 115 and perhaps also 110. It proves for the first time by physical methods that there is a non-zero charge e-o effect for the symmetric mode of $^{236}U^*$. Unfortunately it is not possible to go beyond and give a quantitative value for the effect δ_Z. For the next example, $^{240}Pu^*$, there is at best a vague indication of a 5 amu structure which does not allow to judge whether there is an e-o effect or not. For $^{246}Cm^*$ the symmetric mode has shrunk to a few masses and for $^{250}Cf^*$ the mode is no longer visible. It is questionable whether in these cases it is even in principle conceivable to disentangle symmetric and asymmetric modes for assessing separately e-o effects. Nevertheless the charge e-o effect in the symmetric mode is seen to decrease like for the asymmetric mode with increasing fissility (or Coulomb parameter or charge)

As to theory, neither in the combinatorial nor in the statistical model of charge e-o effects attention is paid to the distinct properties of symmetric and asymmetric modes. Repeatedly it has been stressed that for the standard isotopes in particular of Th and U the symmetric mode is strictly separated from all asymmetric modes. In experiment this independence is clearly established (see Figs. 4.5.3 and 4.5.4). The experimental findings are corroborated by calculations of the potential energy surface (PES) at the saddle point and beyond. The PES of the nucleus ^{232}Th is on display in Fig. 4.6.17 [MOL09]. At the saddle the nucleus faces a lower barrier in case the deformation of the nucleus is asymmetric (pear-shaped) and a higher barrier for symmetric deformations. The difference in barrier height is the obvious reason for the large peak-to-valley ratio in the asymmetric to symmetric mass yield of these nuclei. Beyond the saddle two valleys in the PES evolve, a shorter asymmetric and a longer symmetric valley, both leading down to scission. But for the present discussion most important is the high ridge in the figure which is separating the two valleys. In consequence, once the nuclear fluid is engaged in one of the valleys the ridge will prevent the fluid to spill over to the neighbouring valley.

In the statistical GSI model the basic idea is that for symmetric charge splits of the nucleus the charge e-o effect δ_Z is precisely equal to the survival probability $P_0^Z(U)$ of a fully paired proton system as a function of excitation energy U (see eq. (4.6.13)). The model is quite successful for the interpretation of the GSI experiments [STE98]. However apart from the somewhat higher excitation energies in the GSI experiments compared to thermal induced fission,

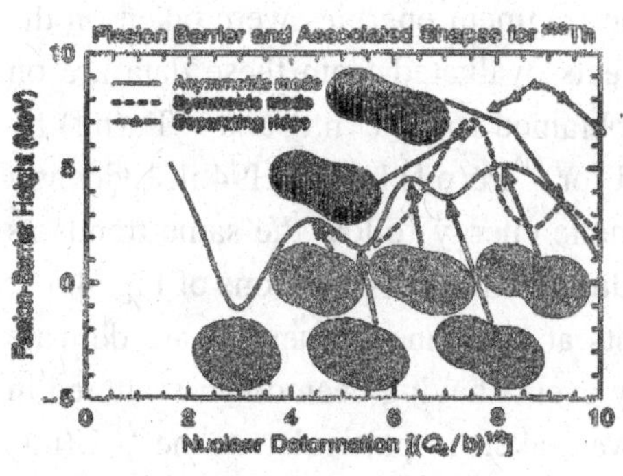

Fig. 4.6.17. PES of the nucleus ^{232}Th.

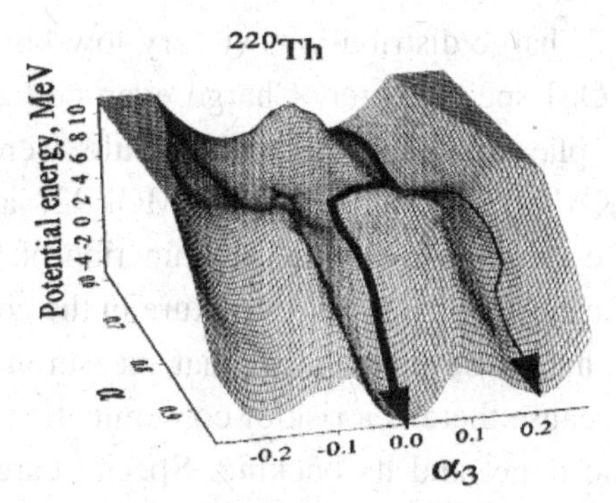

Fig. 4.6.18. PES of the nucleus ^{220}Th.

there is a major difference in the PES for the very neutron-poor Th isotopes under study at the GSI. The PES for the very neutron-poor isotope 220Th is sketched in Fig. 4.6.18 [PAS71, PAS95]. In contrast to 232Th, for 220Th there is no strict separation between symmetric and asymmetric modes. As indicated in the figure, symmetric fission may even evolve from an asymmetric barrier due to a switch-back from the asymmetric to the symmetric valley. For the nuclei investigated at the GSI it appears therefore to be well justified to describe e-o effects in asymmetric fission as proceeding from a basic e-o effect in symmetric fission as proposed in eq. (4.6.13). In thermal neutron fission, however, the mode structure at the saddle point rules out the application of this model. Charge e-o effects develop there independently for the two valleys or modes.

4.6.5 *Charge Even-Odd Effects in Super-Deformed Fission*

Mass-energy correlations of fission fragments were studied in chapter 4.5. A notable result deserving more attention is on display in Fig. 4.5.7. Conditional mass distributions constrained by kinetic energy $Y(A|E_K)$ are shown for the light fragment group in the reactions ^{233}U(n_{th},f) and ^{235}U(n_{th},f). At large kinetic energies there is a pronounced structure with a period of about 5 mass units. This structure was discussed at length in Chapter 4.6.1. It is due to the charge even-odd effect in low energy fission of actinides. In the conditional mass distributions $Y(A|E_K)$ in Fig. 4.5.7 the size of the staggering is not constant. Starting form the highest energies investigated and moving to lower energies the structure first disappears but then comes back at the lowest energies. The question is whether this latter structure in the mass distribution is also linked to an increase of the charge even-odd effect.

Charge distributions at very low kinetic fragment energies were taken on the COSI spectrometer. Charge even-odd effects evaluated from these data are on display in Fig. 4.6.19. The results were obtained for $^{232}U(n,f)$ and $^{239}Pu(n,f)$ by [KAU91a], for $^{235}U(n,f)$ by [MOL92] and for $^{241}Pu(n,f)$ by [SCH94b]. Evidently the charge e-o effects as a function of kinetic energy follow the same trends as the sizes of the 5-amu structure in the conditional mass distributions of Fig. 4.5.7. It has to be pointed out that measurements at low kinetic energies are delicate because there is a risk of contamination by events having been down-scattered in the target and its backing. Special care was taken in particular in the $^{235}U(n,f)$ experiment to identify scattered events. To this purpose (mass, charge, energy) of fragments from the light group were measured by the TOF-energy technique of COSI while the energies of the complementary heavy fragments were obtained by an ionization chamber mounted on a common axis with COSI. From the energies of the two fragments provisional masses were evaluated and compared to the COSI result. The procedure allowed cleaning the COSI results.

To guide the discussion of Fig. 4.6.19 it will be useful to recall an often quoted model, the "Scission Point Model" [THO65]. It is visualized in Fig. 4.6.20. For a scission configuration with two fragments facing each other while their deformations increase, the potential Coulomb and deformation energies V_{Coul} and V_{Def}, respectively, are plotted together with their sum $(V_{Coul} + V_{Def})$. The energy relations were given in equations (4.4.1), (4.4.14) and (4.4.15). Taken together they read

$$\begin{aligned} Q &= TKE^{*}+TXE \\ &= (V_{Coul}+E_K^{sci}) + (V_{Def}+E_X^{sci}) = (V_{Coul}+V_{Def}) + (E_K^{sci}+E_X^{sci}) \end{aligned} \tag{4.6.16}$$

The last term was identified in good approximation as the gain in potential energy from saddle to scission and designated as ΔV in eq. (4.4.11). In the present model it is called the "Free Energy" E_{Free}:

$$E_{Free} = E_K^{sci} + E_X^{sci} \tag{4.6.17}$$

The free energy is the difference between the available energy Q and the sum of potential energies at scission. It is the sum of energies E_K^{sci} and E_X^{sci} which are not bound as potential energies. It should not be confused with the "free energy" from thermodynamics. Attention should be drawn to the fact that in the schematic scission point model the deformations of the two fragments are not further specified. They are tacitly taken to be similar for the two fragments. Later in 1976 the schematic model was put on a quantitative basis allowing for the calculation of mass yields and energy distributions of fragments [WIL76]. As discussed in Chapter 4.5 the predictions from this theory were a cornerstone in the analysis of fission fragment properties.

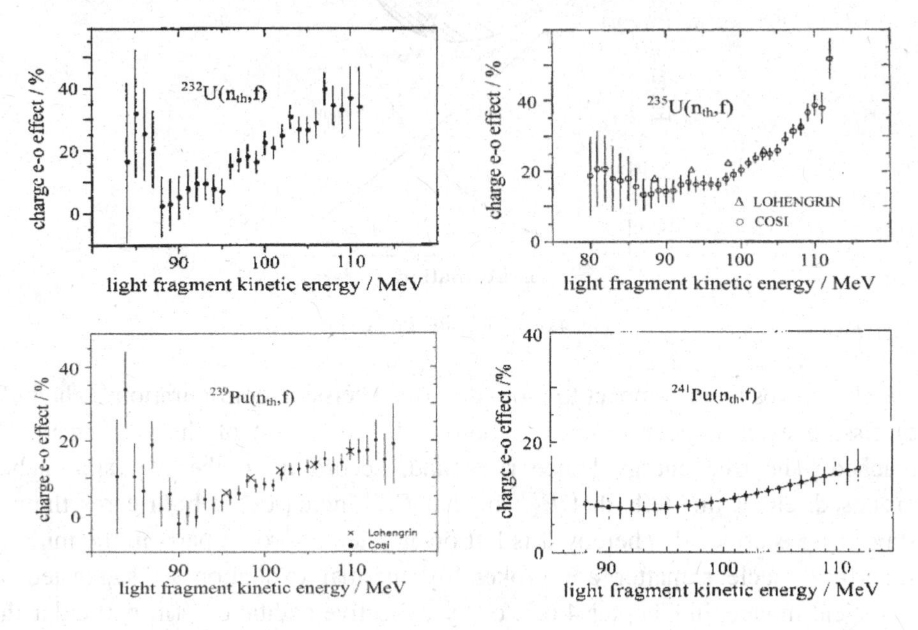

Fig. 4.6.19. Charge even-odd effect δ_Z vs. light fragment kinetic energy / MeV.

In Fig. 4.6.20 two limiting cases labelled 1 and 2 appear. They correspond to the most compact and the most deformed scission configuration with maximum and minimum total kinetic energy *TKE*, respectively. In both cases $E_{Free} = 0$. Vanishing E_{Free} entails $E_K^{sci} = E_X^{sci} = 0$. Apart from these limits of phase space one has in general $E_{Free} \neq 0$. From experiment it is unfortunately not known how the energy is partitioned. One has to invoke theoretical models. As already addressed when discussing charge even-odd effects in Chapter 4.6.2 there are two basic options. In the descent from saddle to scission the potential energy gain $\Delta V = E_{Free}$ may in eq. (4.6.17) either go preferentially into the excitation of intrinsic degrees of freedom as E_X^{sci} or into collective excitations as E_K^{sci}. For the sake of simplicity let us assume that fission is either viscous with the kinetic energy $E_K^{sci} = 0$ or fission is superfluid with the excitation energy $E_X^{sci} = 0$.

Coming back to Fig. 4.6.19 it is observed that from the highest towards lower kinetic energies the e-o effect first decreases. In the Scission Point Model one

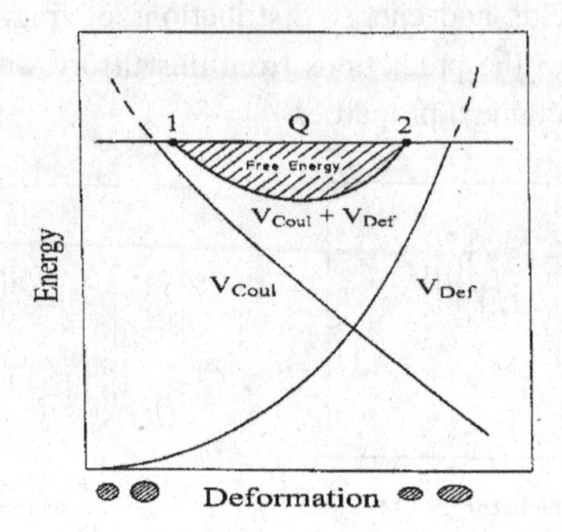

Fig. 4.6.20. Scission Point Model.

thereby moves from compact to more deformed scission configurations. The bulk of fission events undergo scission before the minimum of the free energy is reached. The free energy hence rises and, according to the two approaches proposed, either this tells that E_X^{sci} or that E_K^{sci} increases. In both cases the e-o staggering is affected. Thereby it is left open whether proton pairs in the initially superfluid nuclear matter are broken by thermal excitation as suggested in statistical models in Chapter 4.6.3, or by collective excitations (in particular the downhill fission mode) triggering the breaking of pairs in the very act of scission. In both options pair breaking is enhanced for an increase of E_{Free} and this leads to the decrease of the charge e-o effect δ_Z seen in experiment.

However, what about the constant or even re-enforced e-o effect for very low kinetic energies in Fig. 4.6.19. This phenomenon is best understood in the superfluid model. It concerns those rather rare events having not undergone fission before reaching the minimum of the free energy. We call these events super-deformed fission. Beyond the potential energy minimum of Fig. 4.6.20 the fragments have to climb a rising potential due to the sharp rise of the deformation energy V_{Def} at the expense of kinetic energy E_K^{sci}. Hence the speed of descent from saddle to scission is reduced. For reduced speed the probability of pair-breaking is lower and the charge e-o effect comes back. While formally in super-deformed fission the nucleus may approach the phase space limit with $E_{free} = 0$ labelled 2 in Fig. 4.6.20, it is not evident whether this point is really attained. The discussion will be resumed in Chapter 4.7 on true cold fission.

Historically the first indication for super-deformed fission at very low fragment kinetic energies carrying virtually no intrinsic excitation E_X^{sci} came from studies of neutron correlations emitted from complementary fragments [SIG73]. Neutron evaporation is giving a direct clue to the deformation of fragments at scission. The deformation energy V_{Def} having been accumulated up to scission is transformed after scission into intrinsic excitation. In general V_{Def} is contributing the major part of the total excitation energy $TXE = V_{Def} + E_X^{sci}$ of fission fragments in Eq. 4.6.16. In fact, as argued in connection with eq. (4.4.15), the excitation energy at scission E_X^{sci} only contributes some 30 % to the total excitation energy *TXE*. In the limiting case of super-deformed fission even $E_X^{sci} = 0$ obtains and only the deformation energy V_{Def} contributes to *TXE*. The excitation energy *TXE* is mainly exhausted by neutrons and only to a smaller extent by γ-emission. Hence neutron multiplicity is a measure for fragment deformation at scission. Powerful techniques to detect neutrons are large tanks filled with Gd-loaded liquid scintillators. The tanks are usually hemi-spherical in shape. A central channel allows installing fission sources and fragment detectors. With two tanks mounted to form a sphere with a through-going channel, and with a fission source placed at the centre of the sphere, the total numbers of neutrons $(\nu_L+\nu_H)$ per fission are detected in 4π geometry. Hereby ν_L and ν_H are the multiplicities of the light and heavy fragment, respectively. For the measurement of neutron multiplicity per fragment ν_L or ν_H one of the hemispherical tanks is removed. Neutron detection is hence in 2π geometry. Thereby one fragment is moving towards the interior of the tank while the complementary fragment is moving away from the tank. The neutron data are registered together with the fragment masses and kinetic energies.

Of special interest are correlations between the multiplicities of fragments. The size of the correlation is found by evaluating the covariance $cov(\nu_L,\nu_H)$ of the multiplicities ν_L and ν_H of complementary light and heavy fragments. The covariance is obtained from the experimental variances as

$$2\, cov(\nu_L,\nu_H) = \sigma^2(\nu_L + \nu_H) - \sigma^2(\nu_L) - \sigma^2(\nu_H) \tag{4.6.18}$$

For spontaneous fission of ^{252}Cf, ^{248}Cm and ^{244}Cm experimental covariances are pictured in the top part of Fig. 4.6.21 [KAL00]. They were averaged over all mass splits and are plotted as a function of total kinetic energy TKE. With energy conservation in mind, not surprisingly the covariances are negative, i.e. neutron multiplicities from complementary fragments are anti-correlated. The vanishing of the covariance at the highest TKE_{max} is likewise readily understood since at these energies the neutron multiplicities and hence also variances go to zero. By contrast, the zero covariances for the smallest kinetic energies TKE_{min} deserve special attention.

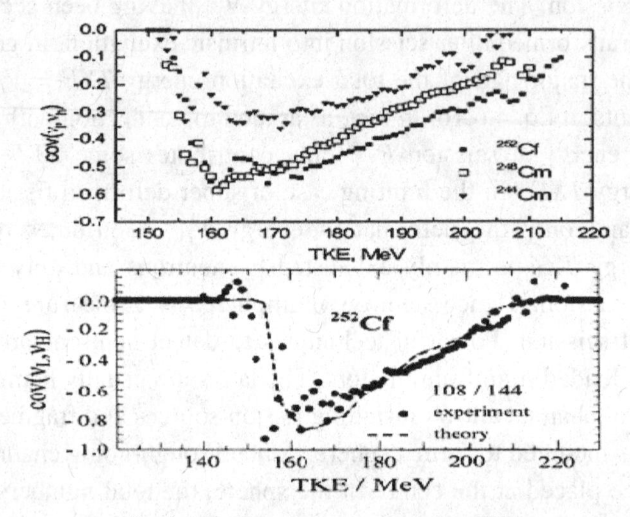

Fig. 4.6.21. Covariance cov(ν_L,ν_H) vs TKE.

Neutron number covariances $cov(\nu_L,\nu_H)$ as a function of total kinetic energy TKE being evaluated with the additional constraint of fixed fragment mass can give a more detailed insight into the process. An experimental example is provided in the lower part of Fig. 4.6.21 [DUE94]. The neutron covariance in spontaneous fission of ^{252}Cf is shown for the ratio of primary masses $A_L^*/A_H^* = 110/144$. For discussion the lines of arguments having been put forward by C. Signarbieux [SIG73] are sketched. At given masses and kinetic energies of fragments the total excitation energies of the fragments $TXE = Q - TKE = E_{XL}+E_{XH}$ are very nearly constant since the Q-value for given masses is almost fixed. With $TXE = E_{XL} + E_{XH} \approx const$ one finds for the variances $\sigma^2(E_{XL}+E_{XH}) = 0$ and $\sigma^2(E_{XL}) = \sigma^2(E_{XH})$. With the assumption that linear regression between multiplicity and excitation energy holds, the neutron multiplicities ν_L and ν_H in eq. 4.6.18 may be replaced by the corresponding fragment excitation energies E_{XL} and E_{XH}, respectively. Note that E_{Xi} (with i = L or i = H) are the excitation energies of the fragments prone to evaporate neutrons. If now, as observed in experiment $cov(E_{XL},E_{XH}) = 0$ at both the maximum and minimum TKE, eq. (4.6.18) tells

$$cov(E_{XL}, E_{XH}) = -\sigma^2(E_{XL}) = -\sigma^2(E_{XH}) = 0 \qquad (4.6.19)$$

Further it is argued that for TKE_{max} and TKE_{min} the most compact and the most deformed scission configurations are reached, respectively. These deformations are unique and therefore their variances and hence also the variances of the corresponding deformation energies V_{Defi} are zero. Besides the deformation energy V_{Def}, excitation energies E_{Xi}^{sci} already present at the scission stage contribute to the final intrinsic excitation: $E_{Xi} = V_{Defi} + E_{Xi}^{sci}$. With $V_{Defi} = const$ for i = L and i = H it follows $\sigma^2(E_{Xi}) = \sigma^2(E_{Xi}^{sci})$. Taking the experimental result $\sigma^2(E_{Xi}) = 0$ from eq. (4.6.19) into account the final result is $\sigma^2(E_{Xi}^{sci}) = 0$. This is only possible for

$$E_{Xi}^{sci} = 0 \qquad (4.6.20)$$

Vanishing covariances of neutron multiplicities at TKE_{max} and TKE_{min} thus corroborate the conclusions from the scission point model in Fig 4. 6.20 that at the limits of phase space the fission fragments are born in a cold state. To the total excitation *TXE* in eq. (4.4.15) only the potential energy of deformation V_{Def} contributes.

For the analysis of experimental data covariances were calculated in a refined scission point model [MAE91]. The result of the calculation for the reaction ^{252}Cf (sf) is on display in Fig. 4.6.21 as a dotted line. The agreement between experiment and theory is excellent.

In the traditional scission point model the deformations of the two fragments are taken to be equal. In a more general approach the deformations may be varied independently of each other. As an example consider the reaction ^{252}Cf(sf). For the mass split A_L/A_H = 110/142 the equi-potential lines of the energy (V_{Coul} + V_{Def}) were calculated in a scission model with two spheroids at a tip distance d. The potential calculated for independent variations of the fragment shape is on display in Fig. 4.6.22 [MAE92]. The deformations of the shapes are given in the ε-parameterization. We recall that ε fixes uniquely the ratio of the minor to the major half-axes a/c of spheroids [HAS88]. Near the centre of the figure the potential energy has a minimum. Away from the minimum the potential increases until for the last equi-potential line shown the Q-value of the fission decay is reached. For equal deformations $\varepsilon_L = \varepsilon_H$ of the light and heavy fragment, respectively, the schematic scission point model of Fig. 4.6.20 is recovered. This is visualized in the lower part of the figure where a cut through the potential along the main diagonal is sketched with the free energy vanishing for compact and super-deformed fission. What is new is the limit of phase space $E_{Free} = 0$ which is predicted for very dissimilar shapes of the two fragments at scission. In the figure these limits are labelled as "cold shape dissimilar". Dissimilar shapes at scission entail dissimilar neutron multiplicities in experiment. Mass distributions for large but dissimilar neutron multiplicities of the two fragments

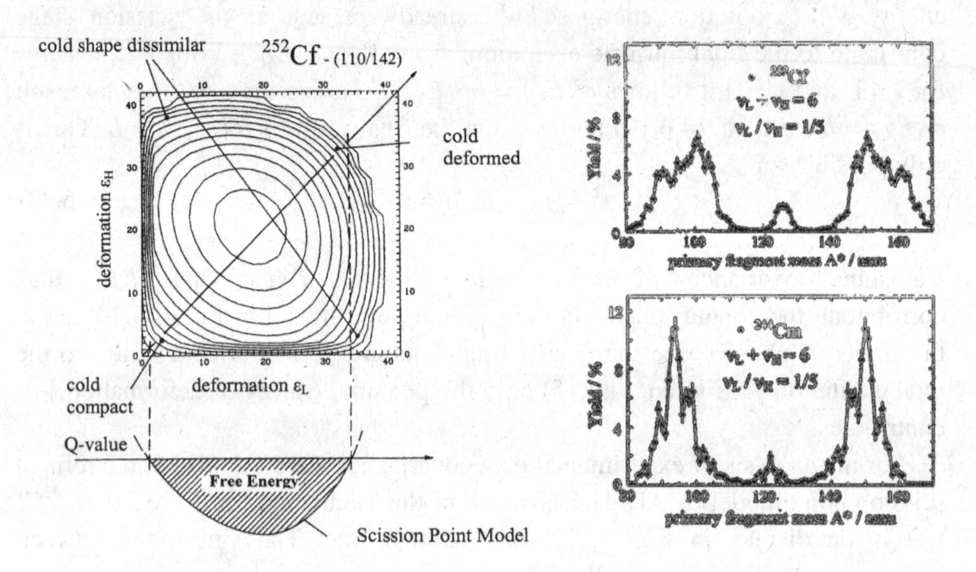

Fig. 4.6.22. Generalized scission point model. Fig. 4.6.23. Y(A) for dissimilar fragment shapes.

are pictured in Fig. 4.6.23 for spontaneous fission of ^{252}Cf and ^{244}Cm [VOR04]. Events have been selected where the total neutron multiplicities $(\nu_L+\nu_H) = 6$ are much larger than average and, even more conspicuous, the ratios of multiplicities $\nu_L/\nu_H = 1/5$ are far away from average. In particular for the ^{244}Cm(sf) reaction the mass distribution exhibits every 5 mass units a prominent structure which in chapter 4.6.1 was traced to a charge even-odd staggering. In chapter 4.6.2 the e-o effect was recognized to be a signature for the intrinsic excitation energy at scission. The salient structures in Fig. 4.6.23 are hence clear evidence for small if not vanishing excitation energies at scission for fragments with dissimilar shapes.

4.6.6 *Kinetic Energy Even-Odd Effects*

General properties of kinetic energy distributions of fission fragments and correlations between fragment energy and mass were considered in Chapters 4.4

and 4.5, respectively. There is a further notable correlation between kinetic energy and fragment charge. In Fig. 4.6.24 the average light fragment kinetic energy is drawn as a function of fragment charge Z [SCH84]. There is a well-defined and regular staggering of the energy. Energies for even charges Z are systematically larger than for odd charges. The obvious definition for the kinetic energy e-o effect is the difference between average total kinetic energies for even and odd charges, respectively. It is designated as δ_{TKE}.

$$\delta_{TKE} = \langle TKE_e \rangle - \langle TKE_o \rangle \qquad (4.6.21)$$

The staggering δ_{TKE} recalls the charge e-o staggering δ_Z and indeed between the two effects there is a close connection. This becomes evident from Fig. 4.6.25 where the charge e-o effect is plotted against the kinetic energy e-o effect. The correlation appears to be linear.

Several models have been proposed to explain the correlation in Fig. 4.6.25 [DJE89].One of these models is sketched in Fig. 4.6.26 [LAN80]. Basically the nucleus is viewed as a superfluid. In the fission process it is assumed that one (and only one) proton pair is broken setting free two quasiparticles (qp). If now

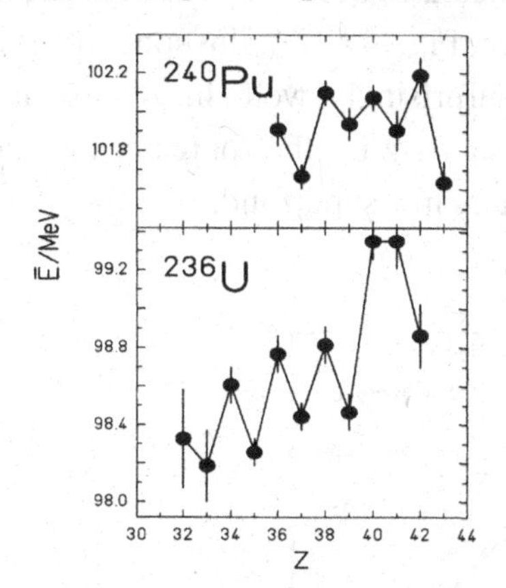

Fig. 4.6.24. Average kinetic energy of light fragment vs charge.

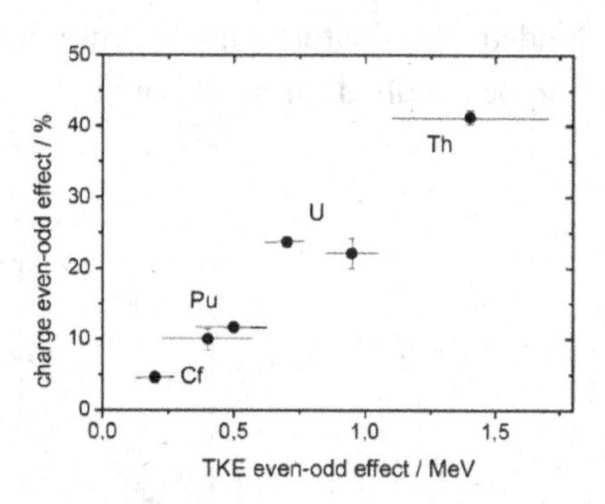

Fig. 4.6.25. Charge e-o effect vs kinetic energy e-o effect.

the energy $\Delta E = 2\Delta$ for breaking the pair is taken from the pre-scission kinetic energy of the fissioning compound there will be a shift in kinetic energy between the superfluid fraction and the broken-pair fraction as sketched in Fig. 4.6.26. In consequence there will be also a shift in kinetic energy between even and odd charge fragments. The shift is calculated as follows. Let p be the fraction of the nuclear fluid with one broken pair. By setting equal the probability that the two qp of the proton system end up in the same or in complementary fragments the yield for the odd charge fragments will be $Y_o = p/2$. To the yield of the even charge fragments Y_e the same broken fraction $p/2$ will contribute but there is in addition the fraction $(1-p)$ of the superfluid. The total even yield is hence $Y_e = (1 - p/2)$. The charge e-o effect is thus $\delta_Z = (Y_e - Y_o) / (Y_e + Y_o)$ or $\delta_Z = (1-p)$. It is equal to the superfluid fraction. As to energies, for even charges one has to add the properly weighted components of the broken-pair and the superfluid fraction while for odd charges only the broken-pair fraction is present. One finds for the relation between δ_Z and δ_{TKE}

$$\delta_{TKE} = 4\Delta\delta_Z / (1 + \delta_Z) \tag{4.6.22}$$

The model does not precisely reproduce the linear relationship between the two e-o effects present in the experimental data. But recalling that at the saddle point the energy gap is $2\Delta \approx 1.7\ MeV$ and that except for ^{230}Th* the δ_{TKE} in comparison is small, the relation in eq. (4.6.22) is not far from linear. To give a numerical example: for ^{236}U* with $\delta Z = 23.4\ \%$ one calculates $\delta TKE = 0.64\ MeV$ while in experiment $\delta TKE = 0.7\ MeV$ is found (see Fig. 4.6.25). In spite of several simplifying assumptions the model works surprisingly well. In a sense it is a downgraded combinatorial model but with an ansatz for the source of the energy having to be supplied for breaking nucleon pairs in a superfluid.

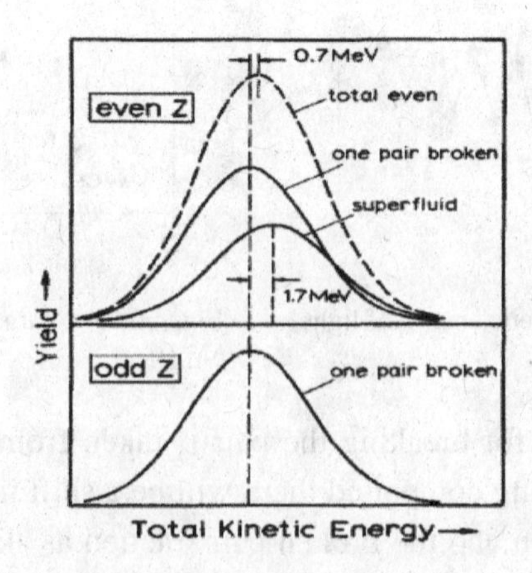

Fig. 4.6.26. Model for the relation between δ_{TKE} and δ_Z.

4.6.7 *Neutron Even-Odd Effects*

In a perfect nuclear superfluid both the proton and the neutron subsystem are fully paired. Phenomena associated to proton pairing have been commented in the foregoing. The question is whether also neutron pairing could not be an equally rich source of information on the fission process. There is however a difficulty. Unlike protons, in general neutrons are evaporated from the primary fragments. For technical reasons neutron multiplicities $\nu(A)$ are usually measured in experiment as a function of the secondary masses A and not the primary masses A* of fragments. Of course $A^* = A + \nu$ and at least average primary masses are to be found from $\langle A^* \rangle = A + \langle \nu \rangle$. Yet, by averaging any primary neutron effect in the fragments will be blurred. Furthermore, the process of neutron evaporation could independently introduce additional e-o effects. Finally, the sawtooth structure of neutron multiplicity as a function of fragment mass and kinetic energy is adding a further complication since from mass to mass the neutron multiplicity can change very rapidly. The idea suggesting itself to correct the fragment neutron yields for the loss by evaporation will therefore not produce reliable results.

In Fig. 4.6.27 charge and isotonic distributions from the reaction ^{233}U(n,f) are confronted for the same choice of energy windows [QUA88]. The data were obtained on the LOHENGRIN spectrometer and cover as usual the light fragment group. Evidently, at lower energies the even-odd staggering is much more pronounced for protons than for neutrons. Note that the fragment neutron numbers are those of the secondary fragments after prompt neutron evaporation.

The e-o effects δ_Z for protons and δ_N for neutrons are compared in Fig. 4.6.28 [QUA88]. For fragment energies near 90 MeV the e-o effect is around 20 % for protons but drops by a factor of 4 to about 5 % for neutrons. At the highest fragment energy of 110.55 MeV in Fig. 4.6.27 the even-odd staggering is getting more pronounced both for protons and neutrons. This is also borne out by the increase of the e-o effect with increasing energy in Fig. 4.6.28. Thereby the factor of 4 between e-o effects for protons and neutrons is maintained up to the highest energies. In Fig. 4.6.27 it is further noteworthy that it is the charge $Z = 40$ and the neutron number $N = 60$ whose yields rise sharply at larger fragment energies. At the high kinetic energy neutron evaporation is not very probable and for the sake of discussion, may be thought to be virtually absent. The heavy fragment complementary to the light fragment in fission of ^{234}U* then has $Z = 52$ and $N = 82$ and is hence very close to a doubly magic nucleus.

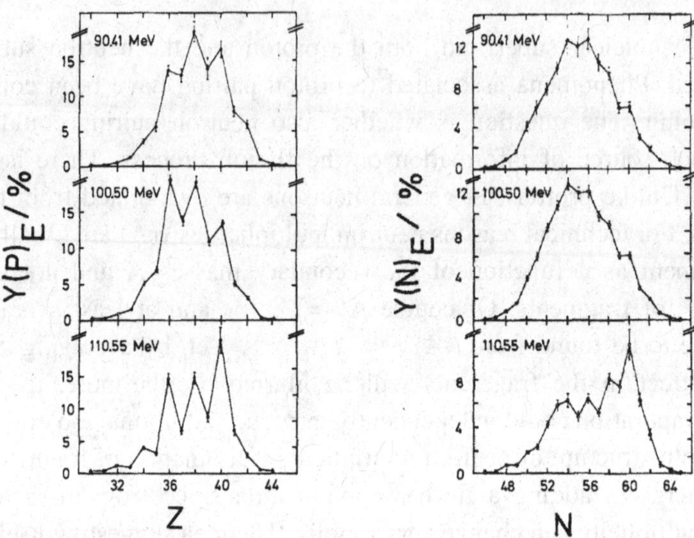

Fig. 4.6.27. Element yields Y(Z | E) (left) and isotonic yields Y(N | E) (right) at given kinetic energy E of light fission fragments from 233U(n,f).

Interestingly, in experiments with the neighbouring isotope 235 in the reaction ^{235}U(n,f) it was observed that at the highest kinetic energy studied of 113 MeV the yield for the mass A = 102 with Z = 40 and N = 62 has a dominant spike [WOH76]. Again the corresponding heavy fragment has Z = 52 and N = 82. These findings could indicate that in specific mass and energy regions shell effects and not even-odd effects are responsible for fine structures in the distributions of protons and neutrons and ultimately mass numbers. So far shell effects have not been taken into account in the models proposed for the interpretation of even-odd effects.

Global neutron e-o data have been published for three thermal neutron reactions, all studied on LOHNGRIN: ^{233}U(n,f) [QUA88], ^{235}U(n,f) [LAN80], and ^{239}Pu(n,f) [SCH84]. The neutron e-o effects δ_N averaged over kinetic fragment energy are very similar. For the 3 reactions they are δ_N = *5.4(17)%, 5.4(7)%* and *6.5(7)%*, respectively. The similarity is a strong indication that only the remnants of evaporation are observed in experiment with any primary e-o effect being masked. Primary neutron e-o effects become only accessible in fission studied at very high kinetic energies when no energy is left for neutron emission. Even at the highest energies on display in Figs. 4.6.27and 4.6.28 neutron evaporation cannot be safely ruled out. Besides, pushing kinetic energies higher and higher towards true cold fission, the mass distributions become fragmentary and the methods for analysis applied so far are no longer appropriate. The approach has to be refined. Neutron e-o effects in true cold fission are discussed in Chapter 4.7.

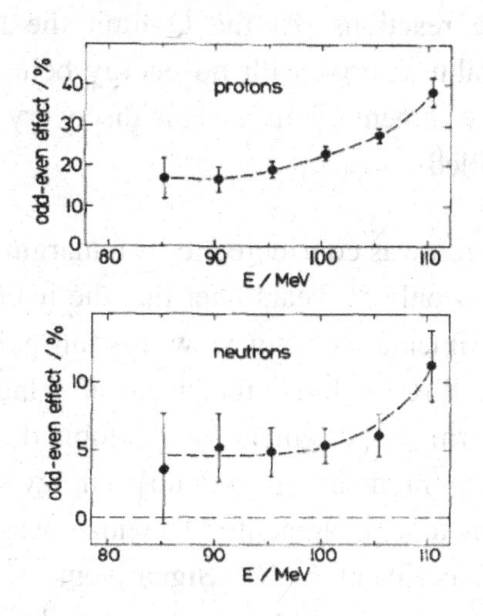

FIG. 4.6.28: proton (top) and neutron (bottom) e-o effect vs fragment energy from $^{233}U(n,f)$

From theory it is predicted that primary neutron e-o effects should be smaller than the corresponding charge e-o effects. In the combinatorial model of eq. (4.6.9) it is the probability ε to break pairs which should be larger for neutrons with $\varepsilon_N \approx N/A \approx 0.6$ compared to protons with $\varepsilon_Z \approx Z/A \approx 0.4$. This will bring down the neutron e-o effect δ_N. Likewise in the statistical model of eqs. (4.6.13) and (4.6.15) the survival probability for a fully paired neutron subsystem should be smaller than for protons. This is traced to the larger single-particle level densities for neutrons which compared to protons reduce the survival probability of the fully paired neutron system when all excitation energy has to be absorbed by the proton system [REJ00]. It should be stressed that these considerations apply to standard fission as opposed to true cold fission where mass by mass the phase space limit at the highest kinetic fragment energies is explored.

4.7 Cold Fission

4.7.1 *Discovery and Experimental Techniques*

Back in 1961 J. C. D. Milton and J. S. Fraser were studying thermal neutron fission of U- and Pu-isotopes by a double time-of–flight technique [MIL61]. In the (TKE, mass) contour diagrams measured they were surprised to see that for a broad mass range the total kinetic energies TKE came very close to the respective Q-values of the reactions. At the Q limit the kinetic energy was exhausting all of the available energy with no energy being left for excitation energy TXE heating up the fragments. This was the discovery of "cold fission" as this process is nowadays called.

For many years cold fission was considered to be a marginal phenomenon and hence just a curiosity. It was only 20 years later that the interest for cold fission was revived. From an experimental point of view it is intriguing that, in case the excitation energies are falling behind the neutron binding energies, the evaporation of neutrons form the fragments is prohibited. This should allow getting access to the primary fragments from double energy or double velocities measurements on complementary fragments. The idea was first tested in an original double velocity experiment by C. Signarbieux at the ILL [SIG81]. Masses in the cold fission regime could be resolved one-by-one. With virtually no excitation energy involved the expectation was that a superfluid mother nucleus should undergo fission into two superfluid fragments. But in the mass distributions of the (e-e) compound nuclei $^{234}U^*$ and $^{236}U^*$ under study even and odd masses were observed with no preference whatsoever for even mass splits. The unexpected finding was a challenge for more detailed investigations.

Searching for rare events is a prerogative of the LOHENGRIN mass separator of the ILL (see Chapter 4.2). The idea of cold fission was in the air at the ILL and it was P. Armbruster to be the first to recognize the new aspects which were hidden and should be disclosed in cold fission [ARM81]. A first experiment focussed on $^{233}U(n,f)$. In neutron-less cold fission the measurement of the mass and energy of one fragment (at LOHENGRIN usually the light fragment) is sufficient to find the mass and energy of the complementary fragment. Mass distributions in the high energy tail corresponding to the fraction 10^{-5} of the energy integrated chain yield were determined. The total energies TKE at this yield level were compared with Q-values averaged over the isobaric charge distribution as measured at high kinetic energies. The difference between Q-values and TKE yielded the excitation energies. For light masses from 80 to 104 the excitation energies were below 7 MeV pointing to neutron-less fission. Like in the above time-of flight experiments odd mass splits were seen also here to compete with even mass splits. Quite systematically even mass chains had actually higher excitation energies than odd mass splits.

Detector development at the ILL led to the finding that carefully designed ionization chambers operated with pure hydrocarbons (preferably methane) as the counting gas had intrinsic energy resolutions δE for fission fragments of better than $\delta E = 100\ keV$ [OED83a]. It is only the energy straggling in the entrance window to ionization chambers which is deteriorating their performance. This gave C. Signarbieux the idea to revive a technique which was already employed in the first experiments on fission physics. There the energies of both fragments were measured simultaneously by a back-to-back or "twin ionization chamber" (TIC). The layouts of two historical chambers are on display in Fig. 4.7.1.

To the left the TIC is shown which was used by Jentschke and Prankl [JEN42] in an experiment proving that fission is a binary process. To the right a more modern version of a TIC as constructed by Brunton and Hanna is sketched [BRU50]. The two chamber halves share a common cathode carrying the target material to be irradiated by neutrons. A major improvement was to install a Frisch grid in front of the electron collecting anodes thereby shielding the anodes from the influence of the positive ions created in the ionization process simultaneously with the electrons. Since that time the construction of TICs has

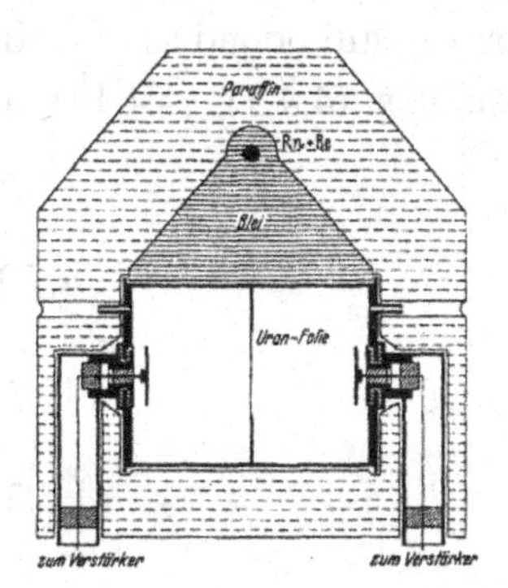

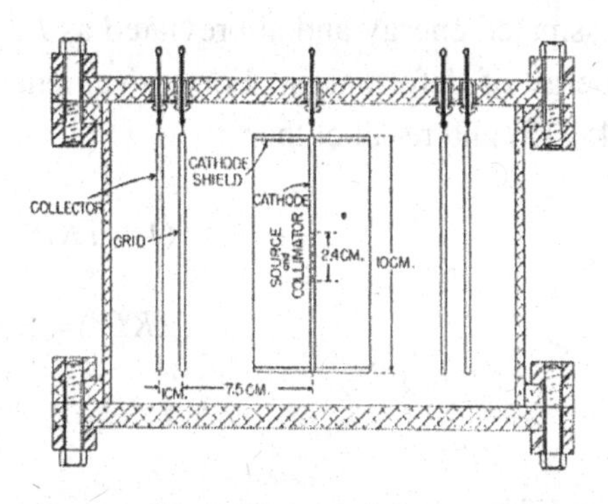

Fig. 4.7.1. Historical twin ionization chamber having allowed proving that fission is a binary process (to the left); version of twin ionization chamber constructed in 1959 (to the right).

not basically changed. Due to momentum conservation the ratio of the kinetic energies E_L/E_H allows to calculate the ratio of fragment masses $A_H/A_L = E_L/E_H$ and from mass conservation the individual masses A_L and A_H are obtained.

Strictly speaking this is only valid in case no neutrons are evaporated, but precisely this condition is fulfilled in cold fission. However, most important for the study of cold fission is the fact that no entrance window is required and that the target-backing can be made extremely thin. This ensures excellent energy resolutions. Very soon it was realized that, besides mass and energy, in studies of cold fission it is crucial to identify also the nuclear charges of fragments [SIG85]. Only with the knowledge of fragment mass and charge a deeper insight into the physics of cold fission is feasible because e.g. the cold fission properties of (e-e) and (o-o) fragments with even mass are very different. The techniques of charge measurements were described in Chapter 4.2.

4.7.2 *True Cold Fission*

The energy liberated in the fission process, the Q-value, is shared between the total kinetic energy TKE^* and the total excitation energy TXE of the fragments. At the very instant of scission the two energies may be further broken down into several contributing terms. At scission, besides a contribution to TKE^* due to the pre-scission kinetic energy E_K^{sci}, the lion`s share of the eventual TKE^* is still tied up as Coulomb potential energy V_{COUL}. Likewise, only part of the eventual TXE of the fragments is already present at scission as intrinsic excitation, often called dissipated energy and abbreviated as E_X^{sci}. Another part is still bound as potential energy of deformation. For convenience we recall the eqs. (4.4.1), (4.4.14) and (4.4.15) and re-label them:

$$Q = TKE^* + TXE \tag{4.7.1}$$

$$TKE^* = E_K^{sci} + V_{coul} \tag{4.7.2}$$

$$TXE = E_X^{sci} + V_{def} \tag{4.7.3}$$

Candidates for cold fission are events where the excitation energy is minimal and, hence, the kinetic energy maximal. These events will come close to the Q-value of the reaction. Since in the kinetic energy of eq. (4.7.2) the potential Coulomb energy is predominant, the corresponding scission configurations will be as compact as possible. This is why the process is called “cold compact fission”.

Energetic relationships for compact scission are visualized in a zoom of the schematic scission point model of Fig. 4.6.20 in Fig. 4.7.2. Q-values and potential energies are plotted as a function of the deformation of fragments at scission. In rather loose sense deformation at scission is figured by two fragments

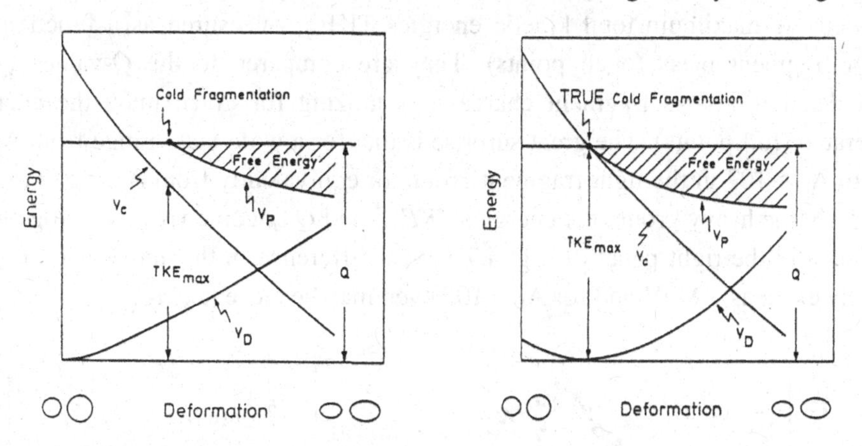

Fig. 4.7.2. Scission point model for cold (left panel) and TRUE cold fragmentation (right panel).

facing each other at a fixed tip distance and with fragment shapes ranging from spherical to deformed. To the left of the figure the more general situation is sketched. Energy conservation $Q = TKE^* + TXE$ imposes a minimum deformation at scission with as a consequence a finite deformation potential energy labelled V_D. At the deformation marked "Cold Fragmentation" the coldest feasible configuration is reached. Since the Coulomb potential decreases faster with deformation than the deformation potential increases, the sum of the potential energies $V_P = V_{coul} + V_{def}$ falls behind the Q-value beyond the cold fragmentation point. The difference $(Q - V_P)$ is hatched in the figure and was called "free energy" in the discussion of Fig. 4.6.20. Evidently one has for the free energy

$$(Q - V_P) = E_K^{sci} + E_X^{sci} \tag{4.7.4}$$

How the energy is partitioned between kinetic and excitation energy has to be found by experiment as outlined in Chapter 4.4. From the figure it is inferred that the maximum kinetic energy release will in general not exhaust the Q-value.

Presently interest will be focussed on "true cold fission" where TKE^* is exhausting Q-values within a few MeV. Inspecting Fig. 4.7.2 one may speculate that true cold fragmentation is achieved in case Q-values are very large and/or fragments have a non-zero ground state deformation. This ideal scission configuration for true cold fission is realized to the right in Fig. 4.7.2.

A first experimental result is depicted in Fig. 4.7.3. Thermal neutron induced fission of ^{235}U was investigated with the TIC technique [SIG91]. To the left are plotted the maximum total kinetic energies TKE_{max} measured as a function of light fragment mass (open points). They are compared to the Q-values Q_{max} calculated for those fragment charges maximizing for given mass the energy liberated (full points). The great surprise is that for a wide mass range from $A_L \approx 80$ to $A_L \approx 107$ in the light fragment group, or equivalently from $A_H \approx 129$ to $A_H \approx 156$ in the heavy group, the energies TKE_{max} and Q_{max} come very close together. As seen in the right panel of Fig. 4.7.3 their difference in the above mass range barely exceeds 3 MeV and for $A_L = 105$ even may be close to zero.

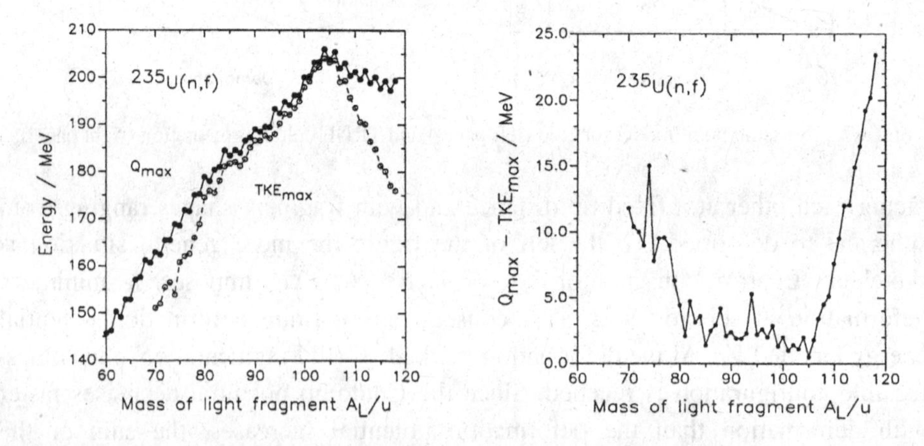

Fig. 4.7.3. Cold fission of $^{235}U(n_{th},f)$.To the left: comparison of TKE_{max} and Q_{max} as a function of light fragment mass A_L.To the right: difference $(Q_{max}-TKE_{max})$ as a function of light fragment mass A_L.

As to the absolute values of the total kinetic energy, it has to be kept in mind that with the TIC method the systematic error of *TKE* is conservatively estimated to be $\Delta TKE = \pm 1$ *MeV*. Hence, in the above mass window neutron emission at TKE_{max} is for sure excluded. In the figure it is further remarkable that towards symmetric fission or super-asymmetric fission the maximum kinetic energies move away from their respective Q-values. This is most pronounced for symmetric fission where the difference $(Q_{max} - TKE_{max})$ becomes 25 MeV. In view of the dip in average kinetic energy at symmetry (see Fig. 4.5.1) the failure to achieve cold fission is not surprising. By contrast, as suggested by the scission point model in Fig. 4.7.2, cold fission could be expected for mass splits with high Q-values and/or for fragments being deformed in their ground states. In fact, the minimum excitation energy for light fragment mass A_L = 104 and the complementary heavy mass A_H = 132 is understood to be associated with the maximum Q-value due to the magic ^{132}Sn and the deformed isotope ^{104}Mo. For masses near $A_L \approx 90$ and equivalently $A_H \approx 146$ it is mainly the strongly deformed heavy Ba isotopes making cold fission accessible. The importance of the deformed neutron shell at N ≈ 88 (corresponding to ^{144}Ba) was already stressed back in 1976 in the scission-point model of Wilkins et al [WIL76]. In the Brosa model this corresponds to the standard II mode[BRO90].

However, the feature in Fig. 4.7.2 with the strongest impact on the general view of nuclear fission is the observation that even and odd masses show up near true cold fission at roughly the same level of probability. At closer inspection fragmentations for an even mass compound nucleus into two odd mass fragments actually appear to come nearer their respective *Q*-values than the even–even mass fragmentations. This was totally unexpected. In view of the sizable charge even-odd effects known from standard fission for e-Z compounds it was conjectured that the preservation of superfluidity being behind the e-o effects should become the more pronounced the colder the system undergoing fission is. With any dissipation being minimized, a result with a 100 % charge even-odd effect in cold fission would not have been surprising. The puzzle becomes even more baffling when besides masses also the charges of fragments in true cold fission are scrutinized.

Before entering the discussion of charge and neutron number distributions in cold fission some more experimental results are presented. In Fig. 4.7.4 two experiments on thermal neutron fission of ^{235}U are compared. Full points are LOHENGRIN data at a mass chain yield level of 10^{-5} / MeV [QUA82]. The dot-dashed line traces yield data obtained with a TIC chamber at the same level of 10^{-5} / MeV [STR87]. The dashed line represents the yields measured in the same experiment but for a yield level of 10^{-4} / MeV. In all cases yields were integrated in an energy window with width 1 MeV. The full line shows Z-weighted *Q*-values. The findings of the two techniques, LOHENGRIN and TIC, agree that in the light mass range from $A_L \approx 88$ to $A_L = 104$ or equivalently in the heavy mass range from $A_H = 132$ to $A_H \approx 148$ cold fragmentation is achieved in the sense that less than 3 MeV of excitation energy is left to the fragments relative to the Z-averaged *Q*-values. The mass range addressed is slightly narrower than in Fig. 4.7.3 but otherwise matches.

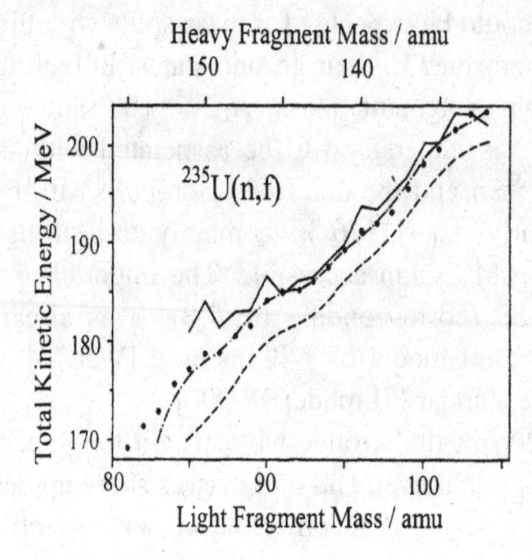

Fig. 4.7.4. Chain yields at the level10^{-5}/MeV For $^{235}U/n_{th},f)$; open points LOHENGRIN, dot-dashed line TIC chamber; dashed line are TIC data at the level 10^{-4}/MeV; full line Q are -value averaged over Z.

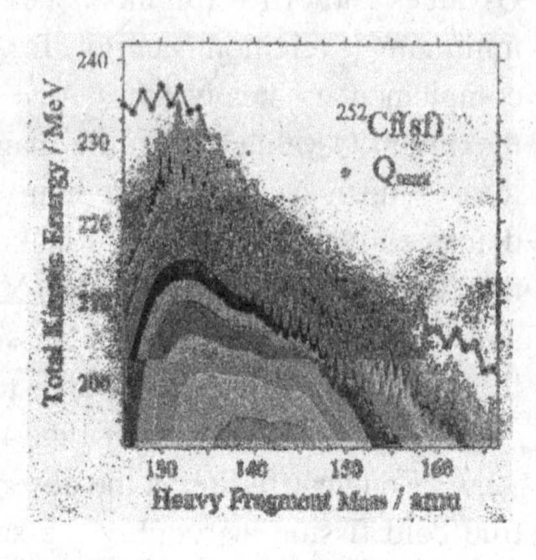

Fig. 4.7.5. Cold fission of ^{252}Cf(sf); for neutron-less fission individual mass lines are resolved in the TIC; total of 10^9 events.

Another fission reaction having been carefully studied is spontaneous fission of ^{252}Cf. The contour plot (*TKE*, A_H) in Fig. 4.7.5 was obtained for a total of 10^9 events accumulated in a TIC chamber with a thin Cf source on a thin C-backing (5μg/cm²) [CRO98]. The experiment had to be run for 1 year. Full points joined by a line are the maximum *Q*-values. Cold fission events with excitation energies not exceeding 1-2 MeV are discerned for $A_H = 132$, $A_H = 134$ and perhaps $A_H = 136$ followed by a broad range of masses from $A_H = 139$ to $A_H = 156$. Again the rule advocated by the schematic scission point model in Fig. 4.7.2 applies: for cold fission to have a chance to materialize either the *Q*-values have to be large (heavy fragments around $A_H = 132$) and/or fragments have to be deformed in their ground states (heavy masses in a window centred around $A_H = 146$). In the plot neutron-less fission is identified as the region where individual mass lines are resolved because as soon as neutron evaporation sets in the mass lines corresponding to a fixed energy or mass ratio are blurred. Surprisingly neutron-less events are to be seen for excitation energies up to 15 MeV. Neutron-less fission is hence not necessarily cold as often argued. On the other hand, from studies of neutron multiplicity it is reported that for ^{252}Cf(sf) the percentage of events with multiplicity $\nu = 0$ is only 0.23% [VOR04]. By contrast, for thermal neutron fission of ^{235}U the multiplicity $\nu = 0$ has a probability of 3.2% [HOL88]. Therefore cold fission studies of ^{252}Cf(sf) require a data base ten times larger than for ^{235}U(n_{th},f). Nevertheless, for the latter reaction a minimum of 10^7 - 10^8 events should be collected to ensure that true cold compact fission can be reliably assessed. Unfortunately, on the COSI spectrometer the statistics was even in the best cases one order of magnitude too low for analysis. Data having nevertheless been published for cold compact fission can therefore not be accepted as being relevant [ASG93].

Upon approaching cold fission significant changes of the fragment mass distributions are observed for small changes in the kinetic energy. In Fig. 4.7.6 mass distributions measured with the LOHENGRIN spectrometer for thermal fission of ^{234}U* and ^{236}U* are on display. The two figures in the upper row are for "standard" fission while the middle and lower figures pertain to cold fission. The yields are integrated for energy windows with 1 MeV in width [CLE86]. In the two lower panels the difference in energy for the light fragment is only 2 MeV yet the mass distributions change dramatically. In the panels in the middle row it is rather peculiar that the peak yields for ^{234}U* near A_L = 100 and A_L = 90 corresponding to heavy masses A_H = 134 and A_H = 144 starts to come into the foreground. They may be associated with the average standard I and standard II Brosa modes, respectively. A similar observation obtains in fission of ^{236}U* for masses A_L = 102 and A_L = 90.

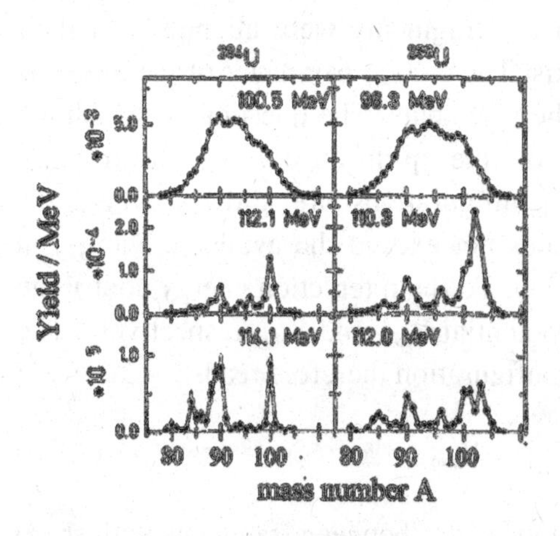

Fig.4. 7.6 Mass distributions for ^{234}U and ^{236}U* taken with LOHENGRIN for E_{KL} windows approaching cold fission.

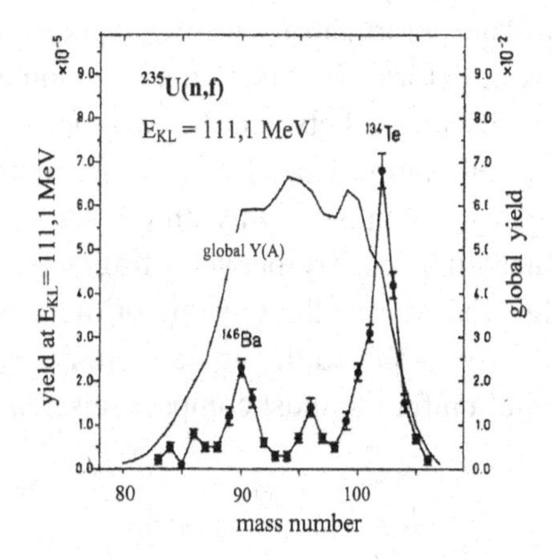

Fig. 4.7.7. Mass distributions integrated over energy (full line) and for E_{KL} = 111.1 MeV (data points) for ^{236}U* LOHENGRIN data.

More detailed results for $^{235}U(n,f)$ from LOHENGRIN are presented in Fig. 4.7.7 [ARM99]. Data points are the yields at the kinetic energy of the light fragment of E_{KL} = *111.1 MeV* in a 1 MeV window (yield scale to the left) while the continuous curve are the energy integrated yields (yield scale to the right). The cold mass distribution comprises the fraction $3 \cdot 10^{-4}$ of all events. The emergence of the characteristic mode masses 102 and 90 in the light fragment group associated with the spherical ^{134}Te (high Q-value) and the deformed ^{146}Ba in the heavy group is striking. It has to be stressed, however, that at the slightly higher energy E_{KL} = *113.9* MeV (not shown) the ^{134}Te yield is fading away [SIM90]. The same observation is made in Fig. 4.7.6 (bottom right panel). This strange behaviour calls for a more detailed analysis.

To summarize the results on cold fission phenomena discussed so far it will be convenient to introduce a model called the "tip model" of cold fission [GON91]. The model allows predicting which fragmentations are liable to lead to cold compact fission. Analogous to the scission point model in Fig. 4.7.2 the scission configuration is parameterized by two fragments almost touching each other. The fragments are taken to be in their ground states with deformations as tabulated by Möller-Nix [MOL81]. In the tables quadrupole and hexadecupole deformations are taken into account. Some nuclei have oblate ground state deformations. In view of the elongation process characteristic for fission, these deformations were considered not being realized. Yet, in most cases besides the oblate also prolate deformations at about the same energy are predicted. These prolate deformations were adopted and the fragments were aligned with their major deformation axis on a common axis. The crucial parameter of the model is the distance d between the two tips of the fragments. The question is: which are the minimum distances? The minimum feasible tip distance was found from an energy argument. Only those scission configurations are allowed where the interaction energy between fragments does not exceed the available energy, in the present case the Q-value of the reaction. To the interaction energy contribute the Coulomb and the nuclear proximity potential V_{coul} and V_{nuc}, respectively. The criterion for the most compact scission configuration therefore reads

$$Q = V_{coul} + V_{nuc} \qquad (4.7.5)$$

From this condition the minimum tip distances d_{min} between fragments with sharp surfaces are found. In keeping with nuclear surfaces being leptodermous and with a diffuseness constant b = 1 fm, tip distances for a valid scission configuration with two touching fragments just about to separate should not be larger than about 2-3 fm. In case the minimum tip distance turns out to be definitely larger than about 3fm the fragments have to be deformed to reach the scission configuration of touching fragments. This means that true cold fission with *TXE* = *0* is ruled out. The *Q*-values are a function of both, fragment masses and charges. For any fixed mass ratio A_L/A_H all experimentally observed charge ratios Z_L/Z_H were considered. Results of calculations for d_{min} are presented in Fig. 4.7.8. For $^{235}U(n_{th},f)$ in the left panel cold compact fission is anticipated for the light fragment mass range from $A_L \approx 80$ to $A_L \approx 108$. This is in perfect agreement with the experimental results shown in Fig. 4.7.2. Surprisingly, whenever according to the tip model a compact scission configuration is accessible the fission process is exploiting it.

In the middle and the right panel of Fig. 4.7.8 minimum tip distances for $^{252}Cf(sf)$ once calculate for spherical and once for deformed fragments are compared. In experiment (see Fig. 4.7.5) cold fission was observed for heavy fragments near $A_H = 132$ and for the range of fragments from $A_H \approx 139$ to $A_H \approx 156$. Evidently, for scission configurations with undeformed spherical fragments the tip model fails to predict cold fission at all. The situation changes dramatically when ground state quadrupole and hexadecupole deformations of fragments are taken into account. Based on the criterion $d_{min} = 3\ fm$ the model rather accurately finds the masses where cold fission is observed, both for the island near $A_H = 132$ and for the more extended mass range from $A_H \approx 139$ to $A_H \approx 159$ (instead of $A_H \approx 156$ in experiment).

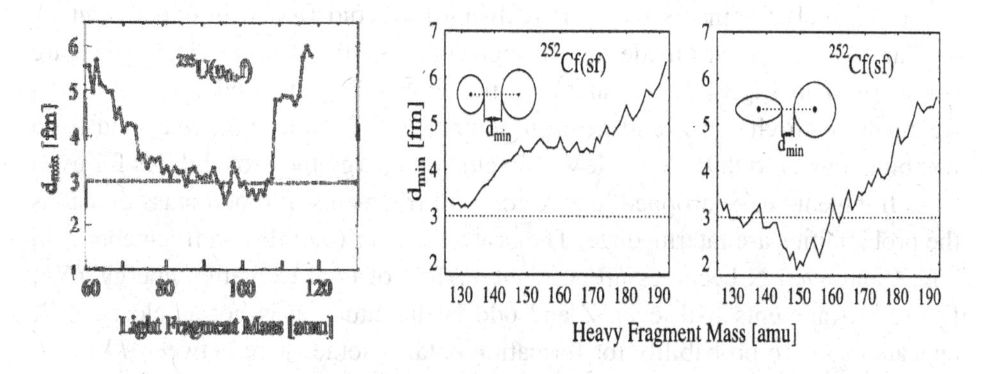

Fig. 4.7.8. Minimum tip distance d_{min} for $^{235}U(n,f)$ vs light fragment mass to the left; minimum tip distance d_{min} for $^{252}Cf(sf)$ vs heavy mass without (middle) and with deformation (right) of fragments.

To assess the properties of true cold fission, plots of mass distributions for given windows of kinetic energy like those in Figs. 4.7.6 and 4.7.7 are not well adapted. This is because yields in a window of kinetic energies depend on whether the *Q*-values allow the windows to be fed or not. Thereby an artificial bias is introduced not telling much on cold fission. Instead one has to switch to a presentation where kinetic or excitation energies are scrutinized for given fragment masses A_L (or A_H) and charges Z_L (or Z_H). Only then is it possible to judge the properties of true cold fission by comparison with the *Q*-values $Q(A_L, Z_L)$ depending likewise on charge and neutron number. The unexpected appearance of even and odd masses in fission of an (e-e) compound turns into the even more surprising evidence that for the even-even compounds under study, in even mass splits the (o,o) fragments are the coldest species while really cold (e,e) fragments are not observed. For odd mass fragments the situation is intermediate. Experiments to demonstrate this are available for $^{236}U^*$, $^{234}U^*$ and ^{252}Cf(sf). In Fig. 4.7.9 results from an experiment with a TIC chamber on $^{236}U^*$ by the Saclay group are on display [TRO89]. The high energy tail of *TKE* for the mass fragmentation (132,104) is inspected. The contributions for the charge splits (50/42) and (51/41) have been disentangled and are plotted separately. Most astonishing is that the yield for the (even, even) charge split (50/42) fades away at TKE_{max} = 203 MeV and fails to reach the *Q*-value at 206 MeV. There is a mismatch of 3 MeV. By contrast, the TKE_{max} for the (odd,odd) charge split (51/41) precisely exhausts the *Q*-value indicated in the figure. The finding tells that for even mass fragmentations of an even-Z compound it is not at all the (even, even) charge and neutron number division but instead the division into two (odd, odd) fragments which is realising true cold fission in experiment. A similar observation was made for all fragment mass distributions. The results are summarised in Fig. 4.7.10 visualizing the probabilities to find in cold fission fragments with charge Z and neutron number N for (e,e), (e,o), (o,e) and (o,o) combinations [TRO89]. At 2 MeV of excitation energy the probabilities for even mass fragments have dropped to zero for (e,e) fragments. For odd mass divisions the probabilities are intermediate. The probability for (o,e) fission fragments with odd Z and even N becomes nil at about 1 MeV of total excitation energy *TXE*. For (e,o) fragments with even Z and odd N the situation is not as clear-cut. It appears that zero probability for formation obtains somewhere between *TXE = 0* and *TXE* = 1 MeV. In any case, for *TXE* below 1 MeV mainly, if not only, (odd,odd) fragments are observed.

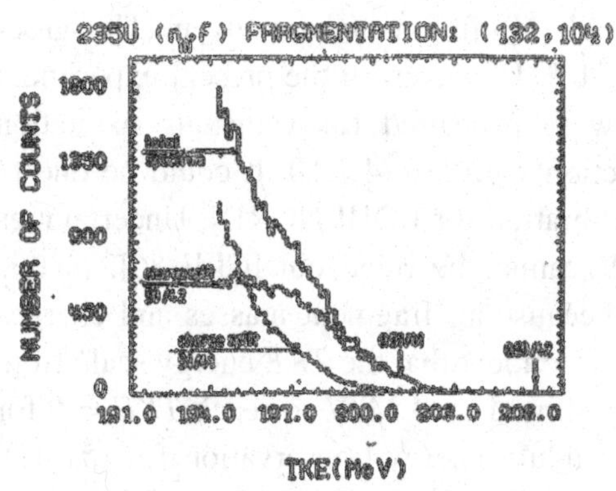

Fig. 4.7.9. High TKE tail for the mass split 132/ 104 resolved into the charge splits 50/42 and 51/ 41.

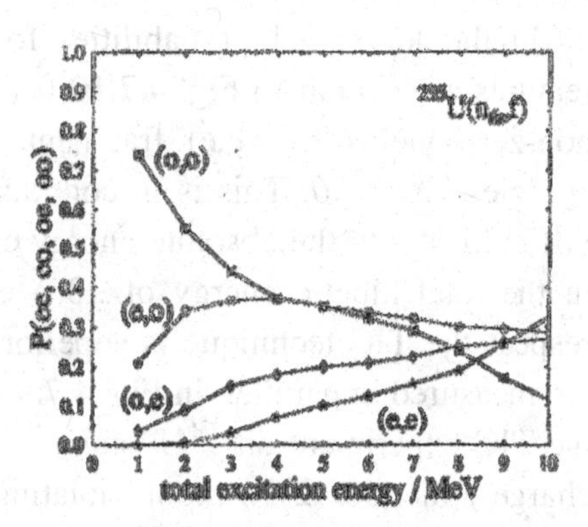

Fig.4. 7.10. Probabilites for (ee), (e,o), (o,e) and (o,o) splits of (charge, neutron number).

These observations are a compelling indication that level densities play a crucial role in cold fission [SIG 91]. When comparing (Z,N) nuclei with different charge and neutron number content the level densities exhibit systematic shifts with excitation energy. In the Fermi gas model effective excitation energies U^* are introduced. For (e,e) nuclei an effective excitation energy $U^* = U - 2\Delta$ is considered to be relevant which is back shifted by the pairing gap 2Δ relative to the physical excitation energy U. For (e,o) and (o,e) nuclei the effective excitation energy is $U^* = U - \Delta$, while for (o,o) nuclei one has $U^* = U$. The physical reason for the back shift is that, due to the pairing interaction, nuclei with even proton and/or neutron numbers are superfluid in the respective subsystem and there are only a few collective levels in the energy gap. For typical fission fragments the energy gap is $2\Delta \approx 2$ MeV. In Fig. 4.7.10 it is precisely at $TXE \approx 2$ MeV that the yield for (e,e) fragments starts to increase. For (o,e) and somewhat less convincing for (e,o) nuclei there is virtually no cold fission below $TXE \approx 1$MeV. Only for (o,o) fragment nuclei true cold fission with $TXE \approx 0$ MeV is attained. This is an interesting new aspect of fission. Cold fission is not steered merely by Q-values and fragment deformations. For true cold fission to become observable ultimately fragment level densities determine the outcome.

On LOHENGRIN cold fission of the isotope ^{233}U, the neighbour to ^{235}U in Figs. 4.7.9 and 4.7.10 was investigated [SCH94a]. Results of interest in the present context are on view in Fig. 4.7.11. The mass range studied was restricted to $A_L = 76$ to $A_L = 91$ and therefore almost ten masses in cold fission could be missing.

In the left panel probabilities for (Z, N) fragments are given. The gross features are similar to Fig. 4.7.10 for $^{236}U^*$. However, in the present experiment non-zero yields for (e,e) fragments were measured for vanishing excitation energies *TXE = 0*. This is in contradiction with Fig. 4.7.10. It could be due to difficulties with the absolute energy calibration for LOHENGRIN. Uncertainties in the total kinetic energy of ±3 MeV cannot be ruled out [CLE 86]. In this respect the TIC technique is superior because all fragment masses and charges are measured in parallel. In Fig. 4.7.9 it is evident that the *TKE* energy scale from the TIC experiment on $^{236}U^*$ cannot be shifted by 3 MeV to reach *TKE = 0* for charge split (50/42) without violating unduly energy conservation for (51/41). Shifting in Fig. 4.7.11 the *TXE* absolute values from LOHENGRIN by about 2-3 MeV to higher energies would bring the results in agreement with the TIC data in Fig. 4.7.10.

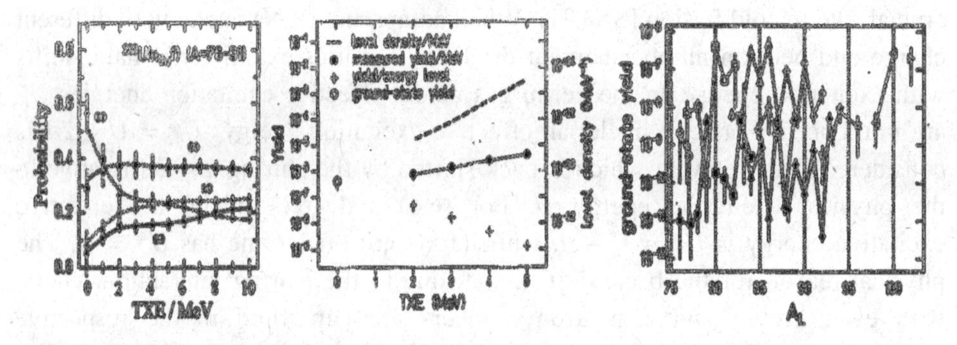

Fig. 4.7.11. Cold fission properties of $^{234}U^*$: probabilities for (Z,N)-fragments (left panel); level density and yield vs TXE for (^{82}Ge+^{152}Nd) (middle panel); extrapolated ground state yield at TXE = 0 vs light fragment mass (right panel).

The middle panel in Fig. 4.7.11 gives the details for the split (^{82}Ge+^{152}Nd). Full points depict the yields as a function of excitation energy *TXE* per MeV (scale to the left). The dashed line traces the level densities vs. *TXE* (scale to the right). Since the level density increases faster with *TXE* than the yields, the yield per energy level (crosses) decreases for increasing *TXE*. Upon extrapolating the yield per energy level to the excitation energy *TXE = 0* the ground state yield is found (open point). The idea of this approach is to make use of data at energies where statistics is still satisfactory to determine yields which are not accessible to experiment. The right panel in Fig. 4.7.11 reviews the extrapolated ground state yields as a function of light fragment mass A_L for up to 3 charges per mass. Open symbols pertain to even charges and closed symbols to odd charges. Even charges exhibit the largest ground state yields. For fixed charges in the cold fission regime with masses $A_L \geq 80$ there is an additional pronounced even-odd staggering: for even charges the ground state yields for even masses (i.e. (e,e) nuclei) are larger than for odd masses (i.e. (e,o) nuclei). For odd charges the inverse is true. This observation discloses a neutron even-odd staggering.

It should be stressed that due to the uncertainty in the absolute energy calibration sizable error bars for the ground state yields shown in Fig. 4.7.11 (right panel) are anticipated. But this should not impair the validity of the general trends discussed for the yields. It has to be emphasised that, in spite of larger probabilities for (e,e) fragments populating individual levels (here the ground state levels), it is the much larger level density for (o,o) fragments which in the end brings the (o,o) fragments to the foreground in cold fission (left panel in Fig. 4.7.11).

A third investigation where measurements were pushed to true cold fission was exploring spontaneous fission of ^{252}Cf [CRO 98]. In experiment the TIC technique was employed. A first result was already presented in Fig. 4.7.5 demonstrating that cold fission was attained for a range of heavy fragment masses from $A_H = 139$ to $A_H = 156$ and, in addition, for some isolated masses near $A_H = 132$. In Fig. 4.7.12 a more instructive view of results obtained is given. Mass distributions for two windows of the total excitation energy *TXE* are presented. It has to be stressed that masses and charges were determined which allowed to calculate the true excitation energy *TXE* as the difference between *TKE* and the corresponding mass and charge dependent *Q*-values. To the left the *TXE* window chosen ranges from 2 MEV to 4 MeV and to the right from 4 MeV to 6 MeV. At the lower *TXE* even masses (black and open bars in the figure) are more abundant than odd ones (shaded bars) while at the higher *TXE* the yields are more equilibrated. For even masses the contributions by (e,e) and (o,o) splits of charge and neutron number are deconvoluted and visualized by open and black bars, respectively. At the higher *TXE* to the right the (e,e) fragmentations (open bars) are clearly seen while at the lower *TXE* to the left virtually no (e,e) splits are registered, except for the mass numbers 132,134 and 136 in a cold fission island and for 140, 142 and 144 in the main agglomeration of cold fission. It is worth noticing that the yields in the spikes for masses A = 146, 150 and 152 are due to La isotopes while the largest kinetic energies are reached for ^{146}Ba, ^{150}Ce and ^{152}Ce. The La-isotopes are strongly deformed in their ground states exhibiting large quadrupole and hexadecupole moments [MOL95]. After integrating the Cf data over fragment mass the data depicted in Fig. 4.7.12 are in perfect agreement with the TIC results for $^{236}U^*$ in Fig. 4.7.10. The overwhelming importance of (o,o) fragmentations close to true cold fission is hence seen both, in thermal neutron induced and in spontaneous fission.

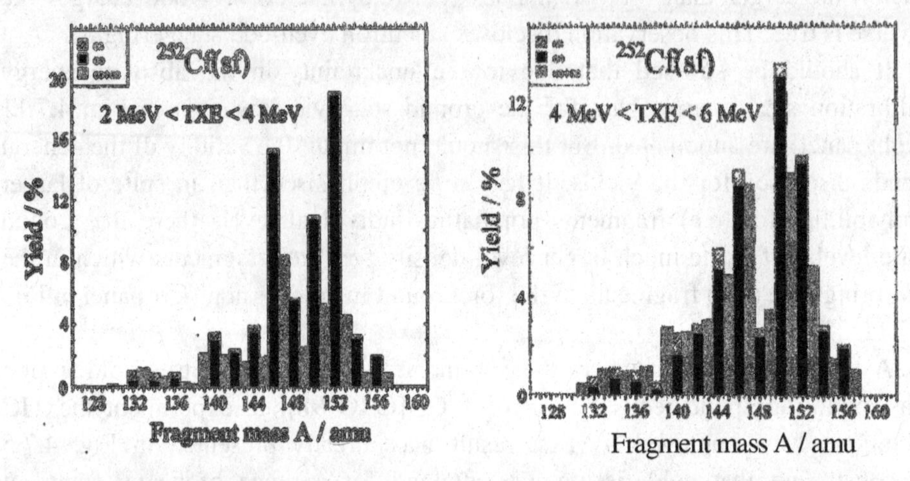

Fig. 4.7.12. Mass distribution with identification of (e,e) and (o,o) splits for even masses: open bars (e,e), black bars (o,o);hatched bars for odd masses. Left panel: 2 MeV < TXE < 4 MeV. Right panel: 4 MeV < TXE < 6 MeV.

Cold fission in the island around the heavy fragment mass A_H= 132 deserves a particular discussion. Similar to Fig. 4.7.9 for thermal neutron fission of $^{236}U^*$, the high energy tails for selected heavy masses 130, 132 and 134 from ^{252}Cf(sf) are compared in Fig. 4.7.13 [CRO98]. In all 3 examples the maximum Q-values for the even charge splits Q_H/Q_L = 50/48 (left and middle panel) or 52/46 (right panel) are not attained. For masses 122/130 about 6 MeV are missing while for 120/132 and 118/134 there is still a gap of about 2 MeV. For masses 122/130 and 118/134 (left and right panel) it is not for sure that true cold fission is reached for the (odd,odd) charge ratios. By contrast, for the mass ratio A_L/A_H =120/132 in the middle panel true cold fission is observed for ^{120}Ag/^{132}Sb, i.e. the (odd,odd) charge split Q_L/Q_H = 47/51. These findings agree with the results reported for thermal neutron fission of $^{236}U^*$ in Fig. 4.7.9 for the mass fragmentation 104/132. Note that for ^{252}Cf(sf) only the odd-odd nucleus ^{120}Ag has in its ground state a sizable but not excessively large prolate quadrupole moment while the odd-odd ^{132}Sb is oblate, though only weakly. True cold fission observed for ^{120}Ag/^{132}Sb in ^{252}Cf(sf) hence confirms that it is less deformation and more the level density of the odd-odd nuclei which is steering the process.

In spontaneous fission of ^{252}Cf there is still a further notable peculiarity. The bump in the *TKE* distribution for energies larger than about 226 MeV in Fig. 4.7.13 is unexpected. It is clearly pronounced for the mass ratio 120/132 and at least indicated for 122/130. In both cases the charge of the heavy fragment is the magic Z_H = *50*, i.e. the nuclei ^{132}Sn and ^{130}Sn are concerned. Both isotopes are spherical. The complementary light fragments are ^{120}Cd and ^{122}Cd, both only slightly deformed. When averaged over the full kinetic energy distribution the main charge contributing to the mass 132 is ^{132}Sb with Z_H = 51 and the ^{132}Te isotope with Z_H = 52 but not the tin isotope ^{132}Sn. This could, however, be an effect due to neutron evaporation by ^{134}Te and ^{133}Sb at low kinetic energies. Though it is tempting to call the pronounced bump structure a magic cluster mode, the term "mode" should better be reserved for characteristic features concerning the fission process as a whole. On the other hand the peculiarity appears not to be related to cluster radioactivity because the ground state to ground state transition typical for cluster radioactivity is virtually absent. Some energy is drained from the available energy, the Q-value, before the peculiar cluster effect is setting in. It appears to be just a manifestation of the compactness of the magic nucleus ^{132}Sn.

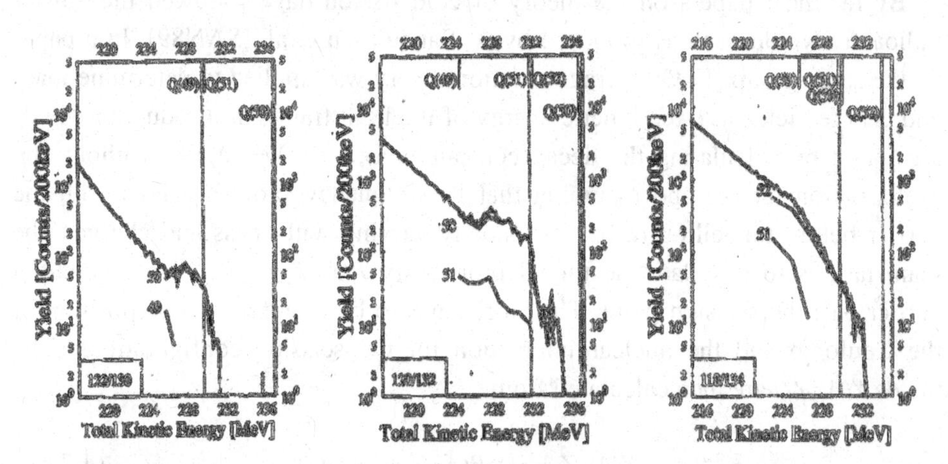

Fig. 4.7.13. High TKE tails for the mass ratios 122/130, 120/132 and 118/134 in ^{252}Cf(sf). The contributions and Q-values for different charge splits labelled by Q(Z_H) are shown.

For the fissioning nucleus ^{236}U* in Fig. 4.7.9 no bump is observed for the mass split 132/104 in the high energy tail of the *TKE* distribution. Yet, decreasing *TKE* by only a few MeV the yield for Z_H = 51 overtakes the yield for

$Z_H = 50$. This is once more the influence of large level densities for the (o,o) nuclei ^{132}Sb and ^{104}Nb in contrast to the (e,e) nuclei ^{132}Sn and ^{104}Zr. The rapid change from (e,e) to (o,o) splits may in the uranium case mask the formation of a bump for fission into (^{132}Sn, ^{104}Zr).

4.7.3 *True Cold Fission: What can be Learned?*

Having surveyed experimental results on neutron-less fission with focus on true cold fission the question arises what can be learned for the fission process in general. In theory two different lines of approach are proposed. On one hand α-decay and/or cluster radioactivity are taken as prototypes for cold fission phenomena. To find the yields of cold fission the task is to calculate, first, preformation factors for clusters in the mother nucleus and, second, tunnel probabilities through the Coulomb and nuclear potential slowing down the decay. Another group of theoretical papers treats cold fission as a limiting case of standard fission. The fragments take shape and are formed, possibly rather early, in the course of the process.

By far most papers on the theory of cold fission have followed the cluster radioactivity approach as proposed by A. Sandulescu et al. [SAN89]. In a paper by the same group [FLO93] fragmentation theory was applied to determine mass and charge yields at fixed kinetic energy of the light fragment. Production yields are found by calculating the decay constant $\lambda(A_L,Z_L) = \nu \cdot P(A_L,Z_L)$ without any preformation factors and assuming that the frequency ν of collisions with the barrier before tunnelling to be very slowly varying with mass and charge. The main task is to calculate the tunnel probability $P(A_L,Z_L)$ through the potential barrier past the scission point. The barrier has to be found as a superposition of the Coulomb and the nuclear interaction for the scission configurations. The yields $Y(A_L,Z_L)$ are then calculated from

$$Y(A_L,Z_L) = P(A_L,Z_L) / P(A_L) \qquad (4.7.6)$$

where $P(A_L)$ is the sum over Z of $P(A_L,Z_L)$. The penetrability $P(A_L,Z_L)$ through the barrier evidently depends on the deformation of the fission fragments at scission. The deformations were adjusted to fit theory to experimental yields. The deformations found are not far away from those elaborated by theory [MOL81].

As pointed out in the preceding Chapter 4.7.2, mass or charge distributions at given kinetic energy of light fission fragments are not revealing much on the properties of cold fission. The more informative mass and charge data as a function of total excitation energy as those e.g. obtained on LOHENGRIN for thermal fission of $^{234}U^*$ and presented in Fig. 4.7.11 were analysed in fragmentation theory [SCH94a, CLE99]. The aim was to reproduce the ground state yields at zero excitation energy which were deduced by extrapolation from measurements at somewhat higher *TXE* (see Fig. 4.7.5). The penetrabilities could be shown to follow the systematic trends of the yields though between penetrabilities and yields there is a factor of twenty orders of magnitude (sic) missing. Hidden behind this factor are preformation probabilities and rates of hitting barrier which are not accounted for. More worrying is the fact that the scission configurations were conceived as two spherical fragments. This is in conflict with the tip model where in Fig. 4.7.8 (middle panel) it is brought to evidence that for undeformed spherical fragments cold fission is not attained.

The same data were analysed as a function of the effective excitation energy *TXE** evaluated in the back-shifted Fermi gas model of level densities which was also advocated for the interpretation of data in Fig. 4.7.9. Differences in level density for even and odd nuclei are thus compensated. Results for effective excitation energies from 2 to 7 MeV were obtained [AVR95]. The dependence on *TXE** was found by assuming that any excitation energy made available near cold fission goes into deformation of fragments. Experiment and theory agree fairly well. However, it must be mentioned again that, first, $^{234}U^*$ data were taken on LOHENGRIN for a restricted range of light fragment masses from $A_L = 76$ to $A_L = 93$ only and that, second, for the specific question of true cold fission the spectrometer data should be discussed cautiously. This was already pointed out in connection with Fig. 4.7.11. In particular the large yields extrapolated to the ground state for (e,e) fission fragments in the mass range 80 to 86 exceeding 10^{-7} per fission in Fig. 4.7.11 are questionable. These yields could not be validated by the TIC chamber technique which for this task appears to be much more reliable. Even-even ground state transitions were so far not observed and, very conservatively, upper limits for their population of 10^{-8} to 10^{-9} per fission are estimated down by two orders of magnitude compared to the LOHENGRIN data.

Fragmentation theory was moreover pushed to scrutinize the dependence of barrier penetrabilities on the relative orientation of two deformed fragments touching each other and prone to tunnel through the barrier [MIS02]. The interaction barrier depending on the relative angles of deformation axes was calculated by double-folding of a heavy-ion potential. The experimental ^{252}Cf(sf) cold fission data discussed in the preceding chapter (Figs. 4.7.12 and 4.7.13) were taken as the test case. Pole-pole orientations of the two fragments facing each other proved to yield the highest penetrabilities. Cold fission both, in the broad range of masses from $A_L = 96$ to 113 (or equivalently $A_H = 139$ to 156) and the narrow range around $A_L = 120$ (or equivalently $A_H = 132$) is correctly predicted. In view of the fact that preformation probabilities were not considered in the paper the success of the calculation is surprising. It tells that the small barrier heights for the pole-pole orientation of fragments prevail over the spectroscopic factors in the formation of cold mass distributions.

An approach very different from fragmentation theory was put forward by J. F. Berger et al. [BER84, BER89]. An ambitious microscopic theory of fission based on constrained HFB calculations has been devised encompassing in particular also cold fission. It describes the fission motion by a few collective parameters like stretching, mass-asymmetry, bending and necking-in among others. The exchange of energy between collective and intrinsic degrees of freedom is neglected based on the argument that the evolution of collective motion is 10 times slower than the internal motion of nucleons. It means that throughout adiabaticity is assumed with the fissioning nucleus remaining in a superfluid state. Remarkably the potential energy surface (PES) calculated for the nucleus ^{240}Pu near the saddle point reproduces the famous double-humped pattern

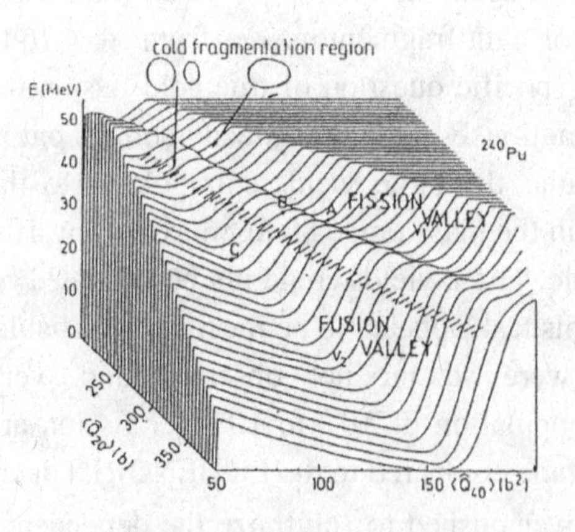

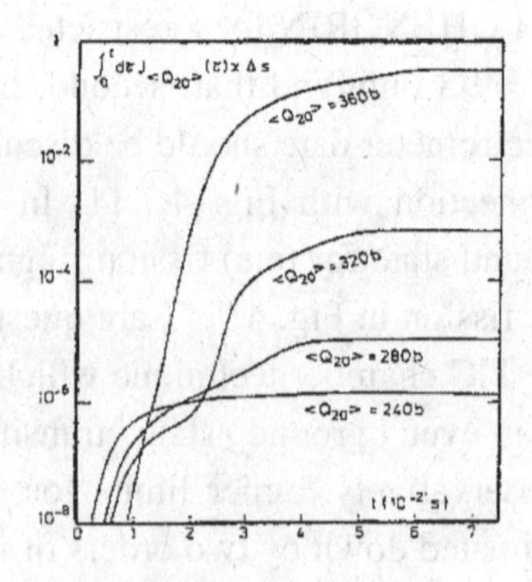

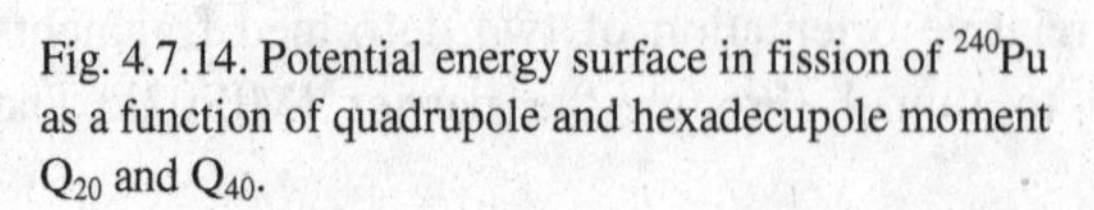

Fig. 4.7.14. Potential energy surface in fission of ^{240}Pu as a function of quadrupole and hexadecupole moment Q_{20} and Q_{40}.

Fig. 4.7.15. Collective flux passing the scission line at four different elongations Q_{20}.

with the second saddle being mass-asymmetric. The PES for ^{240}Pu beyond the saddle point is on display in Fig. 4.7.14 as a function of quadrupole and hexadecupole moments Q_{20} and Q_{40}, respectively. Two distinct valleys appear in the landscape. In fission the nucleus slides down the fission valley as a mono-nucleus. In fusion two heavy ions move up the fusion valley. The two valleys are separated by a ridge whose height is getting lower as the elongation increases. For a nucleus to undergo scission into two fragments, or for two ions to merge into one single nucleus the ridge acts as a barrier. The nucleus will for sure undergo fission when the ridge disappears. The nucleus is getting unstable for elongations expressed by the quadrupole moment Q_{20} at $\langle Q_{20} \rangle \geq 370$ b. This is the so-called exit point at a distance d between the fragment charge centres of d = 17 fm. However, albeit with smaller probability, fission may also occur at smaller elongations on condition that the ridge is tunnelled or overcome. In order to induce scission, energy must then be made available for a motion perpendicular to the stretching motion. With superfluidity being preserved in the downhill motion from saddle to scission, the motion is uniquely slowed down by trnsfer of energy into other collective modes. The mode of importance for the process at issue is a necking-in mode orthogonal to the stretching motion. Cold fission is associated with scission at the smallest feasible elongations. For $\langle Q_{20} \rangle \leq 280$ b (corresponding to d = 15 fm) the fusion valley can only be reached by tunnelling. The fragments emerge in the fusion valley in the hatched zone of Fig. 4.7.14. For the mass split 104 /136 of ^{240}Pu the gain in potential energy calculated at $\langle Q_{20} \rangle = 280$ b is $\Delta V = 3$ MeV. This energy is sufficient for a fraction of the wave packet moving downhill to excite the necking-in mode and pass over the ridge in FIig. 4.7.14. The collective flux crossing the ridge is shown in Fig. 4.7.15 for four different quadrupole moments or elongations. As to be expected the flux increases with elongation. Summing up all flux contributions for elongations up to $\langle Q_{20} \rangle = 280$ b yields $4 \cdot 10^{-4}$ for the fraction of fission events with mass ratio 104/136 from ^{240}Pu. This fraction is clearly in the range of neutron-less fission and even pertains to cold fission in the stricter sense of fission approaching the Q-values. The order of magnitude agrees with experiment. This is an impressive accomplishment. As to even-odd effects it has to be stressed that within the adiabatic approach adopted the appearance of (o,o) fragments in cold fission of (e,e) compounds requires pairbreaking to occur exclusively during the scission process proper.

The comparison of experimental (e,o) effects in cold as opposed to standard fission highlights the intricacy of the problem. As an example the neutron yield distribution for cold fission of ^{252}Cf(sf) [HAM93] is plotted in the left part of Fig. 4.7.16 while in the panel to the right the neutron distribution for thermal neutron fission of ^{234}U* is shown [QUA82]. As judged from Fig. 4.7.5, for Cf the excitation energy of *TXE* = 9 MeV chosen as constraint for the data on display is well inside the range of neutron-less fission and coming close to true cold fission. By contrast, for ^{234}U* the light fragment energy window at E_L = 110.55 MeV is still about 2 Mev below the transition to neutron-less fission. Cold fission has not yet been reached. For both distributions a sizable even-odd effect is visible which at first sight appears to be very similar. But at a closer look there is a striking difference between the two distributions. In cold fission (to the left) yields for odd neutron numbers are larger than for even numbers while in standard fission (to the right) the situation is reversed, i.e. yields for even neutron numbers are favoured compared to odd numbers. Expressed in terms of the even-odd effect for neutrons δ_N which is defined in analogy to the proton even-odd effect δ_Z in eq. (4.6.6) as

$$\delta_N = \frac{Y_e - Y_o}{Y_e + Y_o} \tag{4.7.7}$$

with Y_e and Y_o the neutron yields for even and odd neutron numbers, respectively, in cold fission δ_N is negative while in standard fission δ_N is positive.

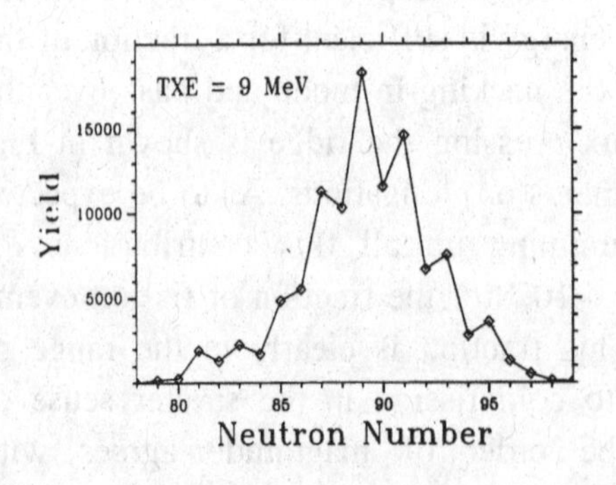

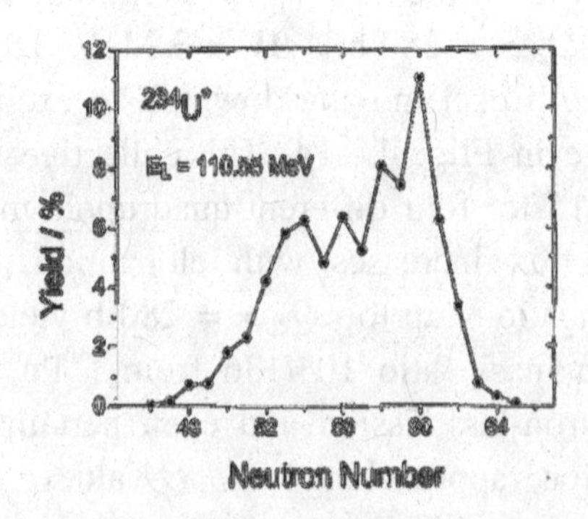

Fig. 4.7.16. Left panel: neutron yield distribution of light fragments for ^{252}Cf(sf) at TXE = 9 MeV; right panel: neutron yield distribution of light fragments for ^{233}U(n_{th},f) at fragment kinetic energy E_L = 110.55 MeV.

Much more detailed information on proton and neutron even-odd effects as a function of total excitation energy *TXE* can be read from Fig. 4.7.10 for the thermal neutron reaction ^{235}U(n,f). The result for the (e,o) effects is plotted in Fig. 4.7.17 [CLE99]. In cold fission with excitation energies smaller than about *TXE* ≈ 8 MeV the two even-odd effects for protons δ_Z and neutrons δ_N are negative. Both are even converging towards $\delta_Z = \delta_N = -100\ \%$ at *TXE* ≈ 0 MeV. There only (o,o) fragments survive. Raising the excitation energy by a few MeV to *TXE* ≈ 4 MeV the proton even-odd effect δ_Z quickly comes close to a vanishing $\delta_Z = 0$ and beyond about 8 MeV turns positive. The neutron even-odd effect rises more gently and nearly linearly with *TXE* and becomes positive at roughly 9 MeV. The Fig. 4.7.17 thus demonstrates the transition between negative even-odd effects in cold fission (Fig. 4.7.16 to the left) and positive effects in standard fission (Fig. 4.7.16 to the right).

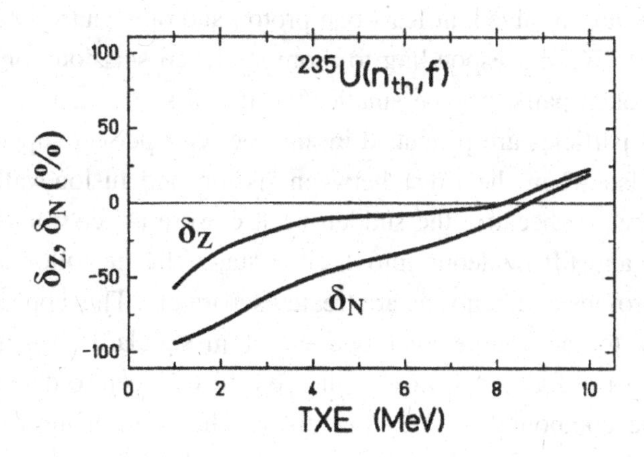

Fig. 4.7.17. Proton and neutron (e,o) effect for 235U(nth,f) evaluated from the yields in Fig. 4.7.10.

These results are thought to be an important contribution of cold fission research with implications for the general theory of fission. Necessarily, in fission of an (e,e) compound the appearance of o-mass or (o,o) fragments requires that somewhere in the process proton and/or neutron pairs have to be broken. For cold fission there is no ambiguity. The energy gained and the time spent on the short path from saddle to scission exclude the coupling from collective to intrinsic degrees of freedom. All available energy will go into the necking-in mode evoked in the above. The nucleus remains superfluid down to scission. From the observation of odd numbers of protons and/or neutrons it has hence to be concluded that nucleon pairs are broken in the very process of scission. As argued already many years ago, the act of scission is a non-adiabatic process [FUL62]. It may be described like the eruption of a volcano inside the nucleus whereby a dividing potential rises in the nuclear potential well cutting the mother nucleus into two fragments. Non-adiabatic scission is commonly thought to be responsible for the ejection of neutrons right at scission [CAR07]. By the same token one has to conjecture that the fast dividing process will affect the pairing of nucleons. For the time being this specific pair-braking process has not been scrutinized by theory. The problem is to find out how pairing correlations may be destroyed when the nucleus disintegrates. In the following a phenomenological model is proposed based on simple and possibly over-simplified arguments. The discussion is restricted to fissioning (e,e) nuclei since only for these nuclei experimental data have been collected for true cold fission.

In true cold fission with $TXE \leq 1$ MeV where only (o,o) fragments show up, two nucleon pairs are involved, at least one proton and one neutron pair must be broken. At higher *TXE* corresponding to more stretched scission configurations the number of broken pairs may be smaller but it is assumed that whenever the neck snaps quasi-particles are produced in an otherwise perfect superfluid. This should be true also when the ridge between fission and fusion valley in Fig. 4.7.14 has disappeared because the sudden neck closure at scission will always be non-adiabatic and lift nucleons into excited states. In the following the two subsystems for protons and neutrons are treated separately. This appears justified since in analogy to the charge even-odd effect in standard fission where in Figs.4.6.9 and 4.6.10 of Chapter 4.6 the charge effect is seen to depend only on the charge of the compound nucleus but not on the neutron number. For the analysis the simplest ansatz is that even in true cold fission at most one proton and/or one neutron pair are broken. With Y_e and Y_o the even and odd yields for either protons or neutrons one has with the normalisation $1 = (Y_e + Y_o)$ from eq. (4.7.7) for the e-o effect $\delta = (Y_e - Y_o)$. Hence $\delta = (1 - 2Y_o)$. The yield Y_o for a fragment with an odd proton or neutron number emerging from the breaking of a nucleon pair is equal to the probability q that indeed a pair has been broken, times the probability p that the two quasi-particles from the pair end up in different fragments. Therefore $Y_o = pq$ and

$$\delta = (1 - 2pq) \tag{4.7.8}$$

Equation (4.7.8) is a special case of the more elaborate combinatorial model of eq. (4.6.9). For pair-breaking due to necking-in splitting the mother nucleus into two halves it is reasonable to assume that the two partners of the pair are separated and go to complementary fragments. It means that throughout the probability p may be set to $p = 1$. The only remaining crucial parameter is the probability q of pair-breaking. In true cold fission with only (o,o) fragments present necessarily both a proton and a neutron are broken for sure and hence experiment tells that $q = 1$ for the two sub-systems. With eq. (4.7.8) one finds for $q = 1$ a maximum negative even-odd effect for both protons and neutrons, $\delta_Z = \delta_N = -100\%$, as shown in Fig. 4.7.17. Yet, one still has to understand why there are only (o,o) fragments for *TXE* ≤ 1 MeV. The answer was already given in the discussion of Fig. 4.7.10 pointing to the influence of level densities. In the back-shifted Fermi gas model of level densities we recall that below the pairing energy $\Delta \approx 1$ *MeV* there are for o-mass nuclei just no states available where the fragments could go (provided the few collective levels in the pairing gap are neglected). A similar reasoning applies for (e,e) nuclei for excitation energies below the pairing gap $2\Delta \leq 2$ MeV. Therefore, in true cold fission with *TXE* ≤ 1 MeV there are only (o,o) fragments and $\delta_Z = \delta_N = -100\%$.

For o-mass nuclei the breaking of either a proton or a neutron pair is required. but not of both. Moving away from true cold fission towards higher excitation energies experiment tells that the probability q for breaking either a proton or neutron pair is no longer 100% or $q = 1$. Paired configurations from the superfluid mother nucleus can survive. For $q < 1$ the even-odd effect in eq. (4.7.8) will start to rise. Remarkably the rise of the even-odd effect for increasing excitation *TXE* is faster for protons than for neutrons. Since there are more neutron than proton pairs in the mother nucleus, breaking a neutron pair should indeed be slightly more probable, i.e. $q_N > q_Z$. This is reflected in the slower rise of δ_N compared to δ_Z in the figure.

The more delicate question still to be answered is why the probability q goes down when the excitation *TXE* is going up. An intuitive answer may be given by recalling that increasing the excitation energy necessarily means decreasing the kinetic energy of fragments. The distances between the centres of charge of fragments are most compact in true cold fission and are getting larger when moving towards standard fission. By the same token the thickness of the neck to be cut at scission will be rather thick in cold fission and become thinner the smaller the kinetic energies are. Due to the short range of the pairing interaction it is therefore to be expected that more nucleon pairs in the neck will be affected by the scission process in cold fission than in standard fission. Hence, when going from cold to standard fission the probability q for breaking will decrease from $q = 1$ to values $q < 1$ and in consequence the even-odd effect in eq. (4.7.8) will rise towards $\delta = 0$ and eventually turn from negative to positive. According to experiment in Fig. 4.7.17 the turnover for the reaction $^{235}U(n_{th},f)$ occurs at $TXE \approx 8$ MeV for protons and at $TXE \approx 9$ MeV for neutrons. At the turnover point $q = 1/2$ in Eq. (4.7.8). Only for $q < 1/2$ the even-odd effects are positive in the present naïve model. As reviewed in Chapter 4.6 positive even-odd effects are common in standard fission.

One may wonder whether the counterpart to the maximal negative even-odd effect $\delta = -1$ in cold compact fission could not be attained with a maximal positive even-odd effect $\delta = +1$ for strongly stretched scission configurations. This would correspond to very low kinetic energies TKE in the laboratory. There would be no pair-breaking at all in this case, q would be nil and $\delta_Z = \delta_N = +1$. This has never been observed in experiment. On the contrary, as shown in Fig. 4.6.11 of Chapter 4.6, for decreasing kinetic energy the positive even-odd effect for charge δ_Z does not increase but decreases. Apparently the pair-breaking probability is controlled by still another parameter which has not yet been addressed. One has to remember once more that small kinetic fragment energies TKE correspond to more elongated scission configurations and hence long downhill paths from saddle to scission. The ensuing large potential energy gain $\Delta V = E_K^{sci} + E_X^{sci}$ in eq. (4.4.11) will then impart the motion of the superfluid a large pre-scission kinetic energies E_K^{sci}. This anti-correlation between the total kinetic energy TKE and the pre-scission kinetic energy E_K^{sci} is startling. It is understood by noting that the main contribution to the total kinetic energy TKE is the Coulomb repulsion between fragments at scission which decreases for elongated scission configurations. The anti-correlation may also be inferred from inspecting the energy diagram of the Scission Point Model in Fig. 4.6.20 for deformations smaller than at the minimum of the free energy. Anyhow, for large pre-scission kinetic energies the Landau-Zener jumps to excited levels will be enhanced and increases the probability for pair-breaking. The decrease of q due to the thinning out of the neck is thus counterbalanced and – as brought to evidence by experiment – even overcompensated by an increase due to the more violent snapping of the neck boosting pair-breaking. The interplay of two parameters controlling the probability q for pair breaking, viz. neck thickness and speed of neck rupture, thus allows to give a coherent interpretation of the complex behaviour of even-odd effects. Evidently, the present discussion of (e,o) effects is closely linked to the problem of freeze-out of charge oscillations in Chapter 4.6.1. The link is due to the fact that the pre-scission kinetic E_{kin}^{sci} playing a role in the (e,o) effects is steering the speed of neck closure at the focus in the discussion of charge variances. Once more it is pointed out that unlike charge variances the problem of pair-breaking at scission has not yet been investigated by theory.

A final remark should address the observation that for very low kinetic fragment energies *TKE* the trend of e-o effects as a function of *TKE* just described is reversed: the size of the e-o effect in super-deformed fission increases again (Chapter 4.6.5). The above arguments on the decrease of e-o effects when kinetic energies are getting smaller due to more elongated scission shapes and pre-scission energies become larger are only valid up to deformations corresponding to the minimum of the total potential energy $V_P = V_{Coul} + V_{Def}$, or maximum of the free energy in the scission point model of Fig. 4.6.20. As to deformations the minimum potential energy should roughly coincide with the scission point in Fig. 4.7.14 where from theory the ridge between valleys for fission and fusion is vanishing. The bulk of fission events will not override this deformation. Yet, there is a small fraction of events where super-deformed states may be reached and where for further deformation the Coulomb potential V_{Coul} and hence kinetic energies *TKE* continue to slope downwards. But in contrast to deformations smaller than the one at the minimum of V_P, the free energy in the scission point model is no longer increasing. Instead it is getting smaller and approaching $E_{free} = 0$ (label 2 in Fig. 4.6.20). In the adiabatic version of the scission point model this is understood to tell that the rise of the deformation energy consumes part or even exhausts the kinetic energy E_K^{sci}. The pre-scission kinetic energy E_K^{sci} will hence become smaller. Though not shown in the potential energy surface from a microscopic theory in Fig. 4.7.14, the interpretation suggested by the scission point model is not in conflict with theory. In line with the model for even-odd effects sketched above, for smaller kinetic fragment energies E_K^{sci} at scission the loss of adiabaticity is less vehement and charge even-odd effects are expected to be re-enforced. This is what is found in experiment (see Fig. 4.6.19). However, the prediction by the Scission Point Model of true cold fission in super-deformed scission with $E_{free} = 0$ (label 2 in Fig. 4.6.20) and hence with no nucleon pairs being broken at all is not confirmed. Even-odd effects of maximum feasible size $\delta_Z = \delta_N = +1$ are by far not reached. To summarize: the limit of phase space where all of the available energy Q is exhausted at scission by potential Coulomb and deformation energies and thus leading to true cold fission is only observed for compact but not for super-deformed scission configurations.

All of the above arguments were based on the assumption that in low energy fission the nuclear fluid remains superfluid and all pair-breaking occurs solely in the scission process. As demonstrated, this approach allows for a comprehensive description of even subtle phenomena. But the assumption will certainly no longer be valid for fission induced at higher excitation energies of the compound nucleus. For these reactions statistical models like the one sketched in Chapter 4.6 with super fluidity being destroyed by thermal excitation are appropriate.

4.8 Ternary and Quaternary Fission

4.8.1 *Ternary Fission Yields*

The discovery of ternary fission was announced in 1947 in experiments with ionisation chambers [FAR47] or with photographic emulsions [TSI47]. The work with ionisation chambers revealed that in coincidence with fragments a third charged particle may be emitted in neutron induced fission of $^{236}U^*$ and $^{240}Pu^*$. It was shown that the ternary particles were α-particles with energies up to 16 MeV. The ratio of ternary to binary fission did not exceed 1/250. The technique of photographic emulsions was often used in the early days of fission studies. For the discovery experiment photographic emulsions were soaked with Uranium solutions. After irradiation and development the photographic plates were scanned under a microscope searching for tracks produced by fission fragments or any other heavy ions. It was found that in rare cases, besides the two fission fragments, a third lighter particle was emitted at roughly right angles to the fission axis defined by the fission fragments. The photo of a ternary fission event is on view in Fig. 4.8.1 [TSI47]. From their ionization density in the emulsion it could be shown that these particles were α-particles. Their range was several times longer than the range of alphas from natural radioactivity and therefore these particles became known as "long range alphas". In the sequel it was discovered that, though α-particles are the most common, there are many other light charged nuclei showing up in ternary fission. Work performed up to the eighties and nineties of last century was reviewed by A. K. Sinha [SIN89], C. Wagemans [WAG91] and by M. Mutterer and J. Theobald [MUT96].

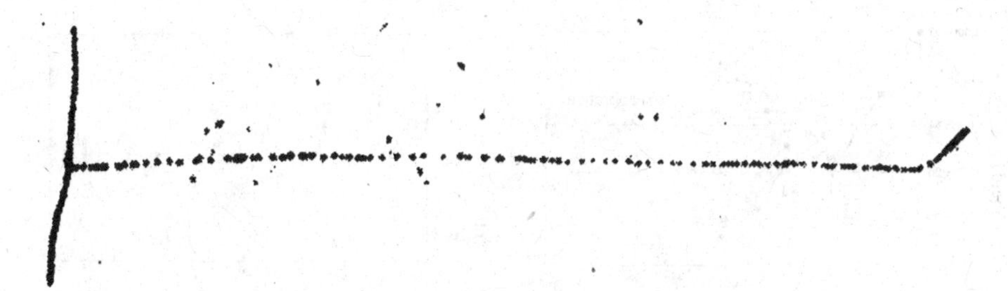

Fig. 4.8.1. Tracks of two fission fragments (to the left) and an α- particle in ternary fission.

A first question to be answered is the type and the rate of ternary particles being emitted in fission. To this purpose the LOHENGRIN spectrometer has been an ideal tool to investigate systematically the yields and energy spectra of these particles down to very low yields. A comprehensive summary of all measurements performed worldwide is given in [KOEa00]. A selection of ternary yields per fission is on display in Fig. 4.8.2 for thermal neutron reactions from ^{234}U* [WOE99], ^{240}Pu*[VOR75, WOE99], ^{243}Am* [VOR75, MUT96, HES96] and 250Cf* [TSE03]. For each element only the yield of the most prominent isotope is plotted. Yields down to an emission rate of 10^{-10} per fission could be determined. This detection limit corresponds to the observation of a few events per day with background events not yet interfering. Evidently, for increasing mass (or charge) of ternary particles their yields decrease very fast. It is further observed that the heavier the fissioning nucleus is, the heavier are also the ternary particles being ejected. For ^{234}U* the heaviest isotopes found were ^{24}Ne and ^{27}Na. For the reaction ^{249}Cf(n,f) ternary silicon and sulphur isotopes with masses up to almost 40 could be identified with yields of 10^{-8} per fission. Still heavier isotopes and elements are to be expected for measurements at still lower detection limits. Generally there is a pronounced even-odd effect in the yields with elements with an even charge number being produced with higher probability. Thus the He-isotopes (above all ^{4}He) exhibit higher yields than the H-isotopes.

A more detailed view of ternary isotope yields is given in Fig. 4.8.3 for the reaction ^{249}Cf(n,f) [TSE03]. Attention should be given to the observation that in general the ternary particles are either stable like ^{4}He or neutron rich. The only exception is the unstable ^{8}Be decaying into 2 α-particles. In a thorough analysis only upper limits for the rather neutron poor isotopes ^{3}He, ^{6}Li, ^{7}Be and ^{10}B could be derived [KOE00b]. On the other hand, for large neutron excess the yields also decrease very fast.

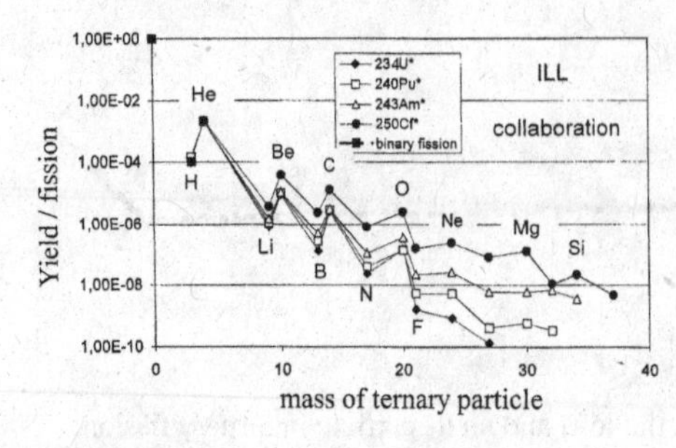

Fig. 4.8.2. Survey of ternary yields.

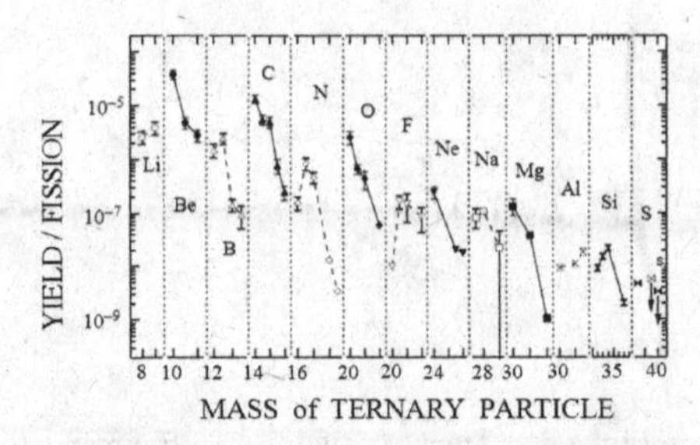

Fig. 4.8.3. Ternary isotopes from 249Cf(n,f).

From Fig. 4.8.2 it is apparent that the dependence of ternary yields on the mass A (or charge Z, or fissility Z^2/A) of the fissioning nucleus is increasingly marked the heavier the ternary particles are. For the lighter ternary particles the dependence is less pronounced. For the long range α-particles LRA and for tritons t this is brought to evidence in Fig. 4.8.4 [WAG08]. Yields as a function of fissility Z^2/A for thermal neutron (full symbols) and spontaneous fission (open symbols) are given separately.

For the range of fissilities indicated the ratio of yields increases by only about a factor of two. The increase of ternary yields with fissility is further interpreted in Chapter 4.8.2 as manifesting their dependence on nuclear deformation at scission. Note that in spontaneous fission the yields for LRA emission in the panel to the left are slightly larger than in thermal neutron fission. The effect is, however, absent for tritons (panel to the right). Helium and hydrogen isotopes are by far the most abundant ternary particles. As a rule of thumb the α-particle ^{4}He carries 90 % and the triton ^{3}H about 6 % of the total ternary yield. He and H isotopes together contribute some 99 % to the total yield. Throughout the actinides the ratio of total yields ternary/binary varies between roughly $2\cdot10^{-3}$ and $4\cdot10^{-3}$.

How and from where the ternary particles are released is an intriguing question. The question was scrutinized back in 1965 by I. Halpern [HAL65]. It was argued that the yields and angular distributions of α-particles being observed cannot be due to evaporation from excited fragments after scission. Much lower yields and isotropic angular distributions in the fragment systems would be expected in this case. But in experiment the bulk of ternary particles is ejected perpendicular to the fission axis as catching the eye in Fig. 4.8.1. This anisotropic emission is traced to the focussing effect of the combined Coulomb field of the two fragments for ternary particles being born in the neck region between the fragments. Compared to the fast separation times of fragments after scission the evaporation times are too long for the focussing of ternary particles to be still effective. The emission mechanism proposed by Halpern considers instead α-particles moving before ejection along the symmetry axis of the fissioning nucleus in the nuclear potential provided by all other nucleons. As sketched in Fig. 4.8.5 the bottom of the potential for nucleons moving just before scission in the neck between the two fragments is to good approximation flat. Just after scission the two stubs of the broken neck recede into the two fission fragments. At the rupture point the potential hence suddenly surges like a volcano. In the sudden approximation the wave function of the α-particle remains unchanged. But the sudden change of the potential imparts energy to the α-particle which allows it to escape [FUL62]. Note that the same mechanism of energy transfer was invoked when discussing the breaking of proton pairs in the very act of scission (see Chapter 4.6.2).

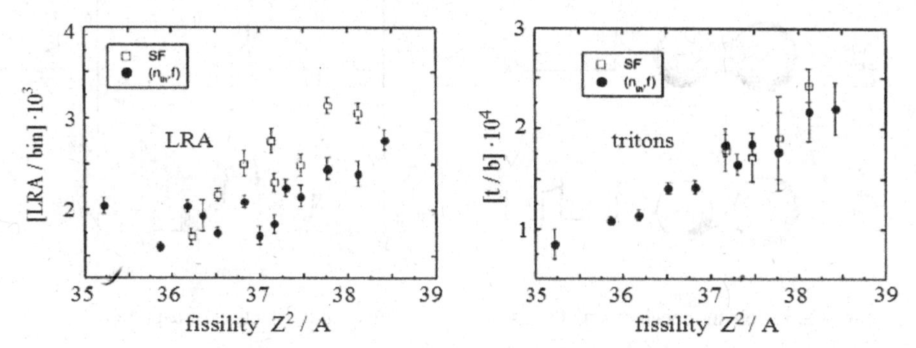

Fig. 4.8.4. Ratio of yields for ternary LRA and tritons relative to binary fission.

In more recent work the sudden approximation was adopted to calculate quantum mechanically the emission probability of LRA particles [SER00]. The wave function along the fission axis z for a symmetric mass division is given in Fig. 4.8.6 together with the potential just after scission (dashed line). The emission probability is largest for ejection points near symmetry. In the Coulomb field of the two fragments these LRAs are focussed perpendicular to the fission axis. Therefore this is called "equatorial" emission contributing 95 % to the total yield. When combined with trajectory calculations the full angular distribution of the ternary particles is obtained [CAR80]. It is found that eventually particles sitting near the wall of the potential appear at angles close to the fission axis. This is called "polar" emission contributing according to the model some 5% of the total yield. From experiment the existence of polar ternary particles is known since long [PIA70]. In a comprehensive review [NOW82] it is confirmed that besides α-particles the three H-isotopes protons, deuterons and tritons are to be detected. Surprisingly, in polar emission the protons are only by a factor of roughly 3 less abundant than α-particles. Among the H-isotopes the protons contribute the largest share, in contrast to tritons in equatorial emission. All polar particles are preferentially emitted in the direction of flight of the light fission fragment. One may therefore suspect that still another mechanism than the one proposed for equatorial emission may be effective, e.g. evaporation of protons from fully accelerated fragments [SCH92]. The issue will not be further discussed here.

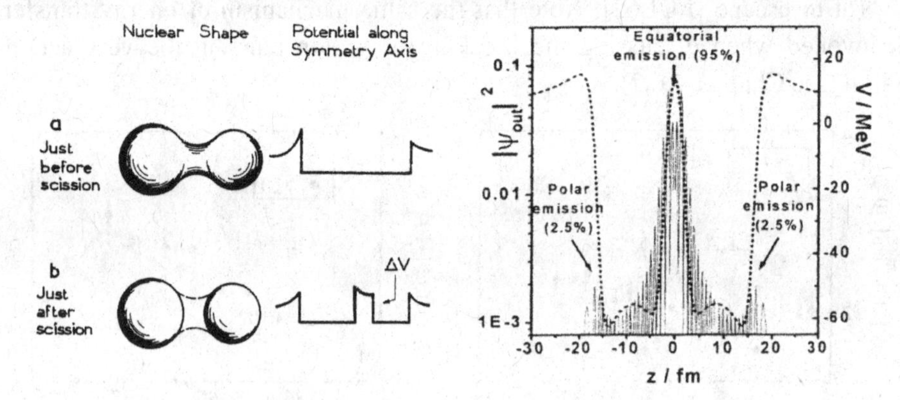

Fig. 4.8.5. Volcano mechanism at scission

Fig. 4.8.6. Ejection of α-particles

In the Serot model outlined [SER00] it is presupposed that in the fission prone nucleus there is a preformation probability for α-clusters. The ratio of LRA to binary yields LRA/ B is therefore parameterized as

$$LRA / bin = S_\alpha \cdot P(TXE>E_{cost}) \cdot P_{trans} \tag{4.8.1}$$

with S_α the spectroscopic factor of α-preformation in the mother nucleus, $P(TXE>E_{cost})$ the probability that sufficient excitation energy TXE is available to overcome the energy E_{cost} required for emission and P_{trans} the probability that the required TXE is transferred to the ejection of an α-particle. The spectroscopic factor is extracted from the analysis of radioactive α-decay. $P(TXE>E_{cost})$ is inferred from the experimental TXE distribution with E_{cost} a parameter to be chosen within reasonable limits (see discussion below). The transfer probability is calculated in the sudden approximation.

The importance of the spectroscopic factor S_α could be demonstrated by experiment. The systematic for the ratio *LRA/bin* for thermal neutron and spontaneous fission as a function of fissility is on display in Fig. 4.8.7 [VER10]. In the left panel the experimental data and in the right panel the same data but divided by the spectroscopic factor are plotted. Evidently, the scatter of *LRA/bin* data is removed when the spectroscopic facto is taken into account. However, except for α-particles, spectroscopic factors for all other ternaries are not known from experiment. In a semi- empirical model spectroscopic factors for exotic cluster decays have been derived [BLE91]. But it is questionable whether it makes sense to consider heavy particles like neutron-rich Si isotopes bouncing around in a nucleus before being set free in ternary fission.

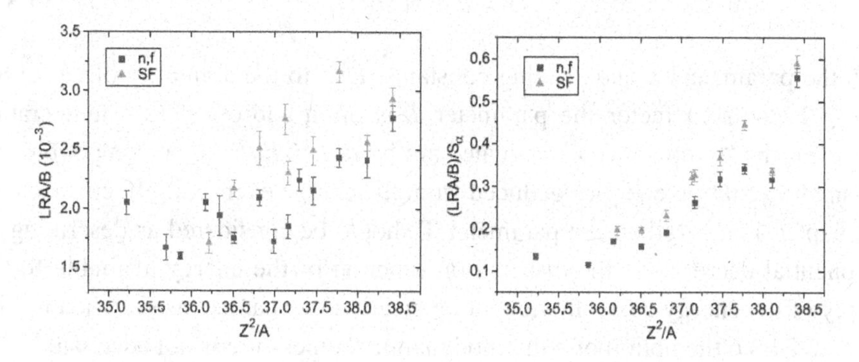

Fig.4.8.7. Systematic of the yield ratio LRA/bin with account for the spectroscopic factor S_α.

Searching for a different approach covering the full spectrum of ternary particles several models have been proposed. In most cases it is the energy having to be provided for the emission of ternary particles which is at the centre of investigation. A first and still much applied model is due to I. Halpern [HAL71], mostly used in a slightly generalized form and called "extended" Halpern model [WOE96]. At the focus are the so-called energy costs E_{cost} that have to be supplied to remove a ternary particle from one fragment and to place it midway between the two fragments. The costs are calculated from

$$E_{cost} = E_B + \Delta V_C^{ter\text{-}bin} + K = \Delta Q^{bin\text{-}ter} + \Delta V_C^{ter\text{-}bin} + K \qquad (4.8.2)$$

with E_B the binding energy of the ternary particle to a mother fragment, $\Delta V_C^{ter\text{-}bin}$ the difference in Coulomb energy for the ternary and binary configuration and K

the kinetic energy of the ternary particle at ejection. Usually this latter term is neglected. There is freedom to choose the specific fragment ejecting the third particle. Most delicate is the proper guess for the binary and ternary scission configurations. Especially for heavy third particles the ternary configuration has to be more stretched than in the binary case. As an example the correlation between measured ternary yields and calculated energy costs is presented in Fig. 4.8.8 for the reaction ^{242}Am(n,f) [HES96]. For the sake of clarity the correlation is shown separately for ternary particles with even (left panel) and odd charges (right panel), respectively. A clear and nearly linear correlation is observed in the logarithmic plot with the yield Y decreasing for increasing energy costs. It is therefore tempting to parameterize the correlation by

$$Y = a\, exp(-E_{cost}/T) \tag{4.8.3}$$

with the parameters a and T being constants. Due to the similarity of eq. (4.8.3) with a Boltzmann factor the parameter T is often addressed as a temperature. However, the "temperatures" evaluated are by far too large to be compatible with the small excitation energies deduced from the charge even-odd-effects discussed in Chapter 4.6.2. Rather, the parameter T should be considered as describing the exponential decrease of the distribution function of the energy available for the supply of the energy costs. In view of an ejection mechanism as sketched in Figs. 4.8.5 and 4.8.6 the notion of a thermodynamic temperature is not adequate.

In variants of the Halpern model either more refined calculations of the energy costs are suggested [PIK94, FAU95] or, in a more complex process, first a proto ternary particle is formed which subsequently grows to the final ternary

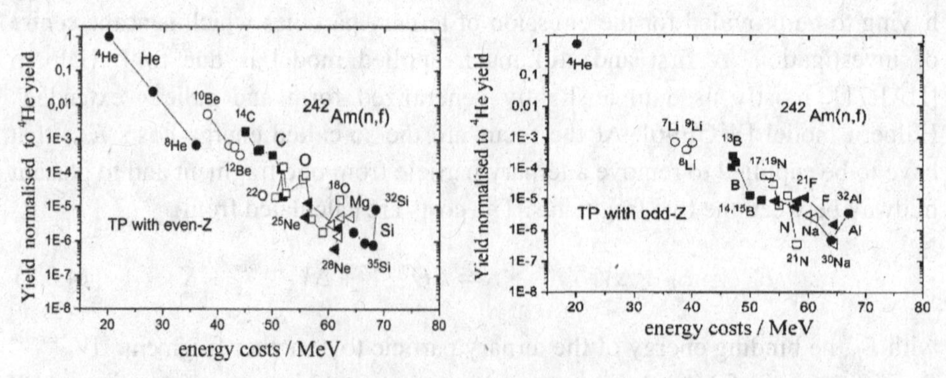

Fig. 4.8.8. Correlation between ternary yields and energy costs for the reaction ^{242}Am(n,f).

particle by nucleon exchange with the fragments [RUB88]. In the latter work it is assumed that light particles are formed as a result of two random neck ruptures. Random neck rupture points to the instability of the neck joining the two fragments once a critical elongation and neck radius of the fissioning nucleus has been reached. It is a generalization of the Rayleigh instability of filaments in hydrodynamics. For nuclear matter the handy formula 2l = 11r for the onset of instability was derived by Brosa et al. where 2l is the total length of the nucleus and r the neck radius [BRO90]. Randomness of neck rupture is at the heart of a theory for nuclear scission with implications for the whole fission process. In the framework of this picture it is natural to invoke for ternary fission a double neck rupture as the initial phase followed by the volcano mechanism illustrated in Fig. 4.8.5 leading to the ejection of ternary particles.

Comprehensive experiments for ^{252}Cf(sf) where fission fragments (twin ionisation chamber), ternary particles (ΔE-E telescopes) and neutrons (4π NaJ crystal ball) were registered in coincidence opened the way to novel approaches. Ternary particles up to carbon could be identified with isotopic resolution being however limited to hydrogen and helium. From these data the total excitation energy *TXE* left to the system in ternary fission was evaluated. In the left panel of Fig. 4.8.9 ternary yields are plotted as a function of the average excitation energy remaining with the system [MUT98]. At scission this energy is present as intrinsic excitation and deformation energy. The energy is eventually exhausted by neutron and gamma emission. Remarkably it is found that, compared to binary fission, for increasing mass of the ternary particle the excitation energy left behind decreases. In the logarithmic plot of the figure the decrease of the yield is close to linear. The yields themselves follow hence an exponential law. From inspecting more closely the figure, it is further tempting to speculate that for ternary masses around 40, which in view of the systematic of yields in Fig. 4.8.2 for californium are expected at yields 10^{-10} of the binary yield, the total excitation *TXE* will have shrunk close to nil. This would then be an example for a true cold ternary fission process similar to those discussed in the preceding chapter 4.7 [GON99].

In the panel to the right of Fig. 4.8.9 the data have been reshuffled to visualize yields as a function of $\Delta TXE = (\langle TXE^{bin} \rangle - \langle TXE^{ter} \rangle)$, the difference between binary and ternary excitation energies *TXE* [MUT98]. The difference ΔTXE represents the excitation energy which has to be taken from the *TXE* reservoir of binary fission to allow ternary fission to proceed. Since at scission most of the excitation energy *TXE* is present as energy of deformation linked to the neck, part of the deformation is removed upon emission of a ternary particle from the neck. The deformation energy having been drained supplies the energy for ejection. The energy consumption should be comparable to the energy costs in Halpern-like models as in Fig. 4.8.8. Yet, comparing Figs 4.8.8 and 4.8.9 there is a mismatch between energy costs and the energy ΔTXE drained, the calculated energy costs being a factor of 2-3 larger than the experimental figures for ΔTXE. As to the energy costs much guess work is required to select and compare for binary and ternary fission simple scission configurations amenable to calculations. The absolute sizes for the costs are therefore not very reliable. Hence the experimental ΔTXE data are considered to give more realistic information on the energy requirements in ternary fission.

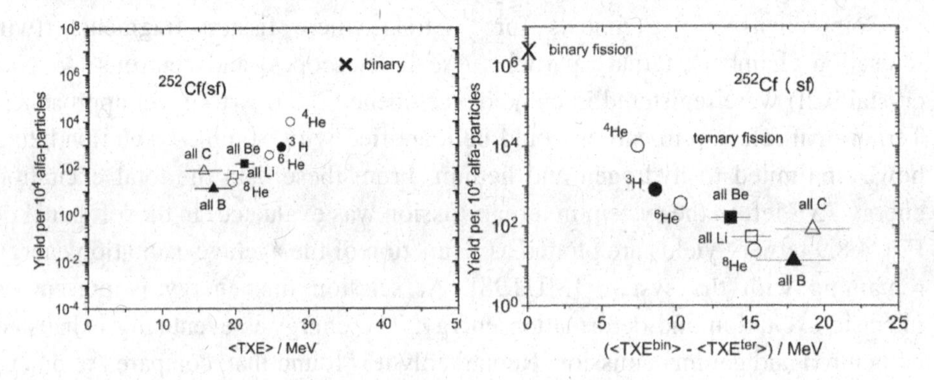

Fig. 4.8.9. Ternary yields as a function of total average excitation energy <TXE> in binary and ternary fission (left). Difference of <TXE> in binary and ternary fission (right).

Unfortunately *TXE* experiments like the one for ^{252}Cf(sf) in Fig. 4.8.9 could not be pushed beyond carbon to search for heavier ternaries, nor could other reactions than the one quoted be investigated. Yet, a heuristic model for yields with features in common with the experimental approach in Fig. 4.8.9 and applicable to any ternary reaction was introduced by Baum [BAU92a]. The starting point is the ansatz suggested by the experiment in the right panel of Fig. 4.8.9:

$$Y \sim exp\,(-\Delta TXE\,/\,T) \qquad (4.8.4)$$

In this ansatz the parameter T is not considered to be a temperature but rather as a parameter which is governing the part of the excitation energy being drained. In a double-neck rupture model T may be imagined to control the locations where the neck is snapping as a function of the distance between rupture points. In turn this distance settles the size of the ternary particle and the corresponding amount of excitation energy being removed. For the evaluation of Eq. (4.8.4) one has to recall the energy relations from eqs. (4.7.1) to (4.73) for binary and analogously for ternary fission

$$Q^{bin} = TXE^{bin} + TKE^{bin} = TXE^{bin} + V_C^{\;bin} + E_K^{\;bin,sci}$$

$$Q^{ter} = TXE^{ter} + TKE^{ter} = TXE^{ter} + V_C^{\;ter} + E_K^{\;ter,sci}$$

with Q^{bin} and Q^{ter} the Q-values, $V_C^{\;bin}$ and $V_C^{\;ter}$ the Coulomb energies and $E_K^{\;bin,sci}$ and $E_K^{\;ter,sci}$ the kinetic energies at scission for binary and ternary scission configurations, respectively. The Q-values are calculated from mass tables and the Coulomb energies are obtained in a model. Note that TKE^{ter} is the kinetic energy of all 3 outgoing particles. Evaluating from these equations ΔTXE one finds

$$\begin{aligned} Y &\sim exp[-(<TXE^{bin}>-<TXE^{ter}>)/T] \\ &= exp[-(Q^{bin} - V_C^{\;bin})/T] \cdot exp[(Q^{ter} - V_C^{\;ter})/T] \end{aligned}$$

In the derivation it is assumed that the difference ($E_K^{\;bin,sci} - E_K^{\;ter,sci}$) is negligibly small. The first factor on the RHS does not depend on the ternary particle parameters and, hence, for ternary fission is a constant. The ternary yields therefore become

$$Y \sim exp[(Q^{ter} - V_C^{\;ter}) / T] \tag{4.8.5}$$

This is the basic formula of the Baum model for calculating ternary particle yields.

Before coming to applications it is worth mentioning that the Halpern model of ternary fission is not at variance with the Baum model. The suggestion in the Halpern model is that the crucial parameter steering ternary fission is the energy which is required to cover the costs for switching from binary to ternary fission. This idea leads to the exponential ansatz in eq. (4.8.3). Halpern prescribed to evaluate the energy costs as given in eq. (4.8.2):

$$E_{cost} = (Q^{bin} - Q^{ter}) + (V_C^{\;ter} - V_C^{\;bin})$$

where the term K, the starting kinetic energy of ternary particles at scission, has been omitted since it is neglected anyhow in all applications. It is readily seen that

$$exp[(Q^{ter} - V_C^{ter}) / T] = exp\,(-E_{cost} /T) \cdot exp[(Q^{bin} - V_C^{bin})]/T$$

The second factor on the RHS of this equation is the same for all ternary events and, hence, the yield calculated from the Baum model in eq. (4.8.5) is proportional to the yield found with the Halpern approach in eq. (4.8.3).

Ternary yields measured for thermal neutron induced fission of ^{235}U on LOHENGRIN fitted with expression (4.8.5) are depicted in Fig. 4.8.10 separately for even and odd charges of the ternaries [BAU92b]. In the logarithmic plot the fit exhibits a smooth and nearly linear correlation between yield and the parameter (Q^{ter} – $V^{ter}{}_C$). The only data points not being reproduced satisfactorily are those for ^4He and ^{20}C. The isotope ^4He is exceptional due to its outstanding binding energy while for ^{20}C the statistical accuracy in experiment was very poor. The close resemblance in the presentation of yield data versus the excitation *TXE* for ^{252}Cf(sf) in Fig. 4.8.9 (left panel) and versus (Q^{ter} – $V^{ter}{}_C$) for thermal neutron fission of ^{235}U in Fig. 4.8.10 should be noted. The excellent quality of the fit with the Baum model becomes even more evident in Fig. 4.8.11 [BAU92b]. Except for single cases the even-odd staggering and even the fine structures in the isotopic yields are fairly reproduced. It should be mentioned that for the fit the partitioning of mass between the two main fragments is a free parameter. The best fit is shown in the figure. It was obtained by choosing for the heavy fragment a mass near the magic ^{132}Sn. Attention should also be drawn to the yield calculated for neutrons in Fig. 4.8.11. These neutrons are not evaporated from the fragments but are ejected at scission. They are usually called "scission neutrons". The yield calculated is some 10^{-2} per fission event. In experiments searching for scission neutrons yields between 1% and 30% are reported.

Yields of the He isotopes from ^{235}U(n,f) are shown in more detail in Fig. 4.8.12 (open points). For comparison the He yields for ^{252}Cf(sf) from the experiment addressed in Fig. 4.8.9 are plotted as full points. In the Cf experiment besides the charged particles also the neutrons emitted in ternary fission were measured. This allowed unraveling in the ^4He and ^6He data the contributions from $^5He \rightarrow {}^4He + n$ and $^7He \rightarrow {}^6He + n$, respectively [MUT00, KOP02]. The lifetimes of the odd isotopes are $\tau = 1.1 \cdot 10^{-21}$ s for ^5He and $\tau = 4.1 \cdot 10^{-21}$ s for ^7He. They are thus short compared to their flight times towards the detectors. They are registered as ^{4}He or ^{6}He by the charged particle detectors, respectively, and are denoted as "secondary" yields. Their contributions are extracted from an analysis

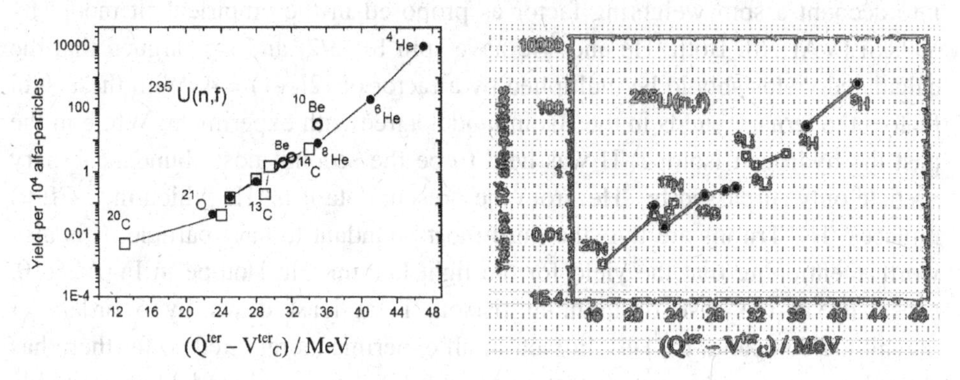

Fig. 4.8.10. Ternary yields from ^{235}U(n,f) vs the parameter (Q^{ter} – V^{ter}_C). Left panel even Z, right panel odd Z of ternary particles.

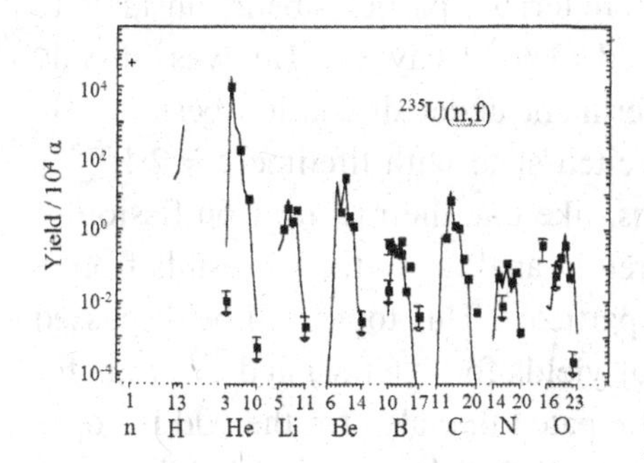

Fig. 4.8.11: Fit of LOHENGRIN data with Baum model for ^{235}U(n,f).

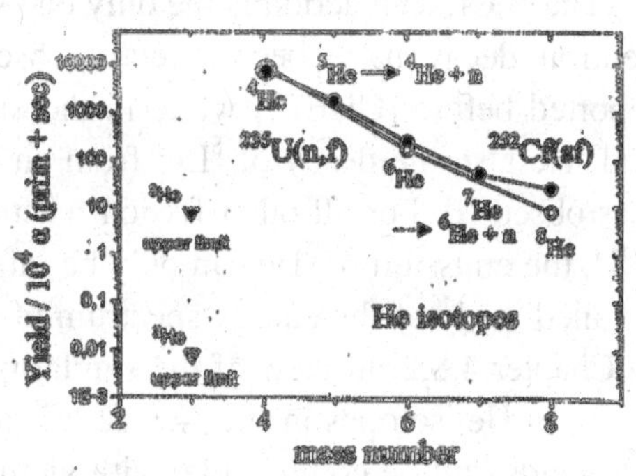

Fig. 4.8.12: Yield of He isotopes. Closed symbols ^{252}Cf(sf), open symbols ^{235}U(n.f).

of the angular correlations between neutrons and charged particles. The yields of the odd isotopes are surprisingly large. The ratios of primary yields ^{5}He/^{4}He and ^{7}He/^{6}He are deduced to be equal within error: ^{5}He/^{4}He = 0.21(5) and ^{7}He/^{6}He = 0.21(5). It means that 17.4(4.0) % of all ^{4}He and ^{6}He particles detected are actually secondary residues from the ^{5}He and ^{7}He break-up reaction. The yields for both, the even and odd He isotopes from ^{252}Cf(sf) are on display in Fig. 4.8.12. They are normalized to the total usually detected yield of primary plus secondary ^{4}He particles $Y(LRA) = 10^4$. In the logarithmic plot the usual even-odd staggering of yields (see Fig. 4.8.11) appears surprisingly to be absent. The decrease of yields for increasing mass number is nearly linear. Theoretical models have difficulties to cope with this particular feature. A way out is to take into account a spin weighting factor as proposed in the empirical fit model by Valskii [VAL76]. Both ^{5}He and ^{7}He have spin I = $3/2^{-}$ and it is argued that the calculated yields should be multiplied by a factor of (2I +1) = 4. With these spin factors the ternary yields in the Baum model agree with experiment. While in the past the hydrogen isotope ^{3}H was held to be the second most abundant ternary particle only surpassed by ^{4}He, from the present data it is ^{5}He which in ^{252}Cf(sf) replaces ^{3}H. Tritons are thus the third most abundant ternary particle. It is also worth mentioning that the yield for the light helium ^{3}He isotope in Fig. 4.8.10, especially in the case of uranium fission, is at least down by 6 orders of magnitude compared to ^{4}He. In fact, in all experiments reported so far there has never been an unambiguous detection of ^{3}He. No theoretical model has been able to predict this extremely low yield. In Fig. 4.8.11 the Baum model overestimates the experimental upper limit of the yield by a factor of 20.

The ^{252}Cf(sf) reaction is the only one where ternary particles being unstable to neutron decay have been directly observed. The decay of ^{5}He was already reported before [CHE72] while in the experiment under discussion besides ^{5}He and ^{7}He also the decay of ^{8}Li* from an excited state with lifetime $\tau = 2{\cdot}10^{-20}$ s was observed. For all other fission reactions, like e.g. thermal neutron fission of ^{235}U, the emission of ^{5}He can only be inferred in analogy to the Cf results from a detailed study of the energy spectrum of α-particles. This topic will be discussed in Chapter 4.8.2. In view of the similarity of yields for ^{252}Cf(sf) and ^{235}U(n,f) for the even He isotopes in Fig. 4.8.12 it is anticipated that also for the odd isotopes the yields will be comparable, with yields for ^{252}Cf(sf) being only slightly larger than for ^{235}U(n,f).

Progress in the identification of ternary particles by ΔE-E telescopes has allowed taking in coincidence with fission fragments ternary particles up to carbon. For helium up to beryllium even the individual isotopes can be resolved. This is a significant step forward because properties of fission fragment distributions for mass and energy may be investigated not only for LRA accompanied fission as before but also for heavier ternaries where shifts of fragment masses and energies become more pronounced. Two examples of yields for Li and Be isotopes measured at a cold neutron beam of the ILL are provided in Fig. 4.8.13 for the reaction ^{235}U(n,f). The Li data from the present measurement [TIS06] (full points in left panel) are compared to former results [VOR72, BAU92] and to models. The yields for ^{7}Li and ^{9}Li with even neutron numbers are clearly enlarged compared to ^{8}Li. The predictions from the Baum model for ^{8}Li and ^{9}Li are satisfactory while the ^{7}Li yield is overestimated. The present data for Be isotopes (full points in right panel) highlight the prominent yield of the ^{10}Be isotope [KOP05]. This is well reproduced by both the Baum and the Halpern model while both fail for ^{8}Be. The interpolation approach by Valskii [VAL76] is also only partly successful. The data are shown to demonstrate that, though the gross trends of ternary yields are very well accounted for in the Baum model (see Fig. 4.8.11), in detail there remain sometimes major discrepancies.

In the above a survey of ternary fission yields has been given. The salient features of their properties are summarized in Fig. 4.8.2. It is tempting to suggest that the dependence of ternary particle yields on the mass of the fissioning nucleus may be related to characteristic featuresof binary fragment distributions. Mass distributions of fission fragments are discussed in Chapter 4.3 for binary fission and in Fig. 4.3.3 the systematic for fissioning nuclei from uranium to californium is on display. In Fig. 4.8.14 the mass dependence of ternary particle andfission fragment yields are brought together for uranium and californium in one single diagram [GOE04]. It is striking that when comparing uranium to californium the extension of sizable ternary particle yields to larger masses finds

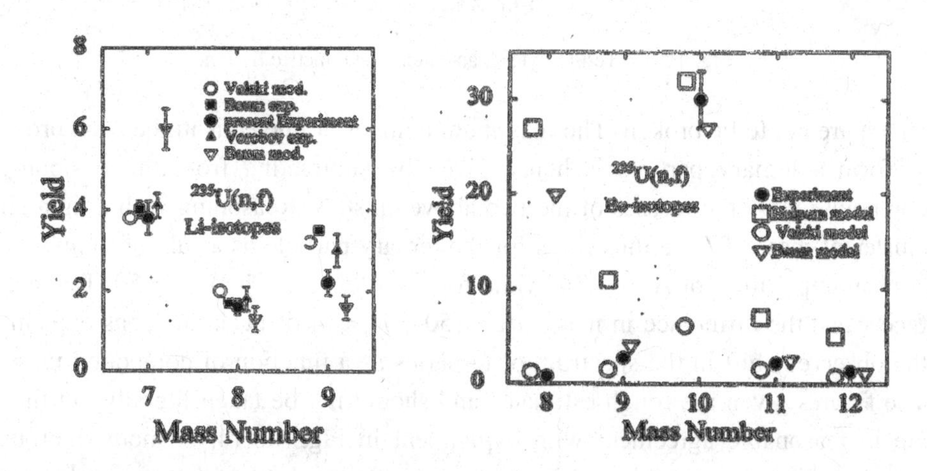

Fig. 4.8.13. Yields of ternary Li (to the left) and Be isotopes (to the right).

its counterpart in the widening of the binary fragment mass distributions. The widening is understood to be due to the extra stability of ^{132}Sn clusters marking the onset of asymmetric fission in the heavy mass group while Ni or Ge clusters with masses from 70 to 80 control the onset of yields in the light fragment group. The latter phenomenon was called super-asymmetric fission in Chapter 4.3 (see Figs. 4.3.3 to 4.3.7). The mirror masses to heavy 132 in the light group and to an average light mass ~ 75 in the heavy group determine the widths of the two groups. For all actinides exhibiting asymmetric fission alike the above clusters mark the lower limiting masses in the corresponding group. It is therefore conjectured that also in ternary fission the cluster nuclei behave as hard cores

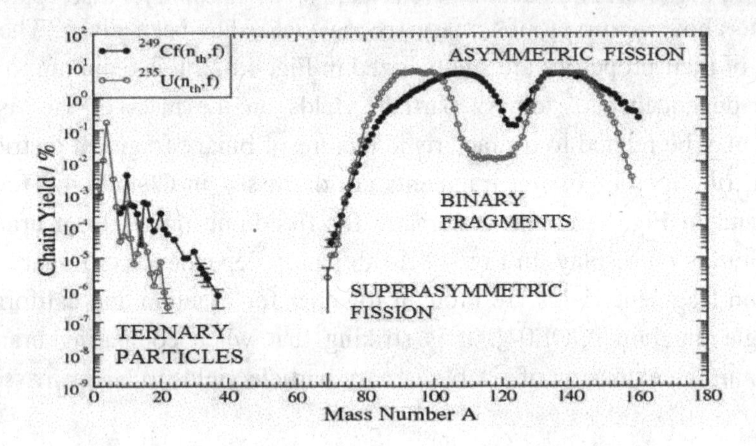

Fig. 4.8.14. Ternary yields and super-asymmetric fission.

which are not to be broken. The maximum number of nucleons in the neck prone to form a ternary particle is hence given by subtracting from the fissioning compound mass the masses of the two above clusters. Reasoning with a light Ni cluster of mass 70, one thus finds for the ternary particle as a rule of thumb for ^{236}U an upper limit of A_{max} = 34, while for ^{250}Cf this mass is A_{max} = 50. Taken at face value the difference in mass ΔA = (50–34) = 14 of neck nucleons explains the observed shift in the spectrum of ternaries as a function of compound mass. The figures given are rough estimates and should not be taken literally but they are in reasonable agreement with experiment in Fig. 4.8.14. A more rigorous proof of the conjecture would be to find an abrupt drop of ternary yields for masses larger than those indicated. Unfortunately, even on LOHENGRIN the detection limit did not allow to push the measurements further into this ternary mass region.

4.8.2 *Energy Distributions and Correlations in Ternary Fission*

First experiments investigating the correlation between ternary particles and properties of fission fragments started in 1979 on an external beam of the ILL [GUE79]. The energies of both fragments and the ternary particles were measured with surface barrier detectors. More comprehensive data with identification of the ternary particles could be collected with the instrument DIOGENES described in Chapter 4.2 (see Fig. 4.2.5). All three outgoing charged particles were analysed as to energy and angle of emission by gaseous detectors. This allows investigating correlations between fragments and ternary particles. The instrument has proven to open new vistas for the study of ternary fission. Some results for thermal neutron fission of ^{235}U(n,f) obtained on an external neutron beam of the ILL are presented in Figs. 4.8.15 through 4.8.18. All data pertain to ternary alphas.

In Fig. 4.8.15 two scatterplots are shown [PAN86, THE86]. For ternary fission of ^{236}U* to the left, the contour lines delineate events with equal yield as a function of the total kinetic energy TKE^{ter} in ternary fission and the mass of thelight fragment. TKE^{ter} describes the total $TKE^{ter} = E_F + E_\alpha$ with E_F the sum of the kinetic energies of the two main fragments and E_α the energy of the α-particle. To the right of the figure the contour lines stand for the yields of binary fission as a function of TKE^{bin} written as $TKE^{bin} = E_F$ and light fragment mass.

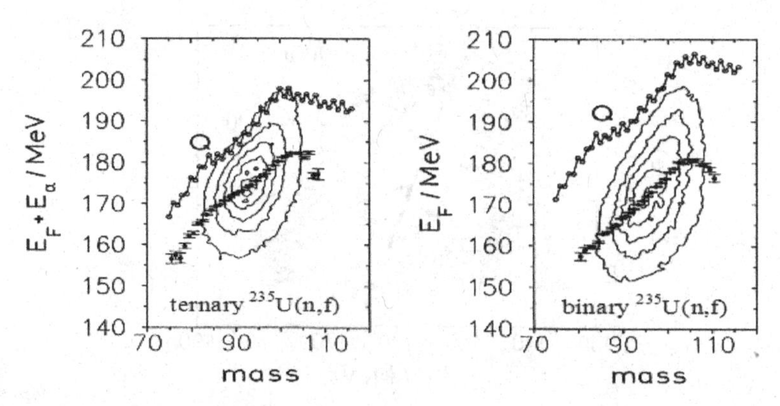

Fig. 4.8.15. Scatterplot (TKE, mass of LF) for ternary and binary fission of ^{236}U*

For both reactions the *Q*-values *Q* and the average <TKE^{ter}> and <TKE^{bin}> are indicated. The comparison in FIG. 4.8.15 of ternary and binary fission is revealing. Compared to the binary reaction, in ternary fission the *Q*-values are lower but the average total kinetic energies as a function of fragment mass are generally larger. The ternary process is thus definitely colder than the binary process. This is also evident from the contour lines for the yields which in the ternary case come close to the *Q*-values for a broad range of masses. On average the total excitation energy *TXE* in thermal neutron fission of $^{236}U^*$ is found to be smaller by about $\Delta TXE = \langle TXE^{bin}\rangle - \langle TXE^{ter}\rangle \approx 10\ MeV$ in ternary contrasted to binary fission [THE89]. This rough figure can be read from FIG. 4.8.15 as the difference between ternary and binary fission for ($Q - \langle TKE\rangle$). For spontaneous fission of ^{252}Cf the corresponding figure is $\Delta TXE \approx 7\ MeV$ (see Fig. 4.8.9). It is underlined that this large drain of energy can only be supplied by an analogous decrease of deformation energy present at scission.

In the scatter plot of Fig. 4.8.16 [PAN 86] the correlation between the α-energy E_α and E_F, the sum of the two main fragment energies, is seen to be negative. The negative covariance is a consequence of energy conservation. Assume for the moment that the excitation energy *TXE* is very small if not nil and consider a fixed mass ratio. The available energy $Q^{ter} = const$ is then virtually found back in the total kinetic energy TKE^{ter}. With $TKE^{ter} = E_F + E_\alpha = const$ the two energies are strictly anti-correlated. For non-vanishing excitation energies *TXE* and averaged over all masses the negative correlation will be partly washed out. The remains of the negative correlation are observed in experiment as demonstrated in Fig. 4.8.16.

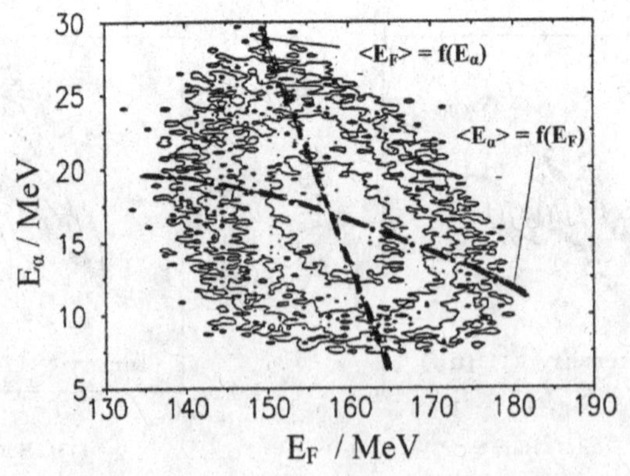

Fig. 4.8.16. Correlation (E_α, E_{FF}) for ^{235}U(n,f).

In Fig. 4.8.17 the total excitation energy <TXEter> averaged over fragment masses in the reaction 235U(nth,f) is on display as a function of the energy Eα of the α-particle [THE86]. For an increase of the α-energy the total excitation energy decreases linearly. At the average α-energy Eα ≈ 16 MeV the average excitation energy is $<TXE^{ter}> \approx 14$ MeV. For ^{252}Cf this energy is about 27 MeV in Fig. 4.8.9. For ^{252}Cf(sf) the decrease of excitation energy for increasing α-energy could less directly be checked by measuring the multiplicity of evaporated neutrons $<\nu>$ [HAN89]. Confirming the result in Fig. 4.8.17 it was observed that for increasing neutron number $<\nu>$ the energy E_α decreases linearly.

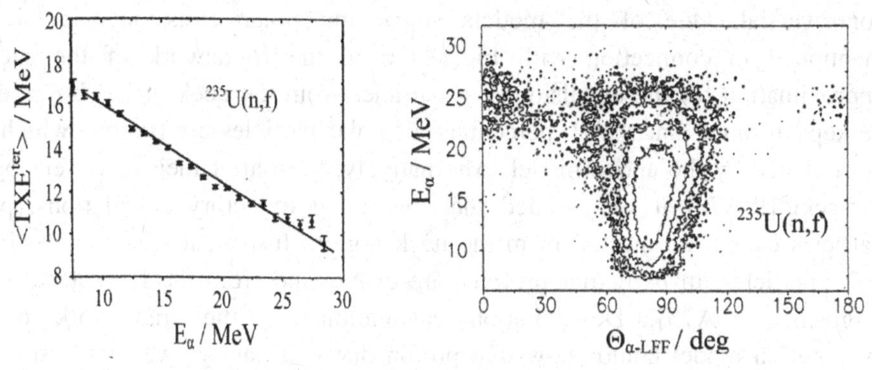

Fig. 4.8.17. Average TXEter vs. α-energy. Fig. 4.8.18. Scatterplot (Eα, Θα-LF).

Angular distributions of ternary particles have repeatedly been addressed in Chapter 4.8.1. Comprehensive information on these distributions for α-particles as obtained in the work under discussion for thermal neutron fission of ^{236}U* is given in Fig. 4.8.18 [PAN86, THE86]. The scatterplot (E_α, $\Theta_{\alpha\text{-}LF}$) of α-energy vs. the angle between the α-particle and the light fission fragment $\Theta_{\alpha\text{-}LF}$ brings to evidence that the overwhelming part of α-particles are emitted at nearly right angles relative to the fission axis. At a closer look the angle for these equatorial alfas is slightly smaller than 90°. This is well understood in the picture put forward for ternary particle emission where due to a double rupture of the neck a ternary particle is born at scission in between the two main fragments. The Coulomb forces exerted by the two fragments will focus the ternary particles at roughly right angles to the fission axis. However the asymmetry of mass and charge division pushes the particles away from the heavy fragment with the higher charge towards the lighter fragment. This is why the angle $\Theta_{\alpha\text{-}LFF}$ is on average slightly less than 90°. In the figure it is also seen that the kinetic energies E_α of equatorial alfas span a wide range of energies up to almost 30 MeV. Energies below about 8 MeV are cut off by absorber foils in front of the α-detectors in order to protect them from radiation damage by fission fragments. In reality the spectrum extends to virtually the energy $E_\alpha = 0$ (see below).

By contrast to the equatorial alfas, the energy distributions for the polar particles in Fig. 4.8.18 flying in direction of the light or heavy fission fragment are narrow with average energies around 25 MeV. It is worth pointing out that by far more polar particles are ejected by the light fragment at angles below $\Theta_{\alpha\text{-}LFF} \approx 50°$ while in the opposite direction for angles $\Theta_{\alpha\text{-}LFF} \geq 150°$ there are much less events. The discussion on the emission mechanism for polar particles is still controversial. One of the models suggesting polar emission was shortly mentioned in connection with Fig. 4.8.6 in the framework of the sudden approximation for the expulsion of α-particles from the neck. It has however to be kept in mind that a high percentage of polar particles are protons which are not covered by the above model. Alternatively, soon after their discovery by E. Piasecki [PIA70] it was argued that, based on trajectory calculations, polar particles cannot be ejected from the neck region. Instead it was proposed that polar particles, in particular protons, are evaporated from the fully accelerated fragments [PIA73]. De-excitation calculations in the framework of an evaporation model could show that proton distributions are well understood in this approach [SCH92].

From an experimental point of view it is none the least the asymmetry in the emission probabilities for polar particles from the light and heavy fragment which is intriguing. One may venture to invoke a connection to the asymmetry in neutron emission, or more basically the asymmetry in fragment deformation between light and heavy fragment. The largest asymmetries in fragment deformation obtain for heavy fragments near the doubly magic ^{132}Sn and light fragments LF with mass number near $A_{LF} \approx 120$. This mass split attracted already attention when searching for the best fit of yield data in fission of ^{236}U* in Fig. 4.8.11. While ^{132}Sn is spherical, the light fragment with $A_{LF} \approx 120$ has at scission the strongest deformation of all fragments. Following relaxation of deformation the light fragment should therefore be a more prolific source for proton evaporation than the heavy fragment. For the much rarer super-asymmetric mass splits the asymmetries of deformation are reversed but due to the low probability for this mass splits their influence on proton evaporation is small in comparison to standard asymmetric fission. Measurements of neutron multiplicities from ^{252}Cf(sf) in coincidence with polar long range α-particles support the above arguments [ALK89, HEE90]. For polar α-emission from the light fragment the ratio of neutron multiplicities from the light and the heavy mass group is $\nu_L/\nu_H = 2.46/1.12 = 2.2$ while in binary fission this ratio is 1.12. It means that polar emission of α-particles shows up for light fragments being much more deformed than in standard binary fission.

In Fig. 4.8.7 the ratio of ternary LRA to binary fission was plotted as a function of the fissility Z^2/A. The general trend of this ratio is to increase with fissility. As proposed in studies of LRA events from spontaneous fission [WIL85] it is more revealing to analyze this ratio as a function of the average total excitation energy $\langle TXE \rangle$ of the fissioning nucleus. To avoid confusion it is stressed that in the present case *TXE* is the excitation energy of binary fission, in contrast to the presentation in Fig. 4.8.9 where the excitation energy is the energy in ternary fission. As usual, the average *TXE* is found as the difference between average Q-values and total kinetic energies of fragments: $TXE = Q - TKE$. The ratio LRA/binary fission is shown in Fig. 4.8.19 for spontaneous fission reactions of Pu-, Cm-, Cf- and Fm isotopes as a function of the binary excitation energy $\langle TXE \rangle$. Isotopes of the same element are connected by lines [SER98]. A clear positive correlation emerges between the probability for emission of ternary LRA-particles with the average excitation energy and hence deformation energy of fragments. This is an interesting aspect telling ternary LRA fission to be more probable the more deformed the nuclei from a fission reaction are. It recalls the correlation between polar α-particle emission and deformation addressed above. The findings are in line with the more general analysis of the spectrum of ternary yields (i.e. not only LRA particles) in the Baum model (see eq. (4.8.5) and Fig. 4.8.10) highlighting the correlation between ternary yields and excitation energies of the ternary system. But it is obvious that excitation or deformation energy is only one of the parameters controlling the outcome of ternary fission. As demonstrated in Fig. 4.8.7, for α-particle emission the spectroscopic factor S_α of the fissioning nucleus is another parameter playing a role.

A very detailed insight into specific properties of ternary particle emission has been obtained in an experiment on spontaneous fission of ^{252}Cf where for all three outgoing particles not only energies but also angles of emission were measured. This allowed a full reconstruction of the kinematics and hence determination of fragment mass distributions in ternary fission [GON05]. Fragment mass distributions in ternary fission are on display in Fig. 4.8.20 for ^{4}He and Be isotopes as ternary particles (full points). For comparison the binary mass distribution is indicated by dashed curves normalized to the respective total ternary yields. It must be pointed out, however, that for experimental reasons the ternary mass distributions pertain to ternary particle energies larger than 8 MeV and 26 MeV for ^{4}He and Be, respectively. Similar results but without ternary particle identification were published by I.D. Alkhazov et al. [ALK89]. In the figures it is striking that in going from binary to ternary fission the mass distributions are not displaced as a whole to smaller mass numbers but are getting narrower the heavier the ternary particle is. In fact the mass number of the lightest fragment in both the light and heavy mass group appears to stay constant while the heaviest fragments in both groups are shifted to smaller mass numbers. It should be recalled that the lightest fragments in one group is complementary to the heaviest fragments in the other group. With the position of the lightest fragments being fixed, the smaller number of nucleons in ternary fission will entail narrower widths of the mass distributions in ternary compared to binary fission. The heavier the ternary particles are the narrower the mass distributions should become.

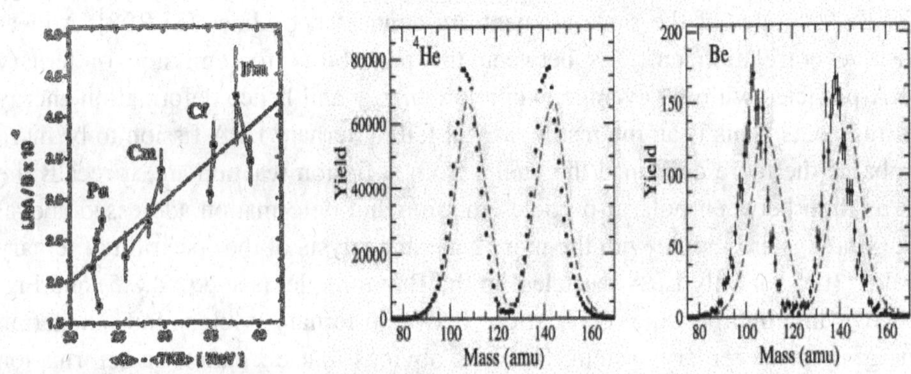

Fig. 4.8.19: Ratio LRA / binary fission vs.total excitation energy in binary fission

Fig. 4.8.20: Fragment mass distributions in ternary fission of ^{252}Cf(sf) accompanied by ^{4}He (left panel) and Be particles (right panel). Closed points: t ernary data. Dashed curves: binary fission with yields normalized to ternary yields

In Fig. 4.8.20 it is conspicuous that the fixed positions of the lightest fragments are near mass number A = 125 in the heavy group and A = 80 in the light group. These mass numbers are sufficiently close to the mass numbers of the magic cluster nuclei responsible for asymmetric and superasymmetric fission (see Chapter 4.3) to venture the statement that also in ternary fission magic cluster fragments determine the positions of the foothills for the onset of sizable yields in the mass distributions. In other words, the magic cluster fragments resist to breakup. This was precisely the argument already advocated in connection with Fig. 4.8.14 where asymmetric and superasymmetric binary fission mass distributions were confronted with ternary yields for the two reactions ^{235}U(n_{th},f) and $^{249}Cf(n_{th},f)$. Discussing these features in a naïve dumb-bell model of the scission configuration (like the one sketched in Fig. 4.8.5) there is a light and a heavy magic fragment approximated by spheres and a neck connecting the two spheres. The number of nucleons left for the neck increases with the mass of the fissioning nucleus. For a given neck radius at scission the length of the neck will hence increase. Thus, the heavier the nuclei are the larger the deformations will be, entailing larger deformation or excitation energies at scission. This is in fact observed in Fig. 4.8.19 where the average excitation is getting larger from Pu to Cf. The correlation between ternary yields and excitation energies in Fig. 4.8.19 is then understood to mean that the probability for ternary fission in the model of a double neck rupture is the more probable the longer the neck is. Intuitively this conclusion appears to be not unreasonable.

Finally the energy distributions of ternary particles are reviewed. Some properties of energy distributions for equatorial α-particles were already addressed in the foregoing in Fig. 4.8.18. Some general rules valid for all ternary particles emerge. All energy distributions of equatorial ternary particles obey three rules:

1) The energy distributions of the ternary particles are Gaussian-like.
2) The energy distributions extend down to energy zero.
3) The average energies and the widths of the distributions increase with the charge of the ternary particles but to good approximation do not depend on the fissility of the fissioning nucleus.

A sample of energy distributions N(E) for the carbon isotopes ^{14}C to ^{17}C from thermal neutron fission of ^{235}U as measured on the LOHENGRIN separator is visualized in Fig. 4.8.21 [BAU92]. Together with the data points the results of trajectory calculations are dawn as histograms. In spite of the cut-off at lower energies imposed by experimental conditions the data are seen to be close to Gaussians. This is corroborated by the trajectory calculations.

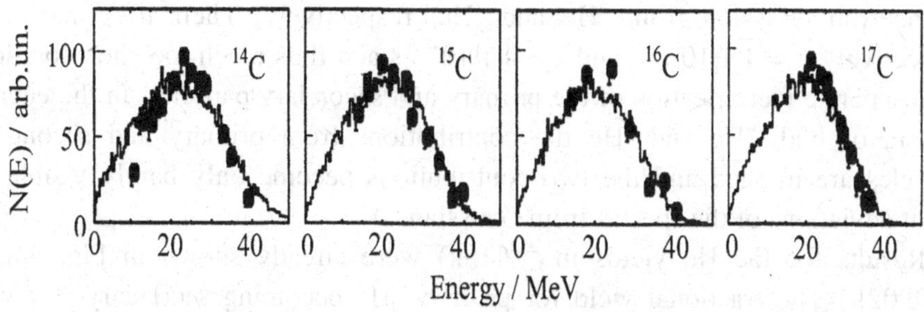

Fig. 4.8.21. Energy distributions of ternary C-isotopes ^{14}C, ^{15}C, ^{16}C and ^{17}C.

For the reaction $^{233}U(n_{th},f)$ average energies and the FWHM widths of energy distributions are plotted for ternary particles from hydrogen to oxygen as a function of mass number of the isotopes in Fig. 4.8.22. Again the experiments were performed on the LOHENGRIN spectrometer [WOE99, GOE99b]. The average energies are observed to increase with the nuclear charge of the ternary particles. This is readily understood as being due to the increasing repulsion of the ternaries by the Coulomb field of the two main fragments. For given charges of the ternary particles their energies decrease for increasing mass since for given charge and hence Coulomb force the acceleration is getting smaller. In comparison to the average energies (to the left in Fig. 4.8.22) the widths are seen to follow very similar trends (to the right in FIG. 4.8.22). Even the figures for averages and widths are very close together with, however, a notable exception for ^{4}He (see the discussion below). The general features outlined are a direct consequence of the fact that the distributions are near-Gaussian in shape with the low energy tail extending down to energy zero. For increasing average energies the widths of the distributions have hence to follow.

Of further interest is to investigate the evolution of average energies of ternary particles as a function of the fissioning compound. The comprehensive study of yields and energies of ternary reaction partners on LOHENGRIN allows compiling results for reactions ranging from $^{229}Th(n,f)$ to $^{249}Cf(n,f)$. For the sake of clarity they are shown in Fig. 4.8.23 separately for even and odd charge isotopes to the left and right, respectively [Woe 99,, Goe 99b]. The energies are rather constant with perhaps a slight increase for the largest fissilities analysed. While discussing ternary yields the special situation for the yields of ^{4}He and ^{6}He was highlighted (see Fig. 4.8.12 and discussion). In the yields observed for these two isotopes there are two contributions: a genuine primary ejection of ^{4}He and ^{6}He, respectively, and an additional secondary side feeding due to the fast decay by neutron emission from ^{5}He and ^{7}He, respectively. Their life times are, respectively, $\tau = 1.1 \cdot 10^{-21}$ s and $\tau = 4.1 \cdot 10^{-21}$ s and thus much too short to allow for a separate identification of the primary and secondary particles. In the energy spectra of both ^{4}He and ^{6}He the contributions from primary and secondary particles are merged and the two contributions become only barely visible as slight deviations of the spectra from Gaussians.

Results for the He yields in $^{252}Cf(sf)$ were already shown in Fig. 4.8.12 [KOP02]. The fractional yield for primary ^{5}He becoming secondary ^{4}He was determined to be 17 % of all ^{4}He particles observed. The averages $<E>$ of the

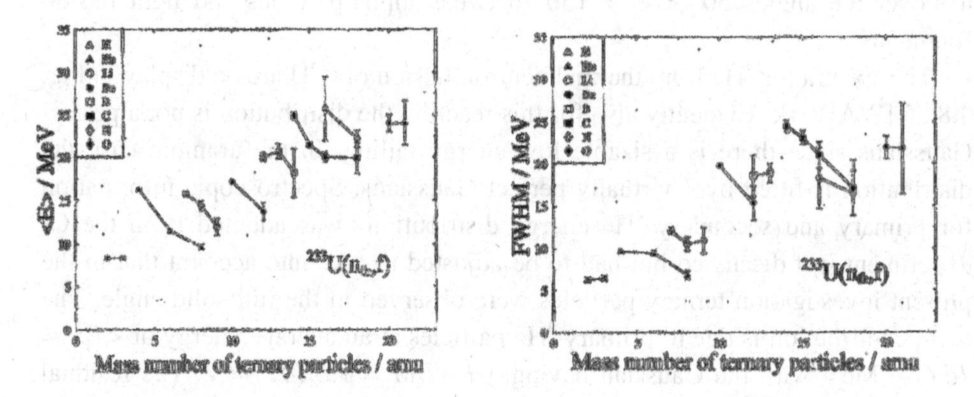

Fig. 4.8.22. Average and FWHM of energy distributions for ternary isotopes from hydrogen to oxygen from the reaction $^{233}U(n_{th},f)$.

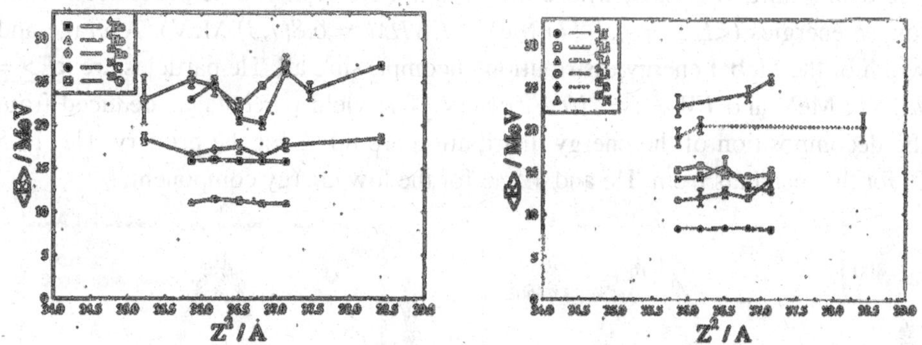

Fig. 4.8.23. Average energies of ternaries for thermal neutron reactions from $^{230}Th^*$ to $^{250}Cf^*$.

energy distributions for primary, secondary and all ^{4}He-particles taken together were measured to be 16.4(3) MeV, 12.4(3) MeV and 15.7(2) MeV, respectively. The corresponding widths (FWHM) were found to be 10.3(3) MeV, 8.9(5) MeV and 10.9(2) MeV, respectively. This spectroscopic information was used to decompose the ^{4}He energy distribution from an experiment where data were taken down to 1 MeV. The decomposition is shown in Fig. 4.8.24 with the dotted curve to the right for primary ^{4}He and to the left for secondary ^{4}He [MUT08]. The sum spectrum is seen to reproduce well the data for energies in excess of 10 MeV. Below 10 MeV, however, there are deviations pointing to a low energy tail. Strictly speaking, when quoting energies it has to be specified whether the data pertain to equatorial or polar or ^{4}He particles taken in the full solid angle. In Fig. 4.8.24 the "equatorial" spectrum is visualized with "equatorial" being meant to cover the angles 50° < Θ < 130° between alpha-particles and light fission fragments.

The spectra for ^{4}He from thermal neutron fission of ^{235}U are on display in Fig. 4.8.25 [WAG04]. Evidently also for this reaction the distribution is not a perfect Gaussians since there is a sizable low energy tailing. In the uranium case the distribution is fitted by 3 virtually perfect Gaussians. Spectroscopic information for primary and secondary ^{4}He energy distributions was adopted from the Cf experiment just discussed but had to be adjusted to take into account that in the present investigation ternary particles were observed in the full solid angle. The main contribution is due to primary ^{4}He particles at an average energy of $<E> = 16.6(4)$ MeV with the Gaussian having a $FWHM = 9.0\ (4)$ MeV. The residual ^{4}He from decay of ^{5}He are centred at an average energy of $<E> = 12.4(4)$ MeV with a Gaussian width of $FWHM = 8.1(6)$ MeV. For a perfect fit to the observed ^{4}He data a third Gaussian, whose origin is not clear, had to be introduced at the lowest energies ($<E> = 6.1(11)$ MeV, $FWHM = 6.8(1.3)$ MeV). Average and width of the global energy distribution encompassing all ^{4}He particles are $<E> = 15.7(2)$ MeV and $FWHM = 10.4(2)$ MeV. The yield percentages deduced from the decomposition of the energy distribution are 80 % for the primary ^{4}He, 15.5 % for the residuals from ^{5}He and 4.5 % for the low energy component.

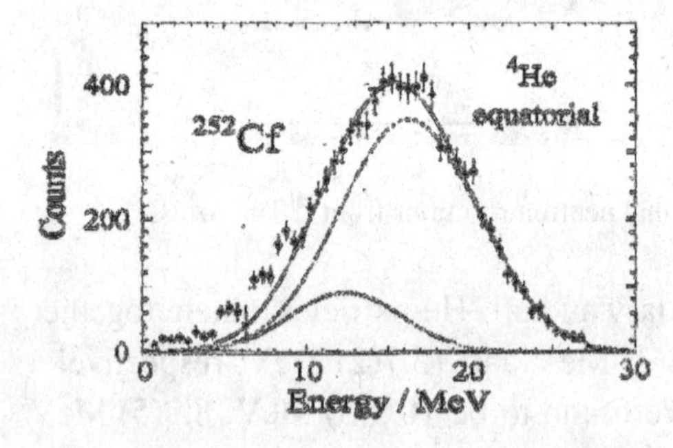

Fig. 4.8.24. Energy spectrum of ^{4}He in spontaneous fission of ^{252}Cf.

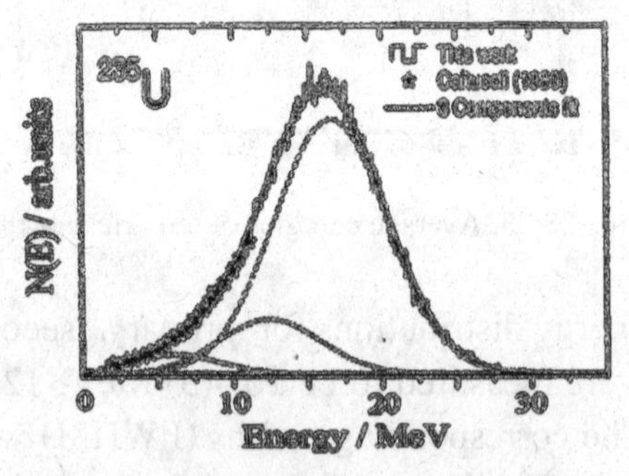

Fig. 4.8.25. Energy spectrum of ^{4}He in thermal neutron fission of ^{235}U.

The question is open whether the energy distributions are nearly perfect Gaussians as assumed for the analysis of the uranium data in Fig. 4.8.25 introducing a third component of unknown origin, or whether the distributions are not perfect Gaussians but exhibit a low energy tailing as suggested for californium in Fig. 4.8.24.

Very briefly it should be mentioned that the DIOGENES detector described in Chapter 4.2 (Fig. 4.2.5) has been used to search for "true" ternary fission into three partners of about equal mass and with angles between them of about 120°. No events of this type were found. Only an upper limit of 10^{-8} per fission could be given [HEE89]. Yet the idea of fission-like tripartition has not ceased to intrigue theoreticians and eperimentalists alike [PYA10].

4.8.3 *Quaternary Fission*

Having discussed at some length ternary fission where the two fission fragments proper are accompanied by a third light charged particle, one may wonder whether still more complex fission-based processes exist. With the ratio for ternary/binary fission of actinides being roughly between $1 \cdot 10^{-3}$ and $3 \cdot 10^{-3}$ the process is rather rare. An even rarer decay mode known is quaternary fission where in addition to the fragments two light charged particles are ejected in one single fission event. In ternary fission the yield of long range alpha-particles (LRA) is dominant (90 % of all cases). With reference to the LRA yield the probability for quaternary fission is down by a factor between $0.3 \cdot 10^{-4}$ for thermal neutron fission of ^{235}U and $3 \cdot 10^{-4}$ for spontaneous fission of ^{252}Cf. Referred to fission the probabilities for quaternary events are ~ 10^{-7} per fission for uranium and ~10^{-6} for californium. These low yields have been a barrier for thorough investigations in the past. Only few experiments have been reported in the literature both, for thermal neutron fission of ^{235}U [AND69, KAP73] and spontaneous fission of ^{252}Cf and ^{248}Cm [KAT73, FOM97]. In the picture of a multiple neck rupture at scission, double neck rupture gives rise to ternary, and triple rupture to quaternary fission.

Quaternary fission is identified in experiment as a coincidence between two heavy fragments and two light charged particles. Up to now only α-particles and tritons were observed in quaternary fission, i.e. precisely those light nuclei which are most abundant in ternary fission. In experiment there is the difficulty to distinguish between simultaneous (or "true") and sequential (or "pseudo") decays into 4 charged particles. The sequential decay is not a genuine quaternary fission process. Very similar to neutron unstable ternary particles ^{5}He, ^{7}He and ^{8}Li* emitting a neutron before arriving at detectors, there exist also particle-unstable ternaries decaying with short lifetimes into charged particle pairs. For sufficiently short lifetimes the ternary process is registered as quaternary fission. The most prominent example for a sequential quaternary process is the formation of a primary ^{8}Be and its subsequent decay into two secondary α-particles. Similarly, pseudo (α, t) quaternary fission has been detected which is attributed to the decay of ^{7}Li* when it is born in its second excited state. For ^{7}Li the ground and the first excited state are particle stable and cannot mimic quaternary fission. In experiment true and pseudo quaternary fission are distinguished by exploiting the different angular patterns between the two light charged particles for the two processes in competition.

In a first experiment quaternary fission was studied at the Institut Laue-Langevin in cold neutron induced reactions with ^{233}U and ^{235}U as targets. The experiments were conducted in parallel to studies of fission induced by polarized neutrons (see Chapter 4.9). Quaternary data were taken irrespective of the orientation of neutron spin [GOE01, JES05]. With better adapted detector assemblies, quaternary fission in ^{233}U(n_{th},f) was investigated in a follow-up experiment. Results from this experiment are reviewed here [GAG03, GOE04]. To identify quaternary fission four independent particle detectors are required. In Fig. 4.8.26 the ^{233}U-target is positioned at the centre of 4 detector assemblies [JES00]. Two position sensitive multi-wire-proportional counters (MWPC) are foreseen for the light and heavy fission fragments, labelled LF and HF, respectively. Two arrays of up to 20 PIN diodes each detect lighter charged particles. The centres of the detectors are arranged on a plane perpendicular to the neutron beam running horizontally through the reaction chamber with the centres of the MWPCs and the diode array at right angles to each other. Since for the majority of ternaries the angles Θ of emission between fragments and light particles is nearly perpendicular (see Fig. 4.8.18), this arrangement optimises the chance to observe ternary or quaternary fission in coincidence with fragments.

The PIN diodes were made from SFH chips (SIEMENS) with thicknesses 380 μm and sizes 3x3 cm² or 1x1 cm². They were mounted with the n+ rear side facing the source and operated at voltages ensuring full depletion. Particle identification was based on the analysis of rise times of the current pulses [MUT00]. With injection of ions from the rear side of the diodes the separation of α-particles from hydrogen isotopes was perfect. However, due to the ranges of H-isotopes exceeding in many cases the thickness of detectors it was not possible in these cases to distinguish between tritons, deuterons and protons. From earlier work [KAT73] it is known that for ^{252}Cf(sf) the ratios of triton / proton yields are virtually the same in ternary and quaternary fission. Apparently due to experimental uncertainties this ratio fluctuates for ternary reactions in thermal neutron fission of actinides between 4 and 6. Assuming that this ratio is also valid for quaternary fission it suggests that at least 80% of all H-isotopes are in fact tritons. In the evaluation the working hypothesis therefore was to consider the observed hydrogen isotopes to be tritons. In Fig. 4.8.26 it is not shown that 12 PIN diodes were of large size (3x3 cm²) and 8 of smaller size (1x1 cm²) with the latter installed in order not to miss too many quaternary events when two particles at close angles to each other hit the same diode. To avoid radiation damage of the diodes by fragments or natural α-particles they were shielded by aluminium foils of appropriate thickness. Inevitably this entails a low-energy cut-off in the energy spectra of ternary and quaternary light charged particles.

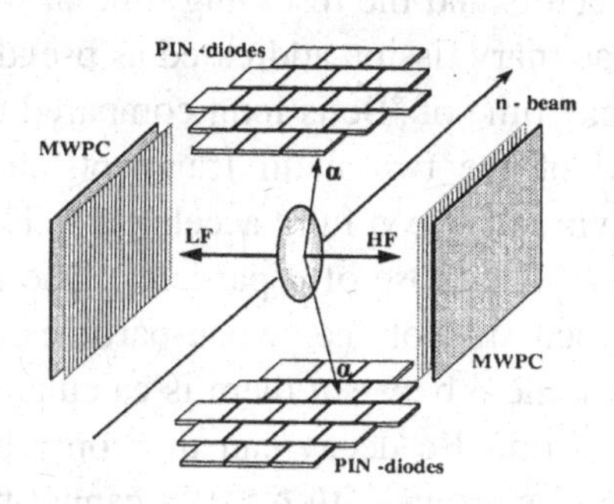

Fig.4. 8.26. Schematic view of detector assembly.

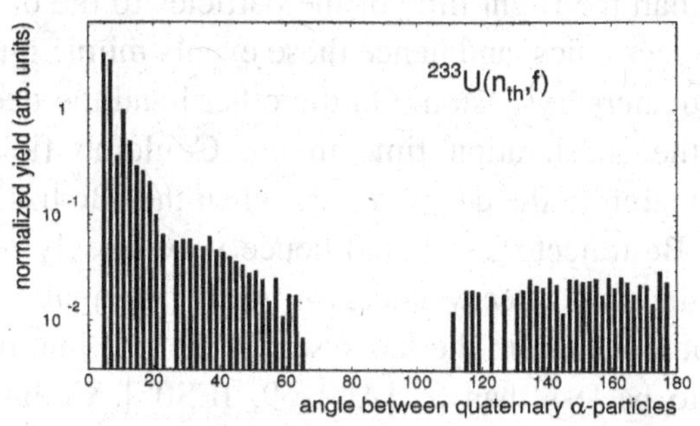

Fig. 4.8.27. Angular correlation between quaternary α-particles.

For the enriched ^{233}U target with about 5 mg of fissile material in total and with a beam intensity of $6 \cdot 10^8$ cold neutrons /cm²·s the number of events registered was 10^6/s for binary events and 10^3/s for ternary particles. Quaternary events are spotted as coincidences between the two MWPCs and two diodes. Small angles between the two light particles are measured for hits in the same diode array while large angles are obtained for hits in opposite arrays (see Fig. 4.8.26). In 6 weeks of measuring time about $6 \cdot 10^3$ quaternary events were accumulated. Particle identification enabled to sort the data into 3 categories: (α,α), (α, H-isotope) and (H- isotope, H-isotope). By far in most cases two alphas in coincidence are observed while the probability for two coincident H-isotopes is down by two orders of magnitude.

The task to differentiate between true and pseudo quaternary fission is solved by evaluating the distribution of angles δ between the two light particles. The distribution for $^{233}U(n_{th},f)$ is shown in Fig. 4.8.27 for quaternary α-particles [GAG04]. For sequential fission the decay in flight of the primary ternary ^{8}Be particle focuses the two quaternary particles kinematically. Hence, for these events small angles $\delta < 90°$, corresponding to the detection of two α-particles in the same diode array, are expected. In Fig. 4.8.27 the yield is observed to increase sharply for small angles δ. The reason for the increase is the contribution from sequential quaternary fission. By contrast, the correlation at large angles $\delta > 90°$ from hits in opposite diode arrays of Fig. 4.8.26 is traced to true quaternary fission. The number of counts is seen here to be independent of opening angles.

To justify the above statements the decay processes have to be inspected more carefully. The ground state of ^{8}Be has a half-life of $T_{1/2} = 6.7 \cdot 10^{-17}$s for the decay into two α-particles with $Q = 0.092$ MeV. The decay time is both, much shorter than the flight time of the particles to the detectors and the resolving time of the electronics, and hence these events mimic quaternary fission addressed as pseudo quaternary fission. On the other hand the decay time of ^{8}Be is long compared to the acceleration time in the Coulomb field of the two main fragments and therefore the decay occurs when the ^{8}Be has virtually been fully accelerated. The ^{8}Be trajectories should hence very closely resemble those of α-particles. Due to the small decay energy a sharp kinematic focussing of the two α-particles is anticipated in the lab system. The maximum angle δ between them is calculated to be less than 8° [AND69, JES05]. Counts from ^{8}Be decay can thus only be found as double hits in one array. Hits in opposite arrays with $\delta > 90°$ cannot be due to ground state decay of ^{8}Be. However, besides ^{8}Be in its ground state also ^{8}Be* in its first excited state may be formed with some probability. The lifetime for the decay of ^{8}Be* at an excitation energy of $E^* = 3.04$ MeV into two α-particles is $T_{1/2} = 3 \cdot 10^{-22}$s and the arguments given above pointing to pseudo quaternary fission are even more compelling here. As yet the ratio of yields for excited to ground state ^{8}Be is not known from experiment. Very cautiously the ratio of ^{8}Be* in its first excited state to the ^{8}Be yield in its ground state is estimated to be 25(10) %. Due to its very short lifetime, ^{8}Be* decays in the acceleration phase. Trajectory calculations for ternary α-particles from fission indicate that at times of $3 \cdot 10^{-22}$s the α-velocity is already 40 % of the final velocity [GUS10]. Simulating excited ^{8}Be* trajectories decaying in the Coulomb field, the opening angles δ to be expected in the lab system yielded a broader distribution than quoted above for ground state decay of ^{8}Be. Taking into account that on average the ternary particles are born slightly off-axis, estimates indicate that α-particles from the decay have virtually no chance to appear in opposite detector arrays of Fig. 4.8.26. Instead, they should be observed in the same diode array but at large opening angles. In the experimental δ-distribution of Fig. 4.8.27 events at angles δ larger than 10° are therefore attributed to α-particles from ^{8}Be* decay. Their probability is seen to decrease for increasing angles δ and becoming at angles of about $\delta = 50°$ equal to the constant probability measured for hits of alphas in opposite arrays. This is considered as a confirmation of the hypothesis that all ^{8}Be* decays are collected in one and the same detector array.

Further the side-feeding of ^{8}Be by the neutron decay of excited ^{9}Be* has to be considered. Besides γ-decay leading to ^{9}Be in its ground state, all excited states of ^{9}Be* are also particle unstable and may eject a neutron to become ^{8}Be. The levels relevant for neutron emission all have lifetimes close to 10^{-21} s and are comparable with the fragment acceleration time of the fissioning nucleus. It is estimated that the ratio of yields for ^{9}Be nuclei produced in these excited states to the primary ground state yield of ^{9}Be is 0.34. Out of these roughly 50% decay back to ^{9}Be and contribute to the measured ternary yield for this isotope. The remaining 50% are registered following neutron emission as ^{8}Be isotopes. Due to the sizable ternary yield of ^{9}Be in ^{233}U(n_{th},f) the side-feeding to the ground state of ^{8}Be accounts for roughly 30% of the measured pseudo (α + α) yield (see Table 4.8.3). All events from the decay of ^{8}Be and ^{8}Be* including side-feeding from the decay of ^{9}Be*contribute to sequential quaternary fission.

Coming back to Fig. 4.8.27 the conclusion of the above discussion is that quaternary events observed for large angles $\delta > 90°$ between the α-particles represent exclusively true quaternary fission. The constant yield as a function of δ for large angles underlines the validity of this conclusion. In fact, in the model of a triple neck rupture the two α-particles are emitted independently but both are focussed by the Coulomb field of the fragments on average closely perpendicular to the fission axis. In a plane perpendicular to the fission axis the distribution of emission angles is expected to be isotropic and due to their independent emission the relative angles between the two α-particles the distribution of δ should be constant for the full range of angles $0° \leq \delta \leq 180°$. This is directly observable in experiment for angles $\delta \geq 90°$ as evidenced in Fig. 4.8.27. For smaller angles true and pseudo quaternary events are mixed. To find the true contribution the constant yield measured at large angles is extrapolated to small angles down to 0°. Subtracting this true quaternary yield from the measured yield one obtains the pseudo quaternary fission yield. By the way, it should not be forgotten to point to a chance background which has to be corrected for in all measured yields.

Besides α-α quaternary fission the α-t quaternary decay was discovered where in coincidence to two main fragments an α-particle and a triton are detected. Also here there are two different types of quaternary events: true quaternary fission with the α-particle and the triton being emitted simultaneously on one hand and on the other hand pseudo quaternary fission with first ^{7}Li* being produced in an excited state and then decaying into an alpha and a triton. The ground and first excited states of ^{7}Li are stable against particle decay. The second excited state of ^{7}Li at an energy of 4.63 MeV decays with a half-life of $T_{1/2} = 4.9 \cdot 10^{-21}$s into (α + t). The production rate of 7Li* in this state is estimated to be about 2% of the rate into stable ^{7}Li. The half-life is comparable to the acceleration time and all events are registered at small angles between alpha and triton in one single diode array. They are hence readily to be distinguished from true quaternary fission showing up as coincident hits in opposite arrays.

The third observed quaternary decay mode with two H-isotopes (mostly tritons) detected in coincidence with two main fission fragments has no pseudo component. All events measured are attributed to true quaternary t-t fission.

For the final evaluation of yields the energy distributions of the quaternary particles have to be known. Examples of the energy distributions of α-particles from ternary, true quaternary α-α and pseudo quaternary α-α fission are on display in Fig. 4.8.28 [GAG03]. The dashed histograms are raw data, the histograms in full are corrected for counting losses in the absorber foils preventing radiation damage to the diodes and the dotted curves are Gaussian fits to the data. Compared to ternary alphas the energies for alphas are down by 2.3 MeV for true and by 4.6 MeV for pseudo quaternary fission (see Table 4.8.1). The smaller energies in case of true quaternary fission may indicate that, on average, the deformation of the scission configuration is larger and the main fragments are already farther apart in quaternary compared to ternary fission. For fragments farer apart the Coulomb forces acting on the α-particles will be smaller. Since in quaternary fission two α-particles have to be accommodated in the neck region between the two fragments, large overall deformations and, hence, lower kinetic energies are quite understandable. As to the α-energies from ^{8}Be and ^{8}Be* pseudo quaternary decay, their still lower energies are in line with an average energy of about 20 MeV to be expected by extrapolation from Fig. 4.8.22 for a ^{8}Be nucleus in ^{233}U(n,f) ternary fission. On average, therefore, following the decay each of the two α-particles should carry some 10 MeV of kinetic energy. This is in reasonable agreement with the experimental findings in Fig. 4.8.28 for ^{234}U*. Similar data were also obtained in the case of ^{252}Cf(sf) [JES05].

The energy characteristics of α-particles and H-isotopes from ternary and quaternary fission are summarized in the Tables 4.8.1 and 4.8.2, respectively. The final results for the yields are summarized in Table 4.8.3 and visualized in Fig. 4.8.29 [GAG03, GOE04]. In the figure the true quaternary yields for simultaneous emission of two light charged particles and two fission fragments are shown. The most abundant reaction is for (^{4}He + ^{4}He) accompanying the fragments. The probability for this reaction is $1.54 \cdot 10^{-7}$ per fission. This figure should be compared to the ternary α-yield which is $2.1 \cdot 10^{-3}$ per fission.

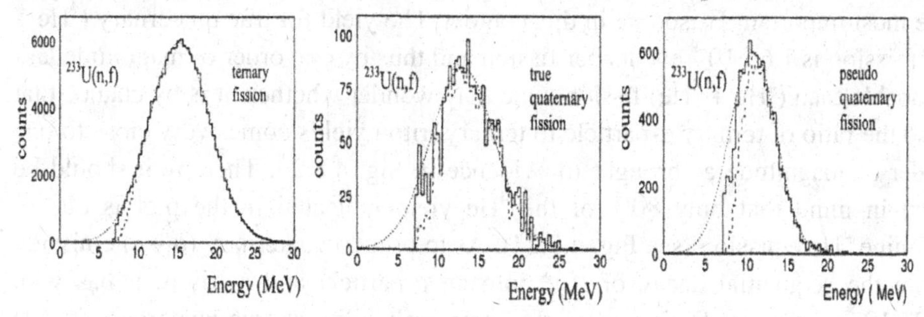

Fig. 4.8.28. Energy distributions of α-particles in ternary and quaternary fission.

Obviously the quaternary α-α yield is definitely smaller than the ternary α-yield squared. Sequential pseudo quaternary fission is labelled ^{8}Be in the figure. The probability $3.36 \cdot 10^{-7}$ per fission comprises the formation and decay of primary ^{8}Be and ^{8}Be* as well as the side-feeding by neutron decay from a primary ^{9}Be*. The individual contributions are $2.15 \cdot 10^{-7}$ events per fission for (^{8}Be + ^{8}Be*) and $1.21 \cdot 10^{-7}$ events per fission for ^{9}Be*. However, the error bars for the partitioning are very large and may come close to 50 %. The probabilities for true and pseudo quaternary fission differ by only a factor close to two. But it is noteworthy that the ternary yield of ^{10}Be is by about a factor of 45 larger than for^{8}Be in Fig. 4.8.29 though on average the Q-values for ternary ^{8}Be and ^{10}Be decay are equal. The comparison shows that predicting ternary yields based merely on Q-values can be entirely misleading.

Table 4.8.1. Energies of α-particles in ternary and quaternary fission of ^{233}U(n_{th},f).

α-particles from	Tern. Fi	True (α-α) Fi	Pseudo (α-α) Fi	True (α-H) Fi	α from ^{7}Li*
Mean Energy / MeV	15.3(3)	13.0(5)	10.7(5)	12.5(10)	11.0(10)
FWHM / MeV	9.2(2)	9.4(5)	6.6(5)	10.6(10)	6.3(10)

Table 4.8.2. Energies of H-isotopes in ternary and quaternary fission of ^{233}U(n_{th},f).

H-isotopes	Tern. t, Tern.d	(t,d,p) from true (α-H-isotope)	t from ^{7}Li*
Mean Energy / MeV	8.4(2)*	5.8(10)	5.8(10)
FWHM / MeV	6.8(2)*	7.0(10)	4.7(10)

* from [VOR69]

The second quaternary reaction identified is for (^{4}He + ^{3}H) as the light charged particles. As the three H-isotopes p, d and t could not be disentangled in experiment the label ^{3}H in the figure is meant to indicate that tritons are by far the most important H-isotope in this context. The yield for true quaternary (^{4}He + ^{3}H) fission is $1.66\cdot10^{-8}$ events per fission and thus by one order of magnitude less probable than (^{4}He + ^{4}He) fission. One may wonder whether it is by chance that also the ratio of ternary α-particle to ternary triton yields comes very close to one order of magnitude as brought to evidence in Fig. 4.8.29. Thereby it should be kept in mind that only 80% of the ^{4}He yield indicated in the plot is due to genuine ^{4}He emission (see Fig. 4.8.24). As to pseudo quaternary (α + t) emission from the sequential decay of ^{7}Li* into an α-particle and a triton, it has with $1.35\cdot10^{-8}$ events per fission about the same probability as true quaternary (α + t) fission. It is tempting to exploit the ratio for the production of excited 7Li* to ground state 7Li in a Halpern-like model of ternary fission (see eq. (4.8.3)) to find in a Boltzmann ansatz the nuclear "temperature" at scission. The temperature evaluated is T = 1.0(1) MeV. As repeatedly stressed, in a ternary fission model based on sudden neck ruptures the quantity T is not a temperature but just a parameter steering the excitation of nascent ternary particles.

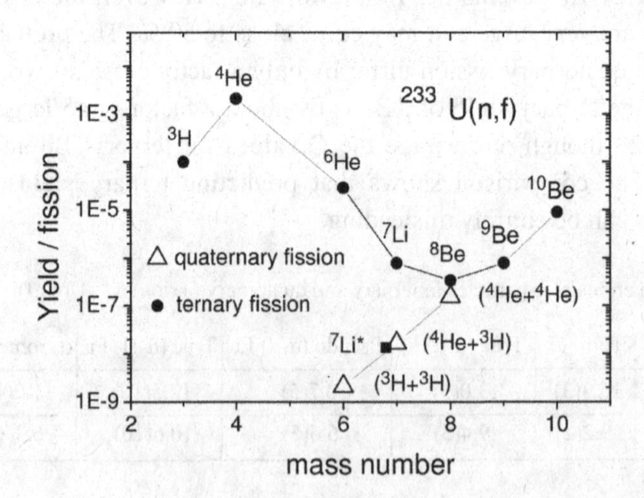

Fig. 8.29. Quaternary and ternary yields in the reaction ^{233}U(n_{th},f).

Finally few events were recorded for the quaternary combination with (^{3}H + ^{3}H) as the light particles. Their probability is $2.1\cdot10^{-9}$ events per fission and thus another order of magnitude smaller than the quaternary (^{4}He + ^{3}H) reaction and two orders of magnitude smaller than the (^{4}He + ^{4}He) reaction. It appears hence that each change from ^{4}He to ^{3}H in quaternary fission decreases the yield by the ternary ^{4}He/^{3}H ratio. This observation is intriguing but could as well be just a chance coincidence. For the (^{3}H + ^{3}H) case it is further surprising that the true one-step and the pseudo two-step decay have vastly different probabilities. The ternary yield of ^{6}He is $3.1 \cdot 10^{-5}$ per fission. In its groundstate ^{6}He undergoes β-decay. The first excited state which decays into (^{3}H + ^{3}H) lies at 14.6 MeV. It is estimated that the probability for its population relative to the ground state is ~ $2 \cdot 10^{-6}$. The expected yield for pseudo quaternary fission ^{6}He* → (^{3}H + ^{3}H) is therefore ~ $6 \cdot 10^{-11}$ per fission. Pseudo quaternary fission is thus way less probable than true quaternary fission. For ^{6}He* only an upper limit $0.4 \cdot 10^{-9}$ per fission could be measured.

The yields discussed are summarized in Table 4.8.3. True yields are given for the three genuine quaternary reactions (^{4}He+^{4}He), (^{4}He+^{3}H) and (^{3}H+^{3}H) and the pseudo quaternary reactions mediated by primary ternary particles ^{8}Be → (α+α) (sum of ground and excited states) and ^{7}Li*→(α+t). For ^{6}He* an upper limit is given.

Table 8.3. Quaternary yields in the reaction ^{233}U(n_{th},f)

Reaction	^{4}He+^{4}He	^{8}Be+^{8}Be*	^{4}He+^{3}H	^{7}Li*	^{3}H+^{3}H	^{6}He*
Yield·10^9 / fission	154(30)	215(100)	16.6(4.0)	13.5(4.0)	2.1(6)	<0.4

For the reaction ^{233}U(n,f) the agreement between the present dedicated experiment and the precursor experiment [JES05] is satisfactory within error bars. In the former experiment the two U-isotopes ^{233}U and ^{236}U were investigated and quaternary yields were found to be virtually identical. By contrast, in experiments conducted at the GSI/ Darmstadt all yields in quaternary fission of ^{252}Cf(sf) were observed to be larger by a factor of ten [MUT02, MUT04]. This figure should be compared to ternary α-yields or alternatively to ternary Be-yields. The ratios of yields for long range α-particles to binary fission increase from $2.1 \cdot 10^{-3}$ for ^{233}U(n_{th},f) to $3.1 \cdot 10^{-3}$ for spontaneous fission of ^{252}Cf, i.e. an increase by a factor of ~1.5 (see Fig. 4.8.4). In contrast, for Be-yields the increase approaches a factor of ten (see Fig. 4.8.2) and is thus comparable to the increase of quaternary yields.

So far, for quaternary fission mainly the systematic of yields and the energy distributions of the light particles have been investigated (see Fig. 4.8.28). As to the energy distributions of the heavy fission fragments in quaternary fission a first experiment has been conducted at the ILL for the reaction ^{235}U(n,f). The facility CODIS [MUT96] already used for the study of the reaction ^{252}Cf(sf) (see Figs. 4.8.9 and 4.8.12) served to investigate the energetics in ternary and quaternary fission. The instrument is sketched in Fig. 4.8.30. It is based on twin ionization chambers for fission fragments and ΔE-E_{rest} telescopes for ternary particles. One of the first results is on display in Fig. 4.8.31 [SPE04]. The energy distributions of the fragments are compared for quaternary α-α (label 1), ternary α-accompanied fission (label 2) and binary fission (label3). As already evident in Fig. 4.8.16 and re-confirmed here, the kinetic energy of the fragments is getting smaller for ternary compared to binary fission (compare fragment energy in ternary fission labelled $E_F \approx 160$ MeV in Fig. 4.8.16 to the secondary total kinetic energy labelled $TKE = 168.6$ MeV in Table 4.4.2) . But, as to be read from Fig. 4.8.15, the total kinetic energy of fragments plus the α-particle is larger in ternary than the fragment energy in binary decays. From the present experiment it is found in Fig. 4.8.31 that, similar to the energy shift in the transition from binary to ternary fission, the fragment energy decreases further from ternary to quaternary fission. It should then be interesting to explore whether the total kinetic energy in quaternary fission as the sum of the energies of both fragments and both α-particles is also larger than in ternary fission. From Table 4.8.1 a sizable contribution by the two α-particles from (^{4}He+^{4}He) decay of 26 MeV to the total kinetic energy release is estimated. The evaluation of the experiment is still in progress.

As to theory, the phenomenon of quaternary fission has been studied in the framework of multi-cluster accompanied fission [POE99, POE05]. To first approximation potential barriers for the decay into two fission fragments and additional light clusters are calculated for an aligned arrangement of all partners with the heavy fission fragments positioned at the two ends of the chain. All nuclei are assumed to be spherical and touching each other. The importance of Q-values is emphasized and probably over-emphasized. For example it is predicted that ^{8}Be and ^{10}Be should have the same probability as ternary light particles. From Fig. 4.8.29 it is however evident that this is not corroborated by experiment. Unfortunately no quantitative predictions are made for quaternary fission. But it is correctly stated that the quaternary (^{4}He + ^{4}He) mode should be the most favourable. On the other hand, the second most favourable mode predicted (^{4}He + ^{6}He) was not seen in experiment. Surprisingly the modes (^{4}He + ^{3}H) and (^{3}H + ^{3}H) observed are not mentioned at all. It appears that ternary particles like ^{3}H which generally are not addressed as cluster nuclei are not taken into account. This shortcoming yields biased predictions for quaternary fission.

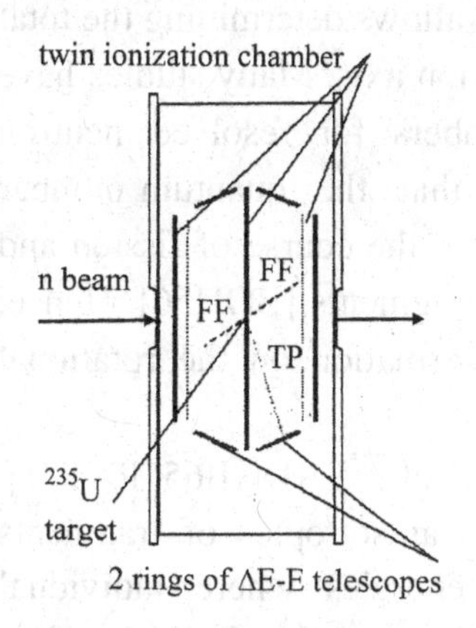

Fig. 4.8.30. CODIS detector for ternary and quaternary fission.

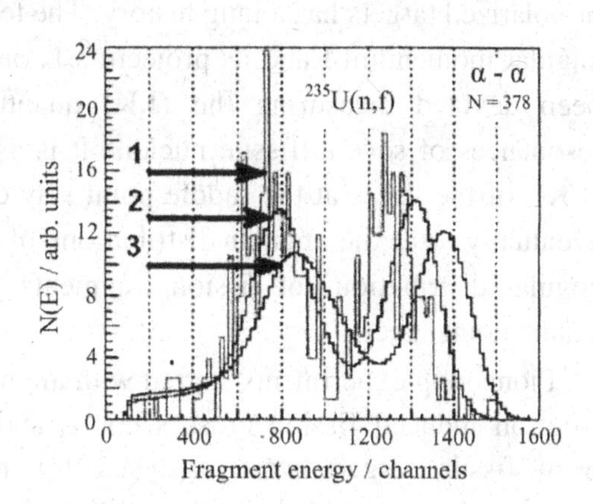

Fig. 4.8.31. Fission fragment kinetic energy 1 α-α quaternary, 2 α-ternary, 3 bin. fission.

Alternatively the challenge is to extend the Halpern model of ternary fission [HAL71] to quaternary fission. In this model the energy costs when going from binary to ternary fission play a central role (eqs. 4.8.2 and 4.8.3). It is suggested to apply the same idea for the transition from ternary to quaternary fission [GOE04]. In the simplest approach only the changes of Q-values are considered while neglecting changes in the scission configurations. For the transition from ternary ^{8}Be to quaternary (^{4}He+^{4}He) scission configurations this is probably an acceptable approximation. The parameter T in eq. (4.8.3) is taken to be T = 1.0 MeV as found in the above from the ratio of excited to ground state yields ^{7}Li*/^{7}Li. Yields for quaternary fission of ^{233}U(n_{th},f) are calculated to be Y(^{4}He+^{4}He)·10^{9} / fission = 235, Y(^{4}He+^{3}H)·10^{9} / fission= 78 and Y(^{3}H+^{3}H)·10^{9} / fission = 0.15. Compared with Table 4.8.3 the yield for quaternary (^{4}He+^{4}He) is within error bars about correctly assessed, while for (^{4}He+^{3}H) the yield is over-estimated by a factor of five and for (^{3}H+^{3}H) it is under-estimated by a factor of 14.

4.9 Fission Induced by Polarized Neutrons

4.9.1 *Parity Violation in Binary and Ternary Fission*

The investigation of neutron induced fission with polarized neutrons and aligned or polarized targets has a long history. The technique allows determining the total angular momentum J and its projection K on the fission axis. Many studies have been devoted measuring the (J,K) quantum numbers for resolved neutron resonances of several fissile nuclei. It is assumed that the quantum number (J,K) of the states at the saddle point stay constant in the course of fission and eventually steer the angular distributions of fission fragments [BOH56]. Hence, angular distributions of fission fragments yield information on the rotational saddle point states.

Pioneering experiments started with aligned targets of ^{235}U [DAB65, PAT71]. Neutron induced fission cross sections and angular anisotropies of fragments from fissile targets were analysed for neutron energies where individual resonances are resolved. Studies with polarized neutrons and polarized targets soon followed [KEY73, MOO78]. At low neutron energies mainly s-wave neutrons are absorbed and in consequence compound spins $J^+ = I + ½$ and $J^- = I - ½$ with I the target spin are populated. A prominent example is ^{235}U with $I = 7/2^-$ and compound spins either $J^+ = 4^-$ or $J^- = 3^-$. Experiments with polarized neutrons and targets enabled to uniquely assign spins to the resonances and to determine spin-separated fission cross sections. In parallel the formal theory of neutron cross sections was further developed taking into account interference effects between resonances [BAR97]. Based on this theory the energy dependent cross sections and anisotropies from an $^{235}U(n,f)$ experiment with aligned ^{235}U were analysed [KOP99]. In this approach the experimental observable is the anisotropy $A_2(E) = \sigma_0(E) / \sigma_2(E)$ of fragment emission in the differential cross section

$$d\sigma(E)/d\Omega = [\sigma_0(E) + f\cdot\sigma_2(E)\cdot P_2(\cos\Theta)] /4\pi \qquad (4.9.1)$$

with E the neutron energy, σ_0 and σ_2 the isotropic and anisotropic part of the cross section, respectively, Θ the angle between fragment emission and the axis of alignment, and f a parameter characterizing the degree of alignment of the target. As a remarkable result the decomposition of cross section and anisotropy into the contribution by the individual (J,K) channels is obtained though as a rule ambiguities remain. In particular the assignments depend strongly on the model adopted for the evaluation [AUC71].

A completely new line of research was opened by the discovery by G. Danilyan [DAN77] of parity violation in neutron induced nuclear fission. Parity Non-Conservation (PNC) in these reactions is disclosed by a correlation between the spin $\boldsymbol{\sigma}_n$ of the incoming neutron and the momentum of one of the two fragments, conventionally the light fission fragment with momentum $\boldsymbol{p}_{LF}$. Owing to the smallness of the PNC effect, experiments require an intense polarized neutron beam. The outstanding characteristics of cold polarized neutron beams at the Institut Laue-Langevin attracted therefore several research groups engaged in studies of PNC effects in fission. For a number of fissile targets detailed data on PNC effects in both binary and ternary fission could be taken.

PNC entails an asymmetry of fragment emission relative to neutron spin described by

$$W(\Theta) \sim 1 + \alpha_{PNC} P_n(\boldsymbol{\sigma}_n \cdot \boldsymbol{p}_{LF}) = 1 + \alpha_{PNC} P_n \cos\Theta \tag{4.9.2}$$

where the neutron polarization $\boldsymbol{\sigma}_n$ and fragment momentum $\boldsymbol{p}_{LF}$ are unit vectors, Θ is the angle between them, P_n is the neutron polarization and α_{PNC} measures the size of the PNC effect. It should be noted that $(\boldsymbol{\sigma}_n \cdot \boldsymbol{p}_{LF})$ is the scalar product of an axial vector with a polar vector and hence is a pseudoscalar. Any pseudoscalar observed in experiment signals parity violation.

Sketching very briefly the theory of PNC in fission, a first remark addresses the surprisingly large size of the effect. Commonly the weak interaction is held responsible for PNC effects. The relative strength of the weak nucleon-nucleon interaction compared to the strong interaction is estimated to be some 10^{-7}. By contrast, in nuclear fission the asymmetry α_{PNC} is typically 10^{-4}. Somewhere in the process of fission a mechanism enhancing PNC must be at work. The theory of PNC in fission was developed by Flambaum and Sushkov [FLA80, SUS81] and Bunakov and Gudkov [BUN83]. Both theories point to the mixing of nuclear states with opposite parity but the same spin at the compound stage of heavy nuclei following neutron capture. In thermal neutron capture the states concerned are $s_{1/2}$ and $p_{1/2}$. The high density of excited levels will favour the mixing of states mediated by the weak interaction. The mixing of those states will entail PNC interference effects. This so-called dynamical enhancement at the compound stage is supported by the experimental evidence that the PNC asymmetry coefficients α_{PNC} fluctuate strongly, and may even change sign when the energy of incident neutrons is shifted by less than 1 eV, the typical distance between levels in heavy nuclei excited by neutron capture. On the other hand, there is no dependence of the asymmetry α_{PNC} on the final states like mass, charge or energy of fragments. This is understood by noting that a fissioning nucleus has to pass over a barrier where the nucleus cools down. Therefore only few and widely spaced transitional states are accessible to the nucleus. Each transition state may encompass many if not all final states. According to A. Bohr [BOH56] the properties of transition states are reflected in the angular distributions of fission fragments which are flying apart along the fission axis. PNC asymmetries in the angular distributions of fragments should be discussed in similar terms. To cover also PNC effects, it is pointed out that the transitional states are in fact closely space doublets differing only in parity [SUS81]. The reason for the splitting is the asymmetric pear-like deformation of the nucleus at the saddle-point leading to the dominant mass asymmetry in fission of actinides [KRA69]. Therefore the mixed parity states of the compound stage have a commensurate chance to pass over the barrier. Interference between transition states of opposite parity will eventually give rise to PNC effects in fission. PNC asymmetries are hence complementary to parity conserving angular anisotropies, both sensing properties of the saddle point.

Experimentally PNC in binary fission is conveniently studied in conventional back-to-back ionization chambers with longitudinally polarized neutrons flying along the chamber axis as shown in Fig. 4.9.1 [ALE94, GRA95, KOE00c]. A very powerful method to detect $(\boldsymbol{\sigma}_n \cdot \boldsymbol{p}_{LF})$ correlations is the spin flip technique. In this technique fragment asymmetries are compared for two opposite spin orientations while the arrangement of detectors remains untouched. Typically the neutron spin is flipped every second between orientations along and against the beam direction. For the two spin directions the polarization P_n in eq. (4.9.2) switches sign and with the count rates $N_\uparrow$ and $N_\downarrow$ for spins along or against beam direction in Fig. 4.9.1 the asymmetry α_{PNC} is found from the normalized difference in count rates $A = (N_\uparrow - N_\downarrow)/(N_\uparrow + N_\downarrow)$ by:

$$A = (N_\uparrow - N_\downarrow)/(N_\uparrow + N_\downarrow) = \alpha_{PNC} P_n \cos\Theta \qquad (4.9.3)$$

At the ILL PNC asymmetries were measured with cold neutrons for the target nuclei ^{229}Th, ^{233}U, ^{235}U, ^{237}Np, ^{239}Pu, ^{241}Pu, ^{241}Am and ^{245}Cm [GAG06]. The results are summarized in Table 4.9.1. Typical sizes are $\alpha_{PNC} \approx 10^{-4}$ but sizes and even the signs of the PNC asymmetry fluctuate strongly. As it appears there is no correlation between the asymmetry and the spin or mass or charge of the target nucleus. This latter statement is strikingly evident when comparing results for the targets ^{234}U and ^{249}Cf which are not enumerated in the table [GAG98a, GAG06a]. The target spins are, respectively, $I = 0^+$ and $I = 9/2^+$, while in both cases the parity violation parameter α_{PNC} is nil within statistical uncertainty. It has also to be pointed out that with thermal neutrons $^{234}U(n,f)$ is sub-barrier fission while for $^{249}Cf(n_{th},f)$ fission proceeds well above the two maxima of the double-humped fission barrier.

An experiment with very thin ^{233}U targets in view of improving mass and energy resolutions was conducted to test the angular dependence of the PNC asymmetry predicted in eq. (4.9.2) and to search for a dependence of the effect on the mass or energy of the fragments [KOE00c]. The targets had a total mass of 5 μg evaporated as UF_4 on carbon backings with thickness 25 μg/cm². The polarized neutron flux was $2\cdot10^8$ n/cm²s and the polarization P_n was larger than 95 %. A total of $2\cdot10^{10}$ events were accumulated. Results are given in Figs. 4.9.2 and 4.9.3. As predicted by theory, the experimental asymmetry A of eq. (4.9.3) is described by a $\cos\Theta$ law in Fig. 4.9.2. In this respect there is perfect agreement between theory and experiment. Taking into account the finite neutron polarization P_n, the PNC asymmetry from the present experiment is $\alpha_{PNC} = +4.00(13)\cdot10^{-4}$.

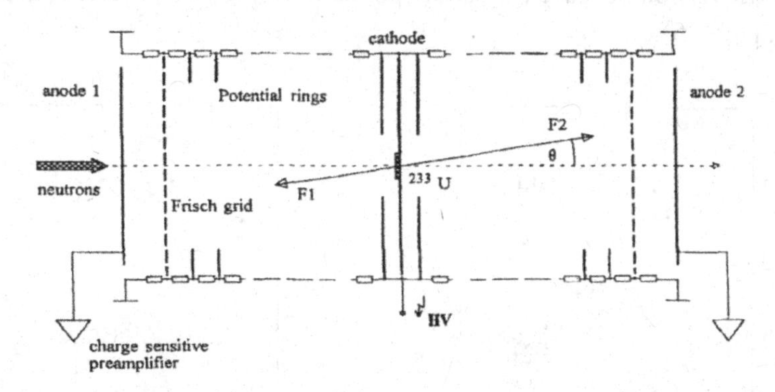

Fig. 4.9.1. Layout of experiments for PNC studies in binary fission.

The dependence of the PNC effect on light fragment mass is sketched in Fig. 4.9.3. The error bars indicate statistical errors. The finite size of the neutron polarization has been taken into account. Unfortunately the measurements could not be extended reliably to the symmetric mass region. In particular for asymmetric mass regions with sizable yield, the PNC effect is seen to be constant within error bars. The same observation was made for the fissioning compounds ^{233}Th, ^{242}Pu and ^{242}Am [ALE94]. This is considered to be a noteworthy result.

Table 4.9.1: PNC asymmetries α_{PNC} for thermal neutron induced fission

target	^{229}Th	^{233}U	^{235}U	^{237}Np	^{239}Pu	^{241}Pu	^{241}Am	^{245}Cm
target spin	$5/2^+$	$5/2^+$	$7/2^-$	$5/2^+$	$1/2^+$	$5/2^-$	$5/2^-$	$7/2^+$
$\alpha_{PNC} \cdot 10^4$	−5.6(2)	+3.6(1)	+0.84(6)	+0.9(4)	−5.0(1)	−0.89(4)	−1.0(2)	+0.26(6)

The mass range covered encompasses in the language of Brosa modes the standard I and standard II modes [BRO90]. The two modes are predicted to bifurcate only once the fission process has moved past the outer saddle-point [BRO90]. The prediction is fully in line with the experimental evidence for a constant PNC effect which according to theory is settled at the saddle-point well before the bifurcation of standard modes I and II. The two modes are hence sharing the same saddle-point and are therefore expected to indeed exhibit the same PNC effect. The more it would be interesting to study the PNC effect for symmetric fission. There the effect should in general be different from the one at asymmetric fission because the saddle points for these two processes are different (see Chapter 4.5).

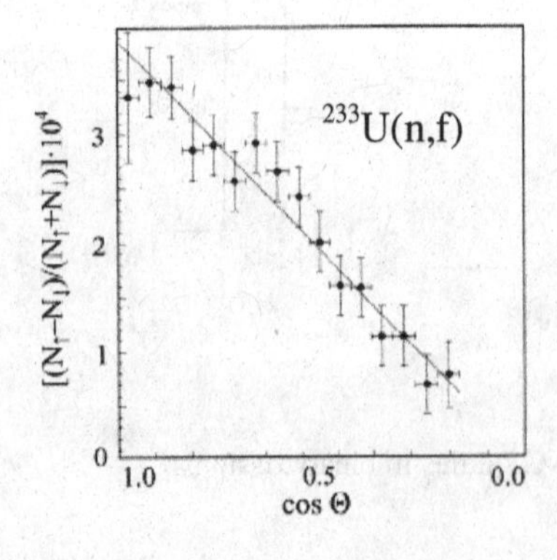

Fig.4. 9.2. Experimental asymmetry A as a function of cosΘ.

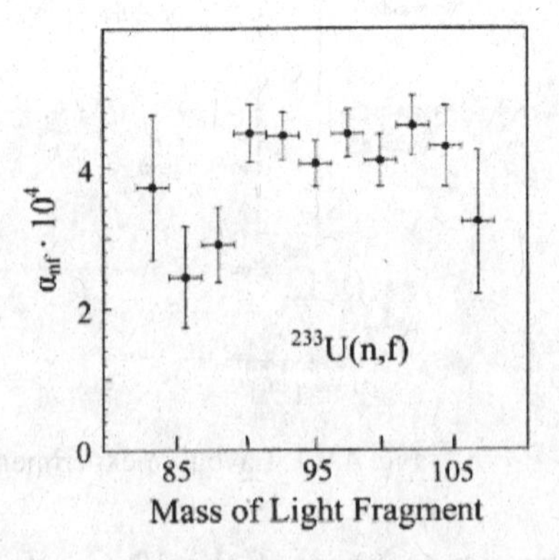

Fig. 4.9.3. PNC asymmetry α_{PNC} as a function of fragment light mass.

Parity violation in neutron induced fission reactions is not the only effect where interference of s- and p-wave neutron capture states are revealed. There are two further parity conserving asymmetries, the so-called left–right (LR) asymmetry and the forward-backward (FB) asymmetry, where interference is essential. These angular distributions are described by

$$W(\Theta) \sim 1 + \alpha_{LR} P_n \boldsymbol{p}_{LF} \cdot [\boldsymbol{\sigma}_n \times \boldsymbol{p}_n] + \alpha_{FB} (\boldsymbol{p}_n \cdot \boldsymbol{p}_{LF}) \qquad (4.9.4)$$

with α_{LR} and α_{FB} the asymmetry coefficients, P_n the neutron polarization and $\boldsymbol{p}_{LF}$, $\boldsymbol{p}_n$ and $\boldsymbol{\sigma}_n$ unit vectors of light fragment and neutron momentum, and direction of neutron polarization, respectively. The LR asymmetry is an asymmetry in fragment emission relative to the oriented plane $[\boldsymbol{\sigma}_n \times \boldsymbol{p}_n]$ spanned by neutron spin and neutron momentum. For maximum sensitivity the neutron polarization is chosen in experiment to be at right angles to the beam direction. Taking data for two opposite directions of neutron spin, the first step in the evaluation is the flip asymmetry $A = (N_\uparrow - N_\downarrow) / (N_\uparrow + N_\downarrow)$ similar to eq. (4.9.3). As it is evident from eq. (4.9.4) the LR asymmetry is conserving parity. To assess the FB asymmetry from Eq. (4.9.4) the neutrons do not have to be polarized. The technique of spin-flip is here replaced by a reversal of the incoming neutron beam direction. In actual practice the measuring device is turned around by 180° relative to the beam.

The ambitious goal in the evaluation of fragment asymmetries is to determine the contribution of p-wave neutrons and to deduce the parameters of p-wave resonances. To this purpose the range of neutron energies to be studied has to be extended from thermal to the resonance region. Except for one isolated example (see Chapter 4.9.2) this type of experiments could not be performed at the ILL. An experimental result on the dependence of PNC and LR asymmetries on neutron energy for the reaction $^{233}U(n,f)$ is shown in Fig. 4.9.4 [PET89]. Neutron energies were scanned with a crystal monochromator. In the PNC asymmetry the change of sign and in the LR asymmetry the minimum at an energy slightly below $E_n = 200$ meV is striking. For this reaction no s-resonances below $E_n = 1$ eV are observed. The structure must hence be due to p-resonances. In fact, theory predicts that at a p-resonance the PNC asymmetry is changing sign [BUN83]. Experiment therefore suggests the existence of a p-resonance at an energy of about $E_n = 200$ meV.

Since the properties of p-resonances cannot be deduced from inspection of fission cross sections, originally the hope was that PNC-, LR- and FB-asymmetries could disclose the position and widths of these resonances. Except for cases like the one just discussed in connection with Fig. 4.9.4, in general this hope was not fulfilled. In the theoretical analysis of asymmetries the complexity of the problem has led to involved formal approaches with the danger that the physics is getting blurred [SUS81, BUN83, BAR04, FUR10]. At low neutron energies a further specific complication arises since resonances at negative energies have to be taken into account whose resonance energies are below the neutron binding energy but being broad enough to contribute to interferences at positive energies under study. Parameters of p-resonances have been tentatively assessed for ^{240}Pu* by [BAR04, FUR10] and for ^{234}U*, ^{236}U* and ^{240}Pu* in [SOK05]. Unfortunately, the parameter sets which can be compared for ^{240}Pu* are in violent disagreement. It is therefore fair to say that the position and widths of p-resonances are not known unambiguously. Due to this ambiguity the fission process is not suited for precision determinations of weak interaction matrix elements. Nonetheless, starting from a weak interaction matrix element of ~ 10^{-7} the dynamical enhancement of ~ 10^3 in heavy nuclei yields the correct average size of the PNC effect of 10^{-4} in Table 4.9.1[FLA80].

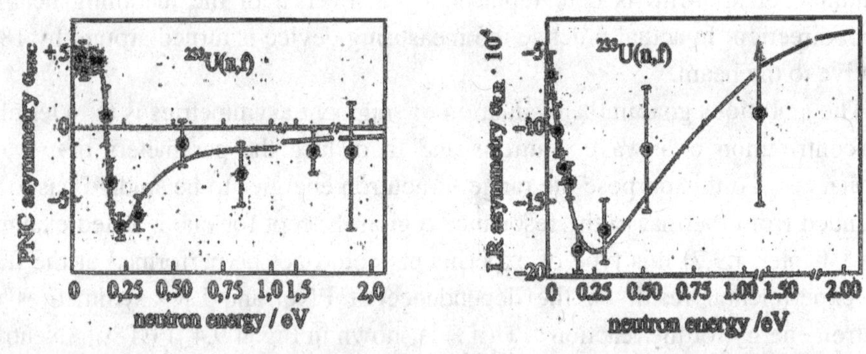

Fig. 4.9.4. PNC asymmetry (left panel) and LR asymmetry (right panel) for ^{233}U(n,f).

Besides this basic understanding of the effect, studying PNC asymmetries has proven being a very useful tool for investigating fine details of the fission process. An example was discussed in connection with the question where the bifurcation points of the Brosa standard I and standard II modes on the potential energy surface of a fissioning nucleus are located. Similarly PNC studies in ternary fission have yielded interesting insight into this rare process. One of the instruments used to this purpose is on display in Fig. 4.9.5. Basically it is the same apparatus as employed for the study of quaternary fission (see Fig. 8.26). Detectors for fragments (MWPC) and ternary particles (PIN diodes) are mounted perpendicular to the neutron beam. But now the neutrons are polarized and pointing either towards the fragment or the ternary particle detectors. In the first case, which is illustrated in Fig. 9.5, the spin flip method allows to investigate the PNC effect for fragments in ternary fission by requiring that in coincidence with the fragments also a ternary particle is observed. Similarly, for neutron polarization in direction of the PIN diodes the PNC effect for ternary particles is measured provided the ternaries are detected in coincidence with the fragments.

In a first experiment for the reaction $^{233}U(n_{th},f)$ the ratio of $\alpha^{ter}{}_{PNC}$ in ternary fission to $\alpha^{bin}{}_{PNC}$ in binary fission was found to be $\alpha^{ter}{}_{PNC} / \alpha^{bin}{}_{PNC} = 1.05(10)$ [BEL91]. Within statistical error this ratio is unity. The experiment was repeated with the instrument shown in Fig. 4.9.5 where for ternary fission the PNC asymmetry was found to be $\alpha^{ter}{}_{PNC} = +3.7(1)\cdot 10^{-4}$ [JES02]. This figure should be compared with the asymmetry $\alpha^{bin}{}_{PNC} = +3.6(1)\cdot 10^{-4}$ reported for binary fission in Table 4.9.1 and also with the value $\alpha^{bin}{}_{PNC} = +4.00(13)\cdot 10^{-4}$ from the experiment described in connection with Figs. 4.9.2 and 4.9.3 making use of a very different type of detectors and geometric conditions [KOE00]. Within statistics the PNC asymmetries for binary and ternary fission are hence seen to be equal. The same conclusion was reached in a study of the reaction $^{239}Pu(n_{th},f)$. There the ratio $\alpha^{ter}{}_{PNC} / \alpha^{bin}{}_{PNC} = 1.12(8)$ was observed [GOE94].

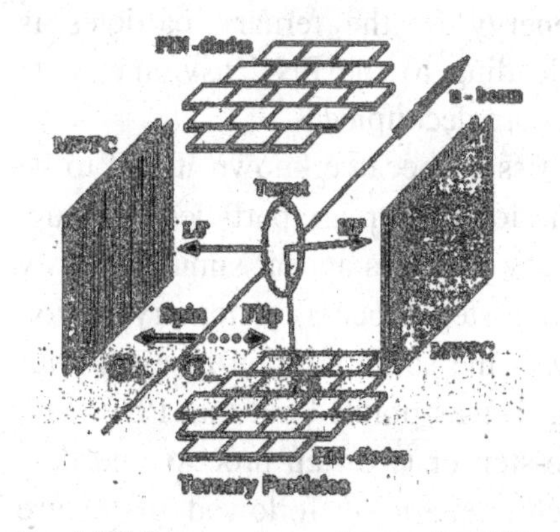

Fig. 4.9.5. Layout of experiments for PNC studies in ternary fission.

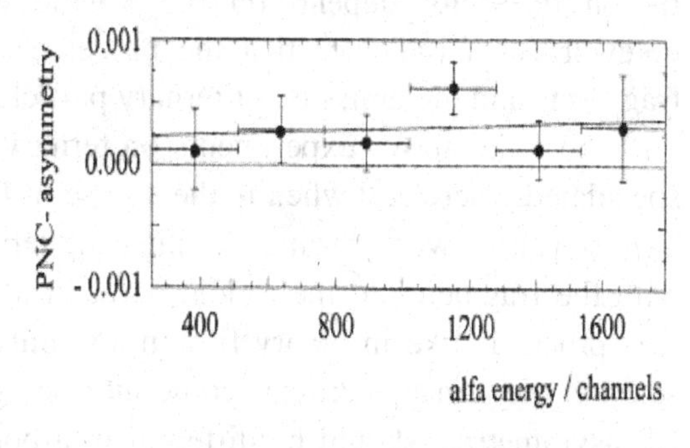

Fig. 4.9.6. PNC asymmetry of fragments in α–accompanied ternary fission) of $^{233}U(n,f)$ as a function of α–energy covering 10 to 30 MeV.

Turning around the spin of incoming neutrons in Fig. 4.9.5 to point in the direction of the PIN diodes for the light charged particles, a possible PNC-effect for the ternary particles can be investigated. The correlation $(\boldsymbol{\sigma}_n \cdot \boldsymbol{p}_{LF})$ for fragments is replaced by the correlation $(\boldsymbol{\sigma}_n \cdot \boldsymbol{p}_{TP})$ for the ternary particles. By contrast to the sizable PNC asymmetry for fragments, both in binary and ternary fission, the PNC asymmetry for ternary particles (TP) is measured to be virtually vanishing. For the reaction $^{233}U(n_{th},f)$ values $\alpha^{TP}_{PNC} = 0.6(4)\cdot10^{-4}$ [PET89] and $\alpha^{TP}_{PNC} = 0.9(13)\cdot10^{-4}$ [JES02] are reported.

It is noteworthy that for fission fragments not only the PNC asymmetry but also the LR asymmetry appears being the same in binary and ternary fission. For the $^{233}U(n,f)$ reaction the LR asymmetries are $\alpha^{bin}_{LR} = (-2.33 \pm 0.25)\cdot10^{-4}$ [ALF95] in binary to be compared to $\alpha^{ter}_{LR} = (-3.1 \pm 1.8)\cdot10^{-4}$ in ternary fission [GAG03]. As to LR asymmetries for ternary particles, similarly to the PNC asymmetry likewise the LR asymmetry for these particles is vanishing: $\alpha^{TP}_{LR} = 0.8(8)\cdot10^{-4}$ [JES02]. For theory it is not too surprising that PNC- and LR-asymmetries behave similarly since both are traced to the interference of s- and p-wave states at the saddle point.

It was further investigated whether in ternary fission the PNC effect in the asymmetry of the fragments depends on the energy of the ternary particles. The PNC effect as a function of the energy of ternary particles is on display in Fig. 4.9.6 for the reaction $^{233}U(n_{th},f)$ [JES01]. The slope of the straight line from a fit to the PNC data is $(1.14 \pm 2.96)\cdot10^{-7}$ per channel. Within statistical uncertainty no dependence is disclosed. The observational fact that the PNC-effect in ternary fission does not depend on the kinetic energy of the ternary particles is noteworthy. It indicates that the processes leading to the PNC asymmetry of fragments and the emission of ternary particles are decoupled.

Before the above experiments on ternary fission became known it had to be speculated where and when in the course of fission the ternary particles show up. Two scenarios were discussed: either the ternary particles appear simultaneously with the fragments at the saddle point in a one-step process, or ternary fission first proceeds like in binary fission and only at the scission point or even after scission the ternary particles come into being. It was pointed out that PNC- and LR-asymmetries should be different in a one-step or two-step process and thus would allow finding from experiment which scenario is followed by nature [BUN85]. In analogy to the discussion of PNC in binary fission which helped to locate the bifurcation point for the standard Brosa modes, the reasoning in the following is based on the notion that PNC- and LR-effects are due to interferences of incoming neutron s-and p-waves at the saddle point stage. The virtual equality of PNC- and LR-asymmetries of fragments in binary and ternary fission proves that binary and ternary fission share the same saddle point or, stated otherwise, the ternary particles are not yet present at the saddle point. The fact that the PNC asymmetry for fission fragments in Fig. 4.9.6 does not depend on the energy of the ternary particles leads to the same conclusion. Likewise the missing PNC- or LR-effects for the ternary particles are in line with the reasoning that these particles are born too late for sensing the subtle s- and p-wave neutron interferences at the saddle. Ternary fission is thus demonstrated to be a two-step process. It is less directly evident from experiment whether in this two-step process the ternary particles are evaporated from one of the pre-fragments, as proposed by Halpern [HAL65], or whether they are ejected from the neck joining the two fragments by a double-neck rupture, as suggested by Rubchenya [RUB82],. Arguments in favour of the double-neck rupture model were given in Chapter 8, in particular in connection with Fig. 4.8.20.

4.9.2 *Tri- and Rot-Effect in Ternary Fission*

Having studied in some detail parity non-conservation in nuclear fission the question arises whether there is also a chance to investigate violation of time-reversal. In particular it was suggested by K. Schreckenbach that time reversal invariance may be probed in ternary fission induced by polarized neutrons [SCH88]. This is in analogy to experiments where in the free beta-decay of polarized neutrons the correlation between neutron spin and the momenta of electron and neutrino are scrutinized. The correlation to be analysed in ternary fission is between the spin of the neutron inducing fission and the momenta of one of the fragments (by convention the light fragment LF) and the ternary particle TP. Formally the correlation is specified by the observable

$$B = \boldsymbol{\sigma_n}\cdot[\boldsymbol{p_{LF}} \; x \; \boldsymbol{p_{TP}}] = \boldsymbol{p_{TP}}\cdot[\boldsymbol{\sigma_n} \; x \; \boldsymbol{p_{LF}}] \qquad (4.9.5)$$

with $\boldsymbol{\sigma_n}$ the neutron spin, $\boldsymbol{p_{LF}}$ and $\boldsymbol{p_{TP}}$ the momenta of LF and TP. All vectors are understood to be unit vectors. The correlation B changes sign under time reversal. It is said to be T-odd. In case a $B \neq 0$ correlation is found it may indicate a violation of time reversal invariance. This statement is indeed valid in the decay of free neutrons. However, in contrast to parity where the measurement of a non-zero P-odd observable proves that parity is not conserved, a non-vanishing T-odd observable is only a necessary but not a sufficient condition for violation of time-reversal invariance. It has in addition to be proved that the reverse reaction is feasible which in the present case of fission is not obvious. Furthermore one has to be aware of reactions which could mimic a violation of time reversal invariance.

The correlation B becomes maximal with B = ±1 when all three vectors in eq. (4.9.5) are orthogonal to each other. For the two particle momenta this condition is rather closely fulfilled by nature since the ternary particles are mostly ejected from the neck and focussed by the Coulomb force between the particle and the fission fragments roughly perpendicular to the fission axis (see Fig. 4.8.18). The neutron spin then has to be oriented perpendicular to the plane $[\boldsymbol{p}_{LF} \, x \, \boldsymbol{p}_{TP}]$. This is achieved by detecting ternary particles and fragments in a plane perpendicular to a longitudinally polarized neutron beam. This arrangement avoids at the same time asymmetries introduced by parity violation (observable $\boldsymbol{\sigma}_n \cdot \boldsymbol{p}_{LF}$ in eq. (4.9.2)) or left-right asymmetries (observable $\boldsymbol{p}_{LF} \cdot [\boldsymbol{\sigma}_n \, x \, \boldsymbol{p}_n]$ in eq. (4.9.4) with $\boldsymbol{p}_n$ the neutron momentum). The detector assembly already in use for the study of PNC-effects in ternary fission and put on view in Fig. 4.9.5 is thus ideally suited for the present purposes provided the neutron beam is polarized longitudinally with neutron spin pointing either parallel or anti-parallel to the beam direction.

The layout of experiments is sketched in Fig. 4.9.7. The detectors for the fragments (MWPC) are positioned to the left and right and the two arrays of PIN diodes are on top and bottom. In the right panel of the figure the diodes have been grouped together. Note that due to the geometric extension of detectors the three vectors in eq. (4.9.5) are perpendicular to each other on average only. The influence of PNC or LR asymmetries is therefore not ideally avoided but it is at least minimized. A series of experiments was conducted. In the most ambitious one the MWPCs were position sensitive and the PIN diodes allowed for particle identification separating He- and H-isotopes.

The triple correlation of eq. (4.9.5) will for given spin orientation yield an asymmetry in the emission of the TP relative to the on average horizontal plane $[\boldsymbol{\sigma}_n \, x \, \boldsymbol{p}_{LF}]$ in Fig. 4.9.7, with count rates in the upper or lower hemisphere being different. The angular correlation is described by

$$W(\boldsymbol{p}_{LF}, \boldsymbol{p}_{TP}) \, d\Omega_{LF} d\Omega_{TP} \sim (1 + DB) W_0(\boldsymbol{p}_{LF}, \boldsymbol{p}_{TP}) \, d\Omega_{LF} d\Omega_{TP} \qquad (4.9.6)$$

with $W_0(\boldsymbol{p}_{LF}, \boldsymbol{p}_{TP})$ the basic correlation being independent of neutron spin and with the coefficient D taken to be a constant measuring the size of the triple correlation B. The correlation is anticipated to be very small and there is only a chance to observe an asymmetry by applying the spin flip technique like in the study of PNC and LR effects. When flipping the neutron spin the observable B

changes sign and therefore the coefficient D is found by evaluating the asymmetry A in the count rates for spin flipped situations at fixed LF and TP momenta. The asymmetry A is given in analogy to eq. (4.9.3) by

$$A = (N_{+z} - N_{-z}) / (N_{+z} + N_{-z}) \tag{4.9.7}$$

with N_{+z} and N_{-z} the count rates for spin in and against the beam direction, respectively, the beam defining the +z-axis. With eq. (4.9.6) the asymmetry is $A = DB$. The sign of A is hence following the switches in sign of B when either $\mathbf{p}_{LF}$ or $\mathbf{p}_{TP}$ are inverted. By convention the sign of the asymmetry coefficient D is defined to be identical to the sign of the asymmetry A for the "reference constellation" with the light fragment in Fig. 4.9.7 flying to the left and the ternary particle being ejected upward. For this constellation the correlator B is positive ($B>0$).

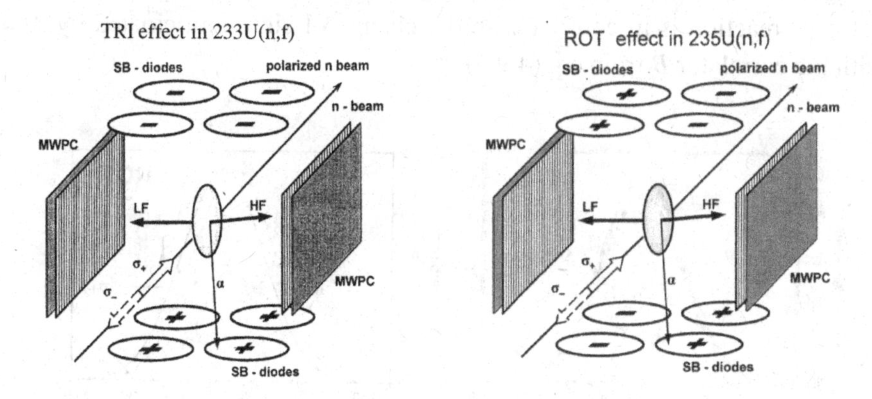

Fig. 4.9.7. Experiments for the study of the triple correlation of eq. (9.5). Signs of asymmetries on the diodes are given for ^{233}U(n,f) in the left panel and for ^{235}U(n,f in the right panel.

The experiments were carried out with the same cold polarized beam of the ILL already employed for the PNC and LR investigations. Recalling the beam specifications, the energy was on average 4 meV, the polarized flux was about 10^9 n/cm² s, and the polarization approached 95%. The reactions having been analysed are the standard fission reactions for the target nuclei ^{233}U, ^{235}U and ^{239}Pu. Up to 4 mg of highly enriched target material was spread as a thin layer on a thin backing. Count rates for binary fission events were typically 10^6 f/s and for ternary events about 10^3 α/s. The flip rate for neutron spin was 1 Hz. The asymmetry in eq. (4.9.7) was evaluated diode by diode which allowed to take data for angles between the momenta $\boldsymbol{p}_{LF}$ and $\boldsymbol{p}_{TP}$ ranging from 40° to 120°.

In a first measurement with a ^{233}U target it became quickly evident that there was a sizable non-zero asymmetry, roughly ten times as large as the PNC or LR effect. The mere size of the asymmetry rules out violation of time as being behind the phenomenon. The sign pattern is on display in Fig. 4.9.7 (left panel). For the sake of clarity the many diodes in Fig. 4.9.5 have been grouped together according to the sign of the asymmetry. The pattern proved to be as anticipated by eqs. (4.9.5) to (4.9.7). By contrast, the neighbouring uranium isotope ^{235}U chosen as target also exhibited an asymmetry, but surprisingly the pattern of signs was completely different. The sign pattern is indicated in the right panel of Fig. 4.9.7. For both Figs. (4.9.7) the light fission fragment LF is taken to be flying to the left. It is likewise baffling that, selecting events where LF is flying to the right, only the signs for the asymmetry in the ^{233}U(n,f) reaction changed while those for ^{235}U(n,f) remained unchanged. This latter property observed for the ^{235}U(n,f) reaction is in conflict with the change of sign aniticipated by $A = DB$ with the correlator B from eq. (4.9.5).

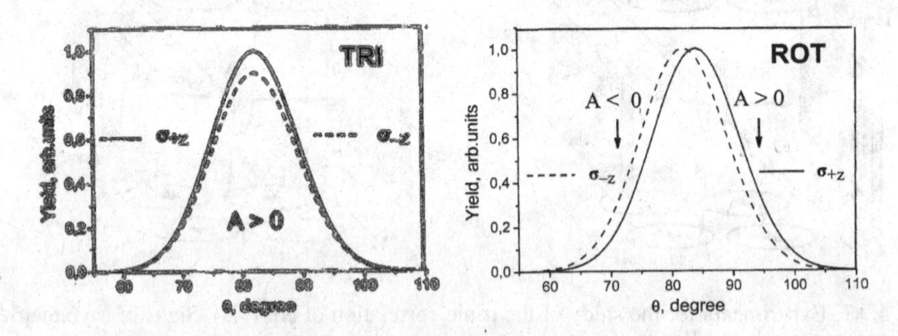

Fig. 4.9.8. Model for TRI and ROT effect.

For the interpretation of the above findings obviously two different phenomena have to be invoked. They are suggested and illustrated graphically in Fig. 4.9.8. In a first effect being supposed to be dominant for the ^{233}U(n,f) reaction, the yields of events depend as anticipated by eqs. (4.9.5) and (4.9.6) on the relative orientations of neutron spin and particle momenta. As already pointed to, for given fragment and ternary particle momenta, the contribution to the yield DB in eq. (4.9.6) is controlled by neutron spin orientation because the observable B of eq. (4.9.5) is changing sign. Between the two distributions of events for spins $\boldsymbol{\sigma}_{+z}$ and $\boldsymbol{\sigma}_{-z}$ there is in eq. (4.9.6) just a scaling factor controlled by the constant coefficient D and the correlation B from eq. (4.9.5). The dependence of the scaling factor on the correlation B is weak since the basic angular distribution $W_0(\boldsymbol{p}_{LF}, \boldsymbol{p}_{TP})$ of α-particles relative to the fission axis is sizable only in a narrow range of angles with $\boldsymbol{p}_{TP}$ nearly perpendicular to $\boldsymbol{p}_{LF}$ (see Fig. 4.8.18). From eq. (4.9.5) the correlator B is hence in most ternary events close to $B = \pm 1$. For the drawing in Fig. 4.9.8 the correlation B has been taken to be $B \equiv +1$ corresponding to the reference constellation introduced above. In consequence the scaling factor is the same for all diodes in one of the two hemispheres separated by the plane spanned by neutron spin $\boldsymbol{\sigma}_{\pm z}$ and fission momentum $\boldsymbol{p}_{LF}$. It was further assumed that the correlation quantified by the constant D is positive leading to a positive asymmetry $A = D > 0$. For the reference constellation there are therefore for neutron spin component $\boldsymbol{\sigma}_{+Z}$ parallel to the beam more ternary particles being ejected into the upper than the lower hemisphere. For the ^{233}U(n,f) reaction, however, it turns out in experiment that the asymmetry A observed for the reference constellation is negative entailing a negative $D < 0$. The distribution of signs for the asymmetries measured diode by diode is shown in Fig. 4.9.7. The ansatz based on the correlator B in eq. (4.9.5) obviously describes well the observed pattern of asymmetries for the reaction ^{233}U(n,f). The asymmetry coefficient D is evaluated to be $D = -3.9 \cdot 10^{-3}$ [GAG04]. The asymmetry D is thus by a factor of ten larger than the asymmetry $\alpha_{PNC} = +3.6 \cdot 10^{-4}$ for parity violation in the same reaction (see Table 4.9.1). Evidently the origin of the D asymmetry has to be traced to another mechanism than violation of time reversal invariance. Nevertheless this new phenomenon has been termed the "TRI-effect" because a non-zero asymmetry A in eq. (4.9.7) signals a non-vanishing TRIple correlation as described by B in eq. (4.9.5).

By contrast, the pattern of asymmetry signs for the fission reaction with the neighbouring ^{235}U isotope to the right in Fig. 4.9.7 does not comply with the reasoning just expounded for the reaction ^{233}U(n,f). There must be a different cause for the distribution of asymmetry signs observed. It is readily seen that the pattern suggests a shift of the angular distribution of ternary particles as a whole to larger or smaller angles depending on the spin orientation $\boldsymbol{\sigma}_{+Z}$ or $\boldsymbol{\sigma}_{-Z}$, respectively. As illustrated to the right in Fig. 4.9.8, the asymmetries A will then switch sign for emission angles of ternary particles smaller or larger than the average emission angle $\Theta = 82°$. This feature is precisely what is observed in experiment to the right in Fig. 4.9.7. As will be argued below in a phenomenological model, and substantiated by trajectory calculations for the ternary particles in the Coulomb field of the fission fragments, the shifts are linked to the non-zero spin J of the compound nucleus. It should be recalled that, following capture of a thermal neutron by a fissile nucleus, the e-e nucleus at the barrier of the saddle-point has not enough energy to break nucleon pairs. All relevant transition states at the barrier are therefore collective in character. In particular this means that the projection K of total spin J on the fission axis will be the head of a band of collective rotational states (see Fig. II-17 in [VAN73]). The axis of collective rotation is thereby perpendicular to the fission axis. The ternary particles are hence ejected from a rotating composite system. After release of ternary particles and fragments at scission the system will no longer be described by a rigid rotation but nevertheless it is straightforward to show (see Chapter 4.9.3) that the observed shifts of angular distributions are a repercussion of the rotation at scission [GOE07]. This has led to the name "ROT effect".

Though for fission of ^{234}U* the change in scale and for ^{236}U* the shift of angular distributions of ternary particles is the dominant effect, it is to be expected that in general both the TRI and the ROT effect play a part in the angular distributions in nature. Both effects are taken to contribute independently to the asymmetry $A(\theta)$ of eq. (4.9.7) which, for the ROT effect, is a function of the angle θ between the momenta of the light fission fragment $\boldsymbol{p}_{LF}$ and the ternary particle $\boldsymbol{p}_{TP}$. For the sake of argument, in the following merely asymmetries for the reference constellation are discussed. Denominating by 2Δ the shift between the two spin-reversed angular distributions in Fig 4.9.8 (right panel), the asymmetry $A(\theta)$ for the ROT effect at the angle θ is found to be $(2\Delta)\cdot[Y'(\theta)/2Y(\theta)]$ where $Y(\theta)$ is the angular distribution of the ternary particles and $Y'(\theta)$ the slope. In the asymmetry for the TRI effect $A = DB$ the correlator was set to $B \equiv +1$. Therefore

$$A(\theta) = (2\Delta)\cdot(Y'(\theta)/2Y(\theta)) + D \qquad (4.9.8)$$

The angular distribution is found by averaging the distributions measured for the two neutron spin orientations taking into account that the efficiency is depending on the mosaic array of diodes. In eq. (4.9.8) it is assumed that corrections for geometry, non-perfect separation between light and heavy fragments, accidental coincidences and finite neutron polarization have been applied [GAG09].

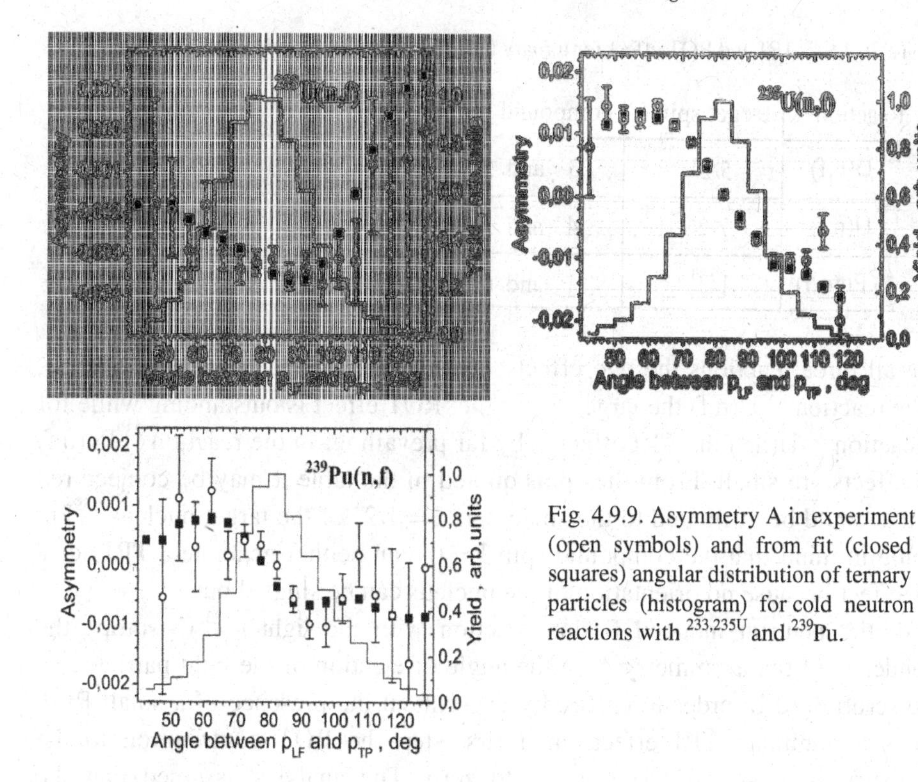

Fig. 4.9.9. Asymmetry A in experiment (open symbols) and from fit (closed squares) angular distribution of ternary particles (histogram) for cold neutron reactions with ^{233,235}U and ^{239}Pu.

Experimental results for the three reactions ^{233}U(n,f), ^{235}U(n,f) and ^{239}Pu(n,f) having been investigated so far are plotted in Fig. 4.9.9 [GAG10]. The measured asymmetry A is marked by open symbols, the fit according to Eq. (4.9.8) by closed squares and the angular distribution as histogram (scale to the right). The changes in scale by a factor of ~20 and ~10 for the asymmetries in fission of ^{234}U* and ^{240}Pu*, respectively, compared to ^{236}U* should be noted. From the figure it emerges that for the three reactions even the shapes of the asymmetry $A(\theta)$ as a function of angle between $\boldsymbol{p}_{LF}$ and $\boldsymbol{p}_{TP}$ are different. For ^{233}U(n,f) the asymmetry A stays throughout negative, for ^{235}U(n,f) the sizable asymmetry A changes sign across the angular distribution while for ^{239}Pu(n,f) the asymmetry also changes sign but is small. Evidently, the fit with the ansatz of eq. (4.9.8) with the correlation B being set to $B \equiv 1$ is perfect. This means that the T-odd correlator B of eq. (4.9.5) which is well established in the theory of neutron decay, is not found to play a role in ternary fission. The ROT shifts 2Δ and the TRI parameters D for the reactions studied are summarized in Table 4.9.2.

Table 4.9.2. TRI and ROT effect for ternary fission induced by polarized cold neutrons.

Reaction	target spin I	compound spin J	$D \cdot 10^3$	2Δ
^{233}U(n,f)	$5/2^+$	3^+ and 2^+	– 3.90 (12)	0.021 (4)°
^{235}U(n,f)	$7/2^-$	4^- and 3^-	+ 1.7(2)	0.215 (5)°
^{239}Pu(n,f)	$1/2^+$	1^+ and 0^+	– 0.23(9)	0.020 (3)°

In all three reactions the two effects TRI and ROT contribute. Nevertheless, for the reaction ^{235}U(n,f) the large size of the ROT effect is outstanding while for the reaction ^{233}U(n,f) the TRI effect is by far prevailing. In the reaction ^{239}Pu(n,f) both effects are small. From the spins quoted in the table it may be conjectured that this could be attributed to the small spin $I = 1/2^+$ of the target nucleus ^{239}Pu, keeping in mind that the compound spin $J = 0^+$ can neither generate a TRI nor a ROT effect because no orientation of the nucleus can be singled out.

For the neutron induced fission reaction with the lighter ^{233}U-isotope the dependence of the asymmetry A on the angle of ejection of the light particle had to be scrutinized in order to ensure by experiment the presence of a small ROT besides a dominant TRI effect. In a first step the ROT contribution to the asymmetry A in eq. (4.9.8) was set to zero. The analysis assumed that the physical TRI effect D is a constant. The TRI effect to be observed in experiment was then found by applying all corrections mentioned in connection with eq. (4.9.8). In particular for ejection angles near the border lines of the angular distribution (histogram in Fig. 4.9.10) the correction is sizable. In Fig. 4.9.10 (left panel) the resulting TRI effect was fitted to the experimental data. The fit is shown in the figure as full squares while the experimental data are displayed as open circles. The comparison is seen to be not very favourable. The mismatch points to the presence of a further effect. In a second step the fit data were subtracted from the experimental data. The difference should bring the ROT effect into the foreground. The difference data are on view in the right panel of Fig. 4.9.10 as open points. A fit with the angular dependence anticipated for the ROT effect reproduces well the difference data (full squares). The analysis hence demonstrates the presence of an albeit small ROT effect [GAG10].

In most experiments ternary particles were registered without identification of the isotope. Due to the dominant α-yield in ternary fission the effects observed are therefore mainly linked to the ejection of α-particles. A few experiments were conducted, however, where particle identification in the diode arrays of Fig. 4.9.7

was operational. In particular, for thermal neutron fission of ^{233}U the TRI effect was measured for α-particles to be $D = -3.92(12)\cdot 10^{-3}$, while for the H-isotopes (i.e. mostly tritons) the asymmetry is $D = -2.94(46)\cdot 10^{-3}$. Within error bars of 2σ the sizes of the asymmetries are identical. It is hence inferred that the TRI effect at most depends weakly on the type of the ternary particle being emitted.

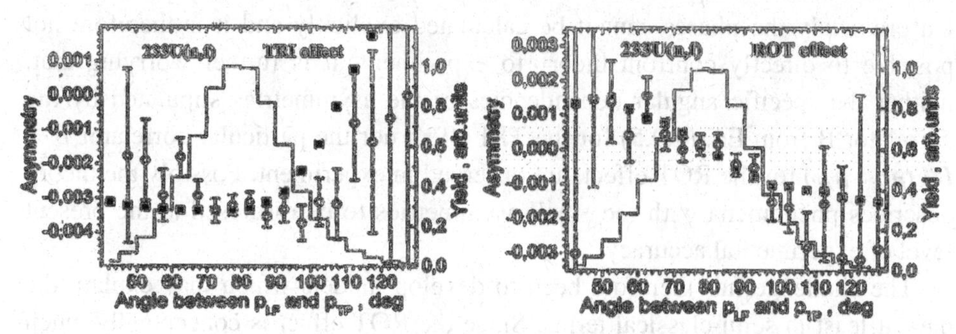

Fig. 4.9.10. Analysis of angular dependence of asymmetry A. See text.

For binary fission of ^{234}U* it was argued in Fig. 4.9.4 that both, Parity-Non-Conservation and the Left-Right asymmetry as a function of incoming neutron energy suggest the presence of a p-resonance at an energy of $E_n \approx 0{,}2$ eV. It was therefore interesting to investigate whether also the TRI effect is markedly changing near this p-resonance. A neutron beam with energy $E_n = 160$ meV is available at the ILL. The asymmetry observed was $D = -2.4(8)\cdot 10^{-3}$ and thus within error bars of 2σ compatible with the value given in Table 4.9.2 for this reaction at cold neutron energies. There is hence no indication for p-resonances to be relevant for the TRI effect. If interferences of resonances are to play a role [BUN08b] these must be s-resonances. It is noted in passing that with the interpretation for the ROT effect given in the above the effect is not contingent on the interference of neutron resonances. The ROT effect should become manifest even for an isolated single resonance.

4.9.3 *Models for the Tri and the Rot Effect*

For both the TRI and the ROT effect a theory has been developed. A strictly quantum-mechanical theory of the TRI effect was proposed soon after its discovery in experiment [BUN03]. The theory is based on the notion of a rotating nucleus where the importance of Coriolis forces in rotating systems is stressed. When it was realized that there are two different effects the theory was generalized to cover also the ROT effect [BUN08b]. While the correlator B of eq. (4.9.5) plays a pivotal role for the TRI effect, for the ROT effect it is the correlator $B` = B \cdot (\boldsymbol{p}_{LF} \cdot \boldsymbol{p}_{TP})$ which is figured out to be relevant. The effects are attributed to the interference between neutron resonances and both the magnitudes and the signs of the two effects depend on the differences of phases. Unfortunately the phases cannot be calculated explicitly and it is therefore not possible to directly confront theory to experiment. It is further worrying that, neither the specific angular dependencies of the asymmetries stipulated by the correlator B from Eq. (4.9.5) for the TRI effect nor the particular correlator $B` = B \cdot (\boldsymbol{p}_{LF} \cdot \boldsymbol{p}_{TP})$ for the ROT effect, are observed in experiment. Possibly the theory describes phenomena with too small asymmetries to be disclosed at the present level of experimental accuracy.

The challenge has therefore been to develop models which may explain the data at least in semi-classical terms. Since the ROT effect is conceptually much simpler than the TRI effect, it will be expedient to start by describing a model for the former effect. This is further advisable because, as will be seen, having a better understanding of the ROT effect is a good basis for a discussion of the TRI effect.

In Fig. 4.9.11 the decomposition of the angular momentum $\boldsymbol{J}$ of the fissioning nucleus into its components $\boldsymbol{K}$ and $\boldsymbol{R}$ on the fission axis andperpendicular to it is illustrated in a classical picture. The z-axis is chosen to coincide with the direction of the incoming neutron beam. Following capture of a polarized neutron with $\sigma_Z = \pm \frac{1}{2}\hbar$ the compound nucleus becomes itself polarized. For fission reactions near the barrier a special situation obtains since the K values of the rotational collective transition states act as filters. The K-numbers characterize in the combination (J,K) the transition states of the fissioning nucleus at the saddle point. They are usually assumed to stay constant down to scission. The polarization for given K is defined as $P(J,K) = \langle J_z(K) \rangle / J$ and derived in quantum mechanics to be [BUN08a].

$$P(J,K) = +\frac{p_n}{2J} \cdot \frac{J(J+1) - K^2}{J} \qquad for\ J = J_+ = I + \tfrac{1}{2} \qquad (4.9.8a)$$

$$P(J,K) = -\frac{p_n}{2J} \cdot \frac{J(J+1) - K^2}{(J+1)} \qquad for\ J = J_- = I - \tfrac{1}{2} \qquad (4.9.8b)$$

In eqs. (4.9.8) p_n is the neutron polarization and I is the spin of the target nucleus. The polarization gives in a semi-classical picture the collective rotation vector $\boldsymbol{R}$ in Fig. 4.9.11 a preferred orientation in space along or against the z-axis of the longitudinally polarized neutron beam. According to eqs. (4.9.8) the polarization is getting larger for smaller K. This is evident from Fig. 4.9.11 since for small K the rotation vector $\boldsymbol{R}$ will become large and hence also the polarization $P(J,K)$. However, in a rigorous quantum mechanical approach the above classical picture has to be corrected. The main point is that the orientation of the nucleus around the z-axis is according to the laws of quantum mechanics indeterminate. The tip of the angular momentum may be anywhere on a circle around the z-axis. While the projections M and K stay constant this is not true for the classical vector $\boldsymbol{R}$. A quantum mechanical operator for angular momentum squared R^2 with eigenvalue $R(R+1) = J(J+1) - K^2$ has to be introduced. A further purely quantum mechanical result is noteworthy: on average the angular momentum along the fission axis vanishes since states with +K and –K are degenerate in energy [BUN08a].

The conjecture behind the model for the ROT effect to be expounded in the following is that the shifts observed in the angular distributions of ternary particles have as their origin the collective rotations described semi-classically by R and in quantum mechanics by the expectation value $\langle J_z \rangle$. The arguments are illustrated in Fig. 4.9.12. Due to the rotational motion the fragments are ejected at scission with a tangential velocity. After scission the Coulomb force will drive apart fragments and ternary particles. There is no longer a rotational motion of the fragments. But due to the starting conditions at scission, after scission the fission axis joining the centres of the two fragments will continue to turn around. The Coulomb field exerted by the fragments on the ternary particles will likewise be turning around and give the ternary particle trajectories an additional twist. Since the Coulomb forces are getting rapidly smaller when the particles separate, the ternary particles will not follow exactly the turning of the fission axis. They will lag behind. As usual the light fragment trajectory is taken as reference and the angle θ of light particle emission is quoted as the angle between the particle and the light fission fragment trajectory. This angle becomes smaller (as shown in the figure) or larger, depending on the sense of rotation at scission. The sense of rotation of the fissioning nucleus is steered by both, the helicity $\sigma_z = \pm \frac{1}{2}$ of the neutron and the spin $J_{\pm} = I \pm \frac{1}{2}$ of the capture state. Attention should be given to this latter feature. For given neutron spin σ_z the spin J of the compound state, and hence the sense of rotation, is either parallel to neutron spin for J_+ or antiparallel for J_-. In Fig. 4.9.12 the z-axis of Fig. 4.9.11 points into the plane, the capture spin is assumed to be $J_+ = I + ½$ and the neutron spin is $\sigma_z = + ½$. On the other hand, for given capture spin J the sense of rotation is turned around when the neutron spin is flipped. Consequently, applying the spin flip technique with neutron spin orientation being reversed periodically, the angular distributions of ternary particles are wobbling back and forth. From measuring the sign of the asymmetry of count rates it can be concluded which capture spin $J_+ = I + ½$ or $J_- = I - ½$ is dominating the capture process leading to fission. Summarising it appears that collective rotations of a fissioning nucleus around an axis perpendicular to the fission axis can explain the shifts of angular distributions invoked for the interpretation of the ROT effect in Fig. 4.9.8.

The sizes of the shifts turn out to be very small in experiment. The displacements 2Δ of the angular distributions for neutron spin reversed situations barely exceed 0.2° for $^{236}U^*$ in Table 4.9.2 and for $^{234}U^*$ and $^{240}Pu^*$ are even an order of magnitude smaller. It is astounding that such small angles of nuclear rotation are reliably measured by the spin flip technique. The reason for these small sizes of angular shifts is understood by taking conservation of angular momentum into account. To simplify the arguments consider symmetric fission with two spherical fragments like in Fig. 4.9.12. To start with, up to scission the fission axis is rotating and the nucleus will carry orbital angular momentum. Shortly after scission only Coulomb forces between the separated fragments are active. They will neither create nor destroy any angular momentum. For

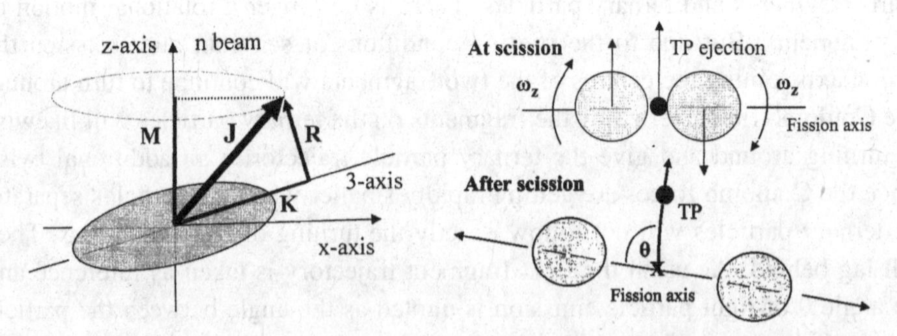

Fig. 4.9.11. Vector model for the de-composition of compound spin J into components **K** and **R** along and perpendicular to the fission axis 3.

Fig. 4.9.12. Model for ROT effect: up to scission the nucleus rotates; this starting Condition gives via Coulomb forces the ternary particle being ejected a twist.

symmetric fission the total angular momentum relative to the scission point will be $2[\boldsymbol{r} \times m\boldsymbol{v}]$ with $\boldsymbol{r}$ the distance of fragments from the scission point, m the mass and v the velocity of each fragment. For the emission geometry in Fig. 4.9.12 the ternary particle does not contribute any angular momentum. The radius vector r is increasing very fast and for constant angular momentum the velocity component $v_\perp$ perpendicular to r in the expression $[r \times mv_\perp]$ for the angular momentum is decreasing rapidly. The angular velocity $\omega = v_\perp/r$ is thus seen to come fast to a virtual stop. In the early stages the typical time constant for the increase in fragment distance r and the decrease of $v_\perp$ is 10^{-21} s. The short effective time constants entail small angular shifts.

However, for proving that the above phenomenological ROT model has a sound theoretical basis, trajectory calculations for the three outgoing particles following scission have been crucial. Even more important has been that the calculations are giving the key to identify the K-numbers of the transition states involved in the reactions. As discussed in Chapter 4.9.1 much effort has gone into the spectroscopy of K-states in fission of oriented nuclei by resonance neutrons. For this task ternary fission induced by polarized neutrons is a new tool.

All calculations of trajectories for fission fragments and ternary particles face the difficulty that the starting conditions at scission are not precisely known. Unfortunately, for the three-body problem at issue, it is not possible to assess unambiguously the initial positions and velocities of particles at start from a measurement of trajectory distributions at infinity. Therefore the input parameters have to be chosen to lie in a range thought to be reasonable with the important constraint, however, that energy and angular distributions calculated comply with experimental evidence. For the present purposes, calculations have to take additionally into account the rotation of the composite system at scission. These calculations were performed by I. Guseva [GUS06, GUS09]. The relevant angular velocity ω_Z at scission around the z-axis of the incoming neutron beam and its polarization is given by

$$\omega_Z(J,K) = \frac{< J_Z(J,K) >}{\Im_\perp} = \frac{J \cdot P(J,K)}{\Im_\perp} \tag{4.9.9}$$

where $P(J.K)$ is the polarization from eq. (4.9.8) and $\Im_\perp$ the moment of inertia around the z- axis perpendicular to the fission axis at the instant of scission. The moment of inertia has to be determined for each of the starting configurations chosen. For an assumed $<J_Z> = 1\hbar$ typical results for the turning angle of the fission axis (solid symbols) and the deflection angle of a ternary α-particle (open symbols) are shown in Fig. 4. 9.13. After a few zs (1 zs = 10^{-21}s) the fission axis and the trajectory of the ternary particle (TP) are no longer deflected. In about 2 zs the fission axis and the TP trajectory have reached 80% of their final orientation. The angular velocity of the deflection is strongly non-uniform. The angles of deflections are as anticipated very small, about 0.2° for the fission axis and 0.15° for the TP trajectory. The TP is indeed lagging behind and does not strictly follow the turning around of the fission axis. The angle of the lagging Δ = 0.05° is indicated in the figure. In experiment the shifts between the angular distributions of ternary particles are measured for opposite directions of neutron spin polarization but always relative to a fission axis which is fixed in space by the detector arrangement. The measured shift is therefore twice the lagging angle Δ. This is why the size of the ROT effect is given in Table 4.9.2 as 2Δ in degrees. It is proportional to the observed asymmetry A which according to Fig. 4.9.8 and Eq. (4.9.8) is further depending on the slope of the angular distributions.

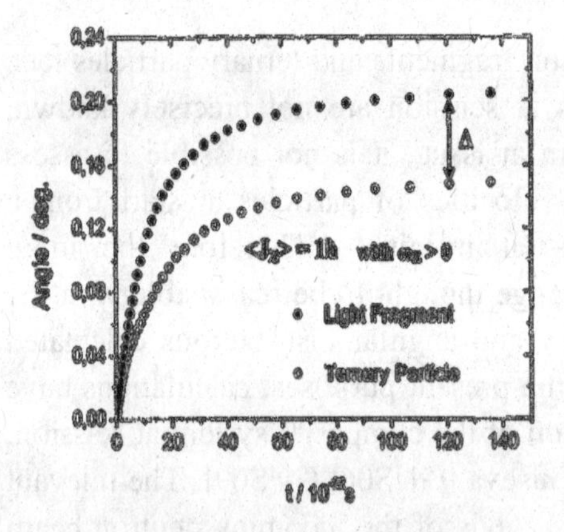

Fig. 4.9.13. Deflection of the fission axis (solid symbols) and the trajectory of ternary particle (open symbols).

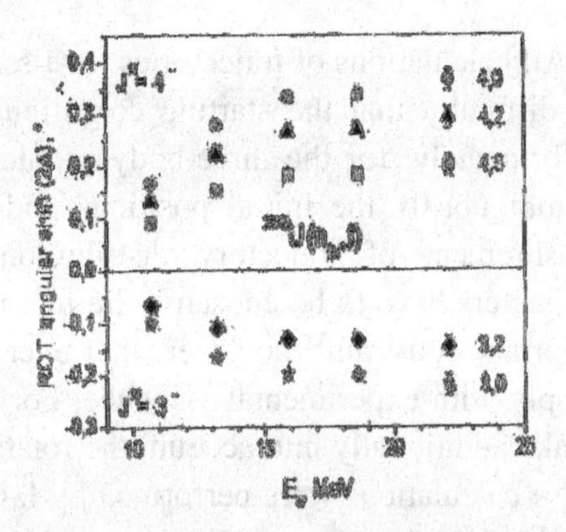

Fig. 4.9.14. ROT effect 2Δ for different (J,K) combinations as a function of ternary α-energy in ^{235}U(n,f).

Trajectory calculations were performed for the two compound spin states $J = I \pm ½$ and for K-values ranging from K = 0 to K = J. For the reaction ^{235}U(n,f) the ROT effect 2Δ calculated for the different (J,K) combinations is on display in Fig. 4.9.14 as a function of the energy of the ternary particle. Large K numbers at the saddle point of fission are thought improbable because the corresponding excitation energies are large. In experiment, however, they are present. The figure demonstrates that the sign of the ROT effect is uniquely correlated with the capture spin $J = I \pm ½$ while the size of the effect is linked to the K-number being effective. The smaller the K-value the larger is the ROT effect. As to be expected and borne out by calculation, the ROT angular shift 2Δ is a function of the kinetic energy of the ternary α-particles. For the identification of K-numbers the average over α-energies has to be evaluated.

Provided the ratio of spin separated fission cross sections $\sigma(J_+)$ and $\sigma(J_-)$ is known from other experiments, the ROT shift data can be exploited to asses K-values. These ratios were adopted from recent evaluations for ^{233}U(n,f) [MAS09], ^{235}U(n,f) [KOP99) and ^{239}Pu(n,f) [MAS09]. They are indicated in the headings of Table 4.9.3. The sizes and signs of the ROT effect were calculated for any combination of (J,K) values of the two spin states, however, under the restrictive assumption that for each of the capture spins J only one K-channel is effective. The assumption would be invalidated in case several K-channels for given spin J_+ or J_- were open with comparable weights. The assumption leads to the following equation for the ROT shift 2Δ.

$$2\Delta = 2\Delta(J_+, K_+)\frac{\sigma(J_+)}{\sigma(J_+)+\sigma(J_-)} + 2\Delta(J_-, K_-)\frac{\sigma(J_-)}{\sigma(J_+)+\sigma(J_-)}$$

Sizes 2Δ calculated for the ROT effect are summarized in Table 4.9.3 [GAG10]. From the table it turns out that for the best studied reaction, viz. ^{235}U(n,f), the assignment of (J,K) saddle point states is quite unambiguously pointing to the combination $(J_+,K_+) = (4,0)$ and $(J_-,K_-) = (3,2)$. For this combination not only the average experimental ROT shift is reproduced by trajectory calculations but also the variation of the shift as a function of both, the energy of the ternary particle and the mass split of the fission fragments. In Fig. 4.9.15 (left panel) experimental data are compared to results from trajectory calculations [GAG09a].The shift is seen in experiment to increase with the energy of the ternary particle and this trend is well accounted for in the trajectory calculations. Only at small energies near 10 MeV this increase is less pronounced in experiment than in the calculations. Very probably the reason is that at low energies there is an admixture of tritons which compared to alphas are less deflected when the Coulomb field of the fragments is turning and therefore the lagging angle Δ in Fig. 4.9.13 and the shift 2Δ become larger. As to the dependence on fragment mass split the near constancy of the ROT shift in experiment is confirmed by the calculations though the slopes as a function of mass split are slightly different. Anyhow, in the detailed analysis of Fig. 4.9.15

the fair agreement of experiment and trajectory calculations gives added confidence that the model proposed is outlining correctly the physics of the ROT effect.

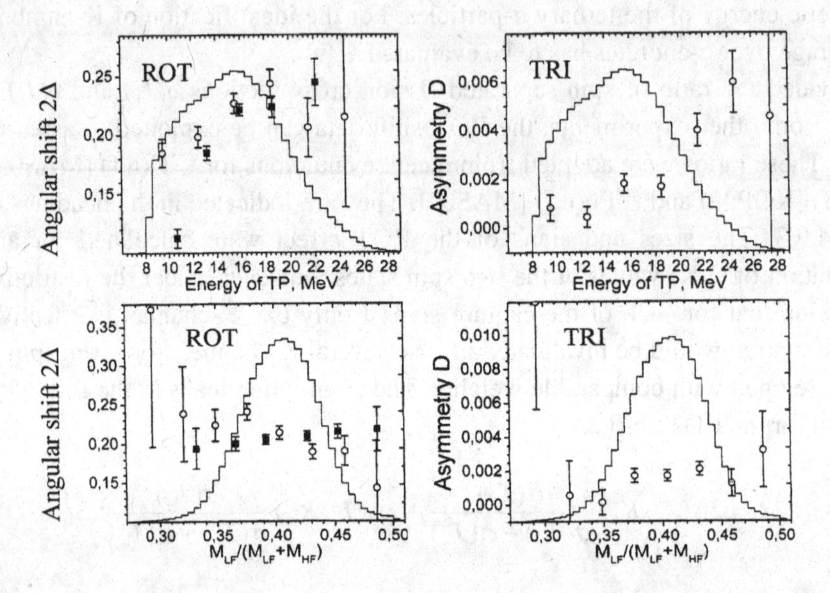

Fig. 4.9.15. Comparison of experimental (open symbols) with calculated ROT asymmetries (closed symbols) as a function of ternary particle energy and mass fraction $M_{LF}/(M_{LF}+M_{HF})$ (left panels) from the reaction ^{235}U(n,f). On the right panels TRI data are shown. The histograms represent energy distribution of ternaries and mass distributions of fragments.

For the reaction ^{235}U(n,f) the TRI effect D contributing to the measured asymmetry A in eq. (4.9.8) is on display as a function of ternary particle energy and fragment mass ratio in the right panels of Fig. 4.9.15, respectively. For small ternary particle energies below ~12 MeV the effect is nearly absent but shoots up at higher energies. For varying fragment masses the TRI effect seems to stay constant. Yet, in an ^{233}U(n,f) experiment with better mass resolution a slight dependence emerged [GAG04].

For the reaction ^{233}U(n,f) the situation in Table 4.9.3 is less clear cut. Partly this is due to the limited accuracy of experiments performed so far. There are two different (J,K) combinations which could be invoked. One combination is (J_+,K_+) = (3,3) in conjunction with (J_-,K_-) = (2,2) and the other is (J_+,K_+) = (3,2) together with (J_-,K_-) = (2,0). Arguments in favour of the latter choice will be given below. The small size of the ROT effect compared to the companion reaction ^{235}U(n,f)

Table 4.9.3. Calculated ROT shifts 2Δ for combinations of (J,K) values.

235U(n,f): target spin I = $7/2^-$

Experiment: **2Δ = 0,215(5)°**; $\sigma(4^-)/\sigma(3^-) = 1.76$

$4^- \backslash 3^-$	(3,0)	(3,1)	(3,2)	(3,3)
(4,0)	0,183	0,191	**0,215**	0,255
(4,1)	0,169	0,177	0,201	0,241
(4,2)	0,128	0,135	0,159	0,199
(4,3)	0,058	0,066	0,090	0,129
(4,4)	–0,040	–0,032	–0,008	0,032

233U(n,f): target spin I = $5/2^+$

Exp.: **2Δ = 0.021(4)°**; $\sigma(3^+)/\sigma(2^+) = 1.27$

$3^+ \backslash 2^+$	(2,0)	(2,1)	(2,2)
(3,0)	0,118	0,131	0,170
(3,1)	0,102	0,115	0,153
(3,2)	0,053	0,066	0,105
(3,3)	– 0,028	–0,015	**0,023**

239Pu(n,f): target spin I = ½$^+$

Exp.: **2Δ = 0.020(3)°**; σ(1+)/σ(0+) = 0.48

$1^+ \backslash 0^+$	(0,0)
(1,0)	0,057
(1,1)	**0,028**

is attributed to a near cancellation of the angular shifts induced by the spin states $J_+ = 3$ and $J_- = 2$ having opposite signs (see Fig. 4.9.14). By contrast, for the reaction ^{239}Pu(n,f) there is a unique solution. Since states with spin $J_- = 0$ are not to be polarized and can hence give rise neither to a TRI nor a ROT effect, the shift is entirely due to the spin state $J_+ = 1$. From the analysis of experiment based on trajectory calculations the channel $(J_+,K_+) = (1,1)$ is determined unambiguously to be relevant for the fission reaction with thermal neutrons. The small size of the ROT effect for this reaction is understood by comparing the angular velocities $\omega_Z(J,K)$ at scission in eq. (4.9.9) for the present channel (J,K) = (1,1) with the dominant channel (J,K) = (4,0) in the reaction ^{235}U(n,f). For ^{240}Pu* this angular velocity $\omega_Z(J,K)$ is by a factor of 5 smaller than for ^{236}U* implying a smaller ROT shift 2Δ.

In summary the ROT effect is shown to be directly linked to the collective orbital rotation of the fissioning system up to scission. It is rather accurately described in simple semi-classical terms. The effect is the first experimental evidence for the rotation of fission prone nuclei in thermal neutron induced fission.

By contrast, the TRI effect is more involved. Evidently, the spin dependent asymmetry of ternary particle yields in Fig. 4.9.8 has to be settled while the nucleus has not yet undergone scission. As proposed in [BUN03] the force, which in the rotating system could steer the distribution of yields, is the Coriolis force acting on the ternary particles moving in the composite nucleus. Phenomenological models based on this conjecture for the TRI effect are proposed. For thermal neutron fission of fissile nuclei studied here, the particularity has thereby to be kept in mind that the transition states are collective. Two cases have therefore to be distinguished. For axially symmetric nuclei there is no collective rotation around the symmetry axis and only $K = 0$ collective channels are allowed. For non-zero $K \neq 0$ the axial symmetry has to be broken. This leads to two types of models. They are presented in Fig. 4.9.16 [GOE09]. To the left of the figure the composite system for asymmetric fission close to scission is sketched. The nucleus is assumed to be axially symmetric, hence only the case $K = 0$ can be handled. The ternary particle TP is located in the neck between the two mains fragments and the nucleus is rotating around an axis perpendicular to the fission axis with angular velocity $\boldsymbol{\omega}$. It is therefore appropriate to introduce a rotating system of reference. Due to the stretching motion of the nucleus on its way to scission, the ternary particle will acquire a velocity $\boldsymbol{v}_{TP}$ provided fission is asymmetric. For a ternary particle TP with mass m and distance $\boldsymbol{r}$ from the centre of mass its equation of motion reads

$$m d\boldsymbol{v}_{TP}/dt = -\partial U/\partial r + 2\, m\boldsymbol{v}_{TP} \times \boldsymbol{\omega} + m\boldsymbol{\omega} \times [\boldsymbol{r} \times \boldsymbol{\omega}] + m\boldsymbol{r} \times d\omega/dt \quad (4.9.11)$$

with U a conservative potential. The different contributions in eq. (4.9.11) are conservative, Coriolis, centrifugal and catapult forces. Note that due to the stretching deformation the rotation is slowed down upon approaching scission ($d\omega/dt < 0$) and hence the catapult force $m\boldsymbol{r} \times d\omega/dt$ for non-uniform rotations appears. The Coriolis and the catapult forces acting on the ternary particle are indicated in the figure. When inverting the sense of rotation by flipping the spin of the neutron inducing fission also the two forces are inverted. In experiment this could indeed give rise to an asymmetry A in the emission yields like the one observed for the TRI effect. From experiment there is one clear example with $K = 0$, viz. the reaction $^{235}U(n,f)$ where a dominant contribution to the ROT effect is due to (J,K) = (4,0). The TRI effect D_{exp} = + $1.7 \cdot 10^{-3}$ in Table 4.9.2 is positive, i.e. in the reference constellation more ternaries are emitted upward (note that in Fig. 4.9.7 the result for the reaction ^{233}U(n,f) is plotted where the TRI effect D is negative). From the K = 0 model it is conjectured that the positive sign of D in the reaction ^{235}U(n,f) may be attributed to the catapult force. But it has to be stressed that the model can only hint to a possible mechanism for the TRI effect. Without a detailed knowledge of the dynamics of the stretching phase towards scission it is not evident that the Coriolis or the catapult force could bring about a sizable and measurable TRI effect.

For channels with non-zero K ≠ 0 the physics is more involved. As repeatedly pointed out a very special situation obtains in near-barrier fission of even-even nuclei as regards the character of the transition states at the saddle point. The collectivity of transition states requires for K ≠ 0 channels that the axial symmetry of the nucleus has to be broken, i.e. the nucleus has to be bent. [BOH75]. An example of axial symmetry breaking is known from the theory of rotational spectra for polyatomic linear molecules. Oscillations perpendicular to the axis of the molecule are considered. There are two such oscillations at right angles to each other being degenerate in energy. They may equally well be described by pure rotations of a bent molecule not possessing axial symmetry.

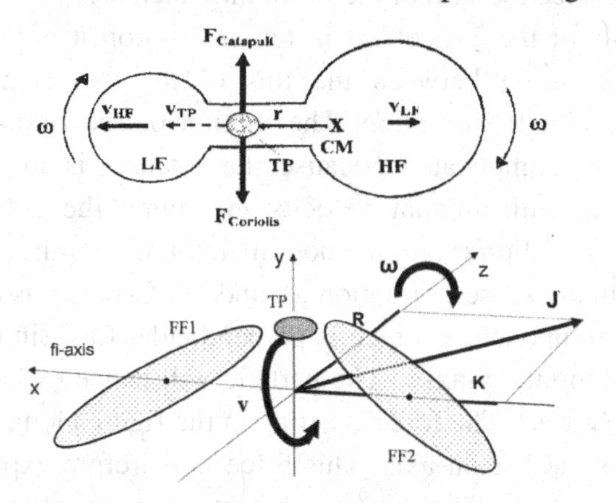

Fig. 4.9.16. TRI model for K = 0 (left panel) and for K ≠ 0 (right panel).

The two rotations with opposite sense of rotation replacing the two oscillations are degenerate in energy. It is interesting to note that the vibration-rotation motion carries angular momentum relative to the original symmetry axis of the molecule. This may be understood from the above equivalence of vibrations and rotations [LAN81]. In fission it is the strongly elongated scission configuration which resembles closely a linear molecule. Fission may indeed be a unique chance to observe bending vibration-rotations of nuclei. It can be shown that generally for a given quantum number N of excitation of the vibration–rotation motion there are (N+1) quantum states N, (N-2), (N-4), …, -N of the angular momentum around the fission axis. These (N+1) states, labelled K in nuclear physics, are degenerate in energy. On average the angular momentum <K> around the fission axis therefore vanishes. For a fissioning nucleus the vibration-rotation motion is sketched in the lower panel of Fig. 4.9.16. It should be noted in passing that bending vibrations of nuclei about to undergo scission have a long history. They have extensively been discussed in connection with the generation of fragment angular momentum [RAS69, ZIE74]. Fragment angular momenta due to bending vibrations are oriented perpendicular to the fission axis and in the $K \neq 0$ model of Fig. 4.9.16 (lower panel) it is worth pointing out that they are either parallel or anti-parallel to the collective rotation vector $\boldsymbol{R}$. The angular momenta of fragments should however not be confounded with the orbital angular momentum of the compound system at issue for the present purpose. Yet, spin-like and orbital-like orbital fragment momenta created at scission have to balance the spin of the fissioning nucleus.

For the analysis of the TRI effect in ternary fission it is presumed that the ternary particle is sitting between the tips of the two fragments and thus participates in the vibration-rotation. The motion of the fissioning nucleus is, however, still more complicated because the nucleus is in Fig. 4.9.16 also rotating as a whole with angular velocity $\boldsymbol{\omega}$ around the z-axis of collective rotation $\boldsymbol{R}$. The vibration-rotation motion imparts to the ternary particle independently from the sense of rotation around the fission axis (in the figure the x-axis) a velocity component $\mathbf{v}$ which is perpendicular to $\boldsymbol{R}$ (in the figure the up or down motion along the y-axis). The particle will hence experience a Coriolis force given by $2m[\boldsymbol{v} \times \boldsymbol{\omega}]$. The force is acting in the figure along the x-axis which in binary fission is the fission axis. This force is therefore expected to have an impact on neck rupture. In the double neck rupture model of ternary fission two ruptures are required to set free the ternary particle. Unfortunately, also in the present model only a full dynamic calculation could indicate where and in which time interval ruptures of the neck occur. From experiment it is observed that not only in binary but also in ternary fission the most stable constituent in the fragment mass distributions is the doubly magic ^{132}Sn isotope (see Fig. 4.8.20). For the sake of argument it is therefore ventured that a first rupture cuts away the heavy fragment. The strongly deformed light fragment with the remnants of the neck eventually becoming the ternary particle is flying in Fig. 4.9.16 to the left along the positive x-axis. As already argued in the K = 0 model of Fig. 4.9.16 (upper panel) the ternary particle will hence experience a Coriolis force. Taking once more the $^{235}U(n,f)$ reaction as an example, the (J,K) channel assignments from the evaluation of the ROT effect yield in Table 4.9.3 (J,K) = (4,0) and (J,K) = (3,2) as the main contributors. In the framework of the presently discussed TRI model the K = 0 states do not contribute to the effect since the velocity **v** perpendicular to the fission axis of the ternary particle is zero. The (J,K) = (4,0) channel is therefore not effective, only the channel (J_-,K_-) = (3,2) is active. For the spin channel $J_- = 3$ the polarization and the sense of rotation is opposite to neutron spin $\sigma_n = +1/2$ along the beam in +z-direction. In Fig. 4.9.16 the angular velocity is therefore pointing in the direction of –z. For an upward vibration-rotation motion oriented parallel to the +y-axis the Coriolis force will then point anti-parallel to the +x-axis. The Coriolis force will thus help to disrupt the ternary particle from the light fragment. It means that the ternary particle will be emitted with higher probability upward into the upper hemisphere than into the lower hemisphere of Fig. 4.9.7. The counting asymmetry A in the spin flip method from eq. (4.9.7) will hence be positive $A > 0$ in the upper hemisphere of diode arrays and negative $A < 0$ in the lower hemisphere. According to the sign convention for the TRI coefficient D, the positive asymmetry A for the reference constellation implies D to be positive. This is in perfect agreement with experiment finding $D_{exp} = +1.7 \cdot 10^{-3}$. It can similarly be argued that for the $^{233}U(n,f)$ and the $^{239}Pu(n,f)$ reactions, where according to the analysis of the ROT effect the channels $J_+ = I + 1/2$ are prevailing, the TRI effect is predicted to be negative. Again this conforms to the signs from experiment in Table 4.9.2. In Fig. 4.9.7 it is precisely this situation which is plotted for the reaction $^{233}U(n,f)$. In hindsight this agreement is taken as an indication that the sequence of neck ruptures ventured in the above is realistic. It is further recalled that in experiment the TRI asymmetry A is changing sign when the light fragment is not flying as in the reference constellation to the left but in opposite direction to the right. This change of sign is correctly predicted by the model without having to advocate for the TRI effect a special correlator like the correlator B in eq. (4.9.5).

Encouraged by these promising results the TRI model under discussion has been further explored with the idea that possibly the K-values assessed in the evaluation of the ROT effect could also be found in the evaluation of the TRI effect [GAG10]. To this purpose a phenomenological ansatz is made which is guided by the obviously well founded conjecture that the Coriolis interaction is the driving force for the TRI effect. In the formula for the Coriolis force the velocity of the ternary particle v and the angular velocity ω of the rotating system enter as factors (see eq. 4.9.11). In the model outlined the up-and-down velocity of the ternary particle is linked to the vibration-rotation motion of the nucleus. The effective size of this velocity is therefore set proportional to the angular momentum of the bent nucleus around the fission axis, i.e. the K-number of the transition states. Taking for the two spin states $J_\pm = I \pm ½$ only one K-channel each, the TRI coefficient D is therefore parameterized by

$$D = const.\cdot [K_+\omega_+ \frac{\sigma(J_+)}{\sigma(J_+)+\sigma(J_-)} + K_-\omega_- \frac{\sigma(J_-)}{\sigma(J_+)+\sigma(J_-)}] \qquad (4.9.12)$$

In eq. (4.9.12) $K_\pm$ and $\omega_\pm$ are the K-numbers and angular velocities ω from eq. (4.9.9) for the spin states $J_\pm = I \pm ½$. It is recalled that for the two spin states the angular velocities are $\omega_+ > 0$ and $\omega_- < 0$ (see eqs. (4.9.8) and (4.9.9)). The weights for the two spin states are proportional to their contributions $\sigma(J_+)$ and $\sigma(J_-)$ to the total fission cross section $\sigma_{fi}(tot) = \sigma(J_+) + \sigma(J_-)$, respectively. The constant in the equation is determined by imposing for the well-studied reaction ^{235}U(n,f) the channel $(J_-,K_-) = (3,2)$ from the analysis of the ROT effect in combination with the TRI coefficient D from Table 4.9.2. Results of the calculations are summarized in Table 4.9.4.

For the reaction ^{235}U(n,f) the combinations (J,K) = (4,0) and (3,2) as determined by the ROT effect are adopted for the TRI effect and therefore the agreement in Table 4.9.4 with this assignment is imposed by the fit. Nevertheless it is remarkable that away from this assignment the TRI coefficients rapidly drift away and are no longer compatible with experimental findings. For the companion reaction ^{233}U(n,f) it is not the combination of (J,K) channels (3,3)and (2,2) favoured by the analysis of the ROT effect in Table 4.9.3 which is found from inspecting the TRI effect. Instead, the combination (3,2) and (2,0) is highlighted by the TRI effect. For the ROT effect this is in Table 4. 9.3 the second-best choice. On the other hand, the (J,K) combination (3,3) and (2,2) put forward by the ROT effect seems to be ruled out by inspecting the TRI effect. This is corroborated by the argument that large K-values require too large excitation energies at the saddle point disfavouring this latter choice. For the time being the ambiguity cannot be resolved. Tentatively combining the results from the ROT and TRI analysis, the (J,K) assignment (3,2) and (2,0) should be the more probable choice for the ^{233}U(n,f) reaction.

By contrast, for thermal neutron fission of ^{239}Pu(n,f) there is no ambiguity. Both, the ROT and the TRI effect identify the channel quantum numbers at the saddle as (J,K) = (1,1) and not (1,0). For this reaction the asymmetry effects thus allow to uniquely assess the quantum numbers of the collective saddle-point states.

In summary, the semi-classical models for the newly discovered ROT and TRI effects appear to be consistent. They can give insight to subtle aspects of ternary fission induced by slow polarized neutrons. Both effects have as their common root the collective character of (J,K) channels at the cold saddle-point with, in particular, the consequence that the rotations of fissioning nuclei are collective. Collective rotations with $K \neq 0$ require the nucleus to violate the usually assumed axial symmetry. The nuclei have to be tri-axial or bent [BOH75]. The collective rotations of bent nuclei suggested are probably a unique feature of nuclei near scission not encountered elsewhere in nuclear physics.

Table 4.9.4. Calculated TRI asymmetries D for combinations of (J,K) values.

235U(n,f): target spin I = 7/2$^-$

Experiment: **D ·10^{+3} = + 1.7(2)**; $\sigma(4^-)/\sigma(3^-) = 1.76$

$4^- \backslash 3^-$	(3,0)	(3,1)	(3,2)	(3,3)
(4,0)	0	1,2	**1.7**	0.96
(4,1)	− 3.5	− 2.4	− 1.8	−2.6
(4,2)	− 6.0	− 4.8	− 4.3	− 5.0
(4,3)	− 6.1	− 5.0	− 4.4	− 5.2
(4,4)	− 3.0	− 1.8	− 1.3	− 2.0

233U(n,f): target spin I = 5/2$^+$

Exp.: **D·10^{+3} = -3.9(1)**; $\sigma(3^+)/\sigma(2^+) = 1.27$

$3^+ \backslash 2^+$	(2,0)	(2,1)	(2,2)
(3,0)	0	0.86	0.69
(3,1)	− 2.4	− 1.5	−1.7
(3,2)	**− 3.5**	− 2.6	− 2.8
(3,3)	− 2.0	− 1.1	− 1.3

^{239}Pu(n,f): target spin I = ½$^+$

Exp.: **D·10^{+3} = -0.23(9)**; $\sigma(1^+)/\sigma(0^+) = 0.48$

$1^+ \backslash 0^+$	(0,0)
(1,0)	0
(1,1)	**− 0.38**

The orbital rotations of the composite system established here for polarized nuclei have to be distinguished from orbital angular moments of fragments generated at scission. The latter momenta are postulated to compensate the mismatch found in experiment between the spins of complementary fragments. But in contrast to the orbital rotation of the polarized nucleus the angular momenta of fragments are in general not correlated with polarization.

Obviously orbital rotations of polarized nuclei should be observable not only in ternary fission accompanied by light charged particles. TRI and ROT effects have indeed been investigated in neutron emission from binary fission of fissile nuclei excited by capture of slow polarized neutrons [SOK04, DAN 06]. On the other hand, the ROT effect has been shown to be present in binary fission by studying angular shifts in the emission of gammas for polarized neutron induced fission (DAN09, GUS10]. ROT and TRI effects have thus become useful tools to investigate phenomena like scission neutrons or scission gammas in binary fission. Especially the search for scission neutrons emitted right at scission very similarly to charged ternary particles should be of interest for future research. The contribution of scission neutrons is one of the major unknowns in the calculation of fission neutron spectra for isotopes not accessible to experiment but of importance in the neutron balance of future power reactors.

4.10 Thermal Neutron Induced Fission Cross Sections

The excellent beam conditions at the curved neutron guides of the ILL are ideally suited for the investigation of small cross sections in fission induced by neutrons in the cold or thermal energy range. As already outlined in Chapter 1, at the end of thermal or cold neutron guides with lengths up to about 90 m intense and very clean neutron beams are available. Neutron fluxes at the end of the guides are typically 10^{10} n/s cm². The ratio of slow neutrons to epithermal or fast neutrons is typically 10^6 and the direct γ-ray flux from the reactor is reduced by the same factor of 10^6. Examples for cross sections obtained by the Geel group (C. Wagemans) are presented.

As known since the early day of fission research, it is the pairing energy of the incoming neutron which for actinide targets with an odd neutron number leads to excitation energies in the compound being sufficient to overcome the fission barrier. By contrast, for actinide targets with an even number of neutrons fission following thermal neutron capture proceeds as a sub-barrier process. The fission cross sections are hence small and their precise determination is a challenge.

The experimental difficulties may be demonstrated by an experiment measuring the thermal neutron cross section for fission of the ^{243}Am isotope. This isotope is produced in sizable amounts in reactors and constitutes an important fraction of the radioactive inventory of burnt fuel elements. The fission cross section is therefore of interest for the incineration process. For ^{243}Am the neutron number N = 148 is even and the cross section is anticipated to be small. Since ^{243}Am decays by α-emission to ^{239}Np and by β-decay further to ^{239}Pu, the Pu with its large fission cross section has to be separated before target preparation. The sample material was highly enriched and contained 99.99 % of ^{243}Am and 0.01 % of ^{241}Am. In spite of the small fraction of ^{241}Am and its small fission cross section a difficulty arises due to (n,γ) capture producing the isotope ^{242}Am in equilibrium with ^{241}Am. The fission cross section for the odd-n isotope ^{242}Am is very large and amounts to 6950(280) b. The ^{242}Am(n,f) process contributes 8.4(8) mb to the fission count rate measured for the above target composition.. The corrected fission cross section for the reaction ^{243}Am(n,f) is reported to be σ_{fi} = 74(4) mb at the thermal neutron energy E_n = 25.3 meV [WAG89].

Among the waste of minor actinides from nuclear power plants there are also several Cm isotopes. For transmutation scenarios the knowledge of reliable (n,f) fission cross sections is therefore required. Thermal neutron induced fission cross sections were determined for the isotopes $^{243,\,246,\,248}$Cm. The element curium has the charge number Z = 96. Hence ^{243}Cm should be fissile while ^{246}Cm and ^{248}Cm undergo sub-barrier fission. The two e-e Cm isotopes decay spontaneously by α-emission and fission. The spontaneous fission decay has to be assessed in parallel to fission induced in the neutron beam. This is all the more important in view of fission cross sections expected to be small. Measurements were performed at a cold neutron beam with average energy $\langle E_n \rangle$ = 5.38 meV. Normalized to the standard thermal neutron energy of E_n = 25.3 meV, the (n,f) cross sections found are 572(25) b for ^{243}Cm, 12(25) mb for ^{246}Cm and 316(43) mb for ^{248}Cm. While the results agree for ^{243}Cm and ^{248}Cm with former work, the cross section for ^{246}Cm is much lower than reported previously [SER05].

An interesting series of measurements was performed for several uranium isotopes. With B_n the neutron binding energy and E_f the height of the fission barrier, it is the sign of the difference ($B_n - E_f$) which determines whether fission proceeds above or under the barrier. For the e-e isotopes this difference decreases for increasing neutron number. Remarkably, for the lighter isotopes ^{230}U and ^{232}U the difference is still positive and turns negative only from ^{234}U onward to heavier isotopes. Therefore only for ^{234}U and heavier (e,e)-isotopes sub-barrier fission obtains with the characteristic small cross sections. Their measurement requires highly enriched samples in order to keep the contribution to fission from the omnipresent odd isotopes ^{233}U and ^{235}U as low as possible. With a sample material enriched to 99.868 % of ^{234}U the thermal neutron cross section for fission could for the first time be determined to be 300(20) mb [WAG98]. Even more demanding as to the purity of the target was the measurement of the fission cross section for the ^{238}U isotope. To further reduce the content of ^{235}U in an uranium batch which already contained only 12 ppm of this isotope, a mass-separator was used which decreased the ^{235}U by a factor of 10^3 down to 0.012 ppm. With this target a fission cross section at thermal neutron energies for ^{238}U(n,f) of 11(2) μb was determined. The cross section for ^{238}U(n,f) at thermal neutron energies is hence by more than 7 orders of magnitude smaller than the cross section for the standard ^{235}U(n,f) reaction. The comparison of the cross sections for the two sub-barrier reactions ^{234}U(n,f) and ^{238}U(n,f) brings to evidence that for ^{238}U the reaction is deeper sub-barrier than for ^{234}U. In terms of the energy difference ($B_n - E_f$), for ^{238}U the difference must be stronger negative than for ^{234}U [DHO84].

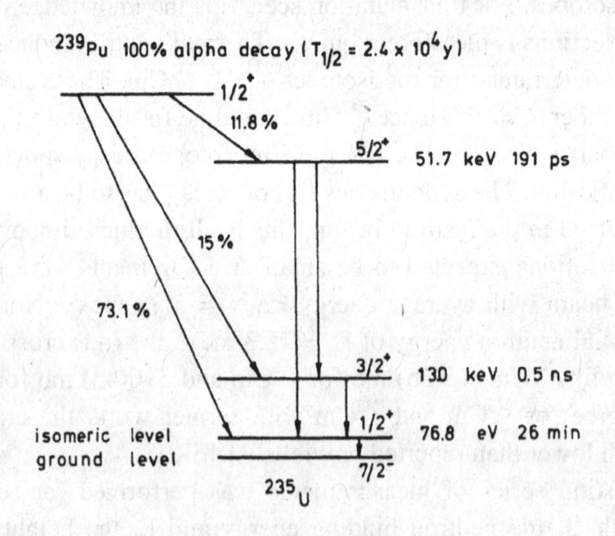

Fig. 4.10.1: Decay scheme of ^{239}Pu

An exceptional cross section measurement was performed for the isomeric state of ^{235}U. The ^{235}U^m state with spin $I^\pi = 1/2^+$ has a half-life of $T_{1/2} = 26$ min. The ^{235}U^m target for cross section measurements has therefore to be produced in the vicinity of the neutron beam position. The isomer is formed in the α-decay of ^{239}Pu. As shown in Fig. 4.10.1 all decays lead in times shorter than 1 ns to the isomeric state $^{235}U^m$ at an excitation energy of 76.8 keV. It decays to the ground state of $^{235}U^g$ by an E3 transition. As plotted in the appliance of Fig. 4.10.2, to collect the isomer $^{235}U^m$ the inner surface of a sphere with diameter 40 cm was covered with a thin layer of ^{239}Pu. Following α-decay the recoiling isomers were caught on a target backing positioned at the centre of the sphere. Due to the short lifetime of the isomer a series of experiments with fresh targets had to be foreseen. Cross section measurements were performed for two neutron energies of ≈ 5 meV and ≈ 56 meV. In the evaluation both the γ-decay of $^{235}U^m$ and the growing in of $^{235}U^g$ contributing fission events had to be accounted for. At the above energies the ratios of cross sections for isomeric to ground state were found to be 1.61(44) and 2.47(45), respectively. The isomeric state has hence a slightly larger fission cross section than the ground state [DEE88].

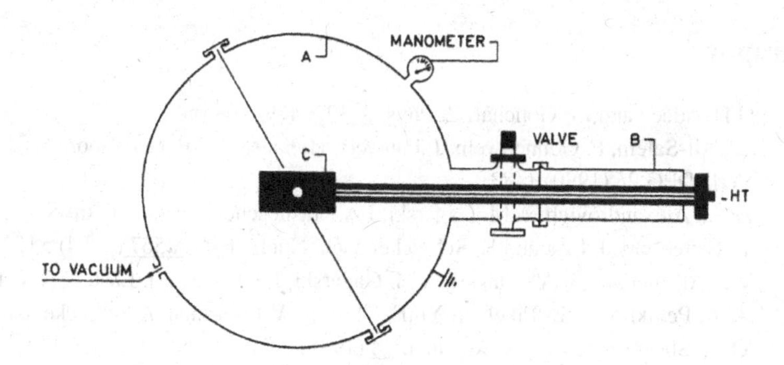

Fig. 4.10.2. Appliance for the production of $^{235}U^m$.

Final remark

In the present survey of results from 35 years of fission research at the high flux neutron source of the Institut Laue Langevin, it should have become apparent that the properties of fragments from fission reactions induced by thermal neutrons have been scrutinised with in many cases unprecedented accuracy and detail. Several new fission phenomena were disclosed. It has, however, to be underlined that many outstanding properties of fragments from low energy fission have at the ILL so far only been studied occasionally or not addressed at all. There are still many topics in fission where the facilities of the ILL could make essential contributions. An evidently still open task for the LOHENGRIN spectrometer is the measurement of mass distributions in the heavy fragment group though, as explained in Chapter 4.2 , nuclear charges in this group are for the time being not accessible. A further very challenging topic is the systematic study of angular momentum carried by the fragments which could give insight into the scission process for a broad range of masses and fission modes. More attention should also be given to the de-excitation of fragments after scission. The intense and clean beams at the exit of neutron guides could be exploited in studies of neutron and gamma emission from fragments as a function of other fragment properties like mass and energy. Finally it should not be forgotten that at excitation energies of the fissioning compound slightly beyond those attainable with thermal neutrons many phenomena like isomeric fission or angular distributions of fragments are carrying crucial information on the fission process. There is obviously still interesting work at low excitation energies ahead.

Bibliography

[ADE85] G.D. Adeev and I.I. Gonchar, Z. Phys. A 322, 479 (1985)S

[AIT90] M. Ait-Salem, F. Gönnenwein, J .Pauwels, H. Faust, and P. Geltenbort, Verh. DPG 25 (1990) 1483

[ALE94] A.Ya. Alexandrovich, A.M. Gagarski, I.A.Krasnoschekova, G.A Petrov, P. Geltenbort, J. Last and K. Schreckenbach, Nucl. Phys. A 567 (1994) 541

[ALF95] V.P. Alfimenkov, G.V. Valskii, A.M. Gagarski, I.S. Guseva, J. Last, G.A. Petrov, A. K. Petukhov, L.B. Pikelner, Yu.E. Sokolov, V.I. Furman, K Schreckenbach and O.A. Shcherbakov, Phys. At. Nucl. 58 (1995) 737

[ALK89] I.D. Alkhazov, B.M Alexandrov, A.V. Kuznetsov, S.I. Lashaev and V.I. Shpakov, Int. Conf. "Fiftieth Anniversary of Nuclear Fission", Leningrad, USSR, 1989 p 229

[AMI73] S. Amiel and H. Feldstein, IAEA Conf. "Phys. and Chem. of Fission" 1973, Vol. II, p 65

[AND69] V.N. Andreeev, V.G. Nedopekin, V.L. Rogov, Sov. J. Nucl. Phys. 8 (1969) 22

[ARM76a] P. Armbruster, K.Sistemich, J.P. Bocquet, Ch. Chauvin and Y. Glaize, Nucl. Instr. Meth. 132 (1976) 129

[ARM76b] P. Armbruster, M. Asghar, J.P. Bocquet, R. Decker, H. Ewald, J. Greif, E. Moll, B. Pfeiffer, H. Schrader, F. Schussler, G. Siegert and H. Wollnik, Nucl. Instr. Meth. 139 (1976) 213

[ARM81] P. Armbruster, U. Quade, K. Rudoph, H.-G. Clerc, M. Mutterer, J. Pannicke, C. Schmitt, J.P. Theobald, W. Engelhardt, F. Gönnenwein and H.. Schrader, CERN 81-09, Vol. II, p 675

[ARM87] P. Armbruster, M. Bernas, J.P. Bocquet, R. Brissot, H.R. Faust and P. Roussel, Europhys. Lett. 4, 793 (1987)

[ARM99] P. Armbruster, Rep. Progr. Phys. 62 (1999) 465

[ASG81] M. Asghar, F. Caitucoli, B. Leroux, P. Perrin and G. Barreau, Nucl. Phys. A 368 (1981) 328

[ASG84] M. Asghar and R.W. Hasse, J. Physique C6, 455 (1984)

[ADE85] G.D. Adeev and I.I. Gonchar, Z. Phys. A 322, 479 (1985)S

[AIT90] M. Ait-Salem, F. Gönnenwein, J .Pauwels, H. Faust, and P. Geltenbort, Verh. DPG 25 (1990) 1483

[ALE94] A.Ya. Alexandrovich, A.M. Gagarski, I.A.Krasnoschekova, G.A Petrov, P. Geltenbort, J. Last and K. Schreckenbach, Nucl. Phys. A 567 (1994) 541

[ALF95] V.P. Alfimenkov, G.V. Valskii, A.M. Gagarski, I.S. Guseva, J. Last, G.A. Petrov, A. K. Petukhov, L.B. Pikelner, Yu.E. Sokolov, V.I. Furman, K Schreckenbach and O.A. Shcherbakov, Phys. At. Nucl. 58 (1995) 737

[ALK89] I.D. Alkhazov, B.M Alexandrov, A.V. Kuznetsov, S.I. Lashaev and V.I. Shpakov, Int. Conf. "Fiftieth Anniversary of Nuclear Fission", Leningrad, USSR, 1989 p 229

[AMI73] S. Amiel and H. Feldstein, IAEA Conf. "Phys. and Chem. of Fission" 1973, Vol. II, p 65

[AND69] V.N. Andreeev, V.G. Nedopekin, V.L. Rogov, Sov. J. Nucl. Phys. 8 (1969) 22

[ARM76a] P. Armbruster, K.Sistemich, J.P. Bocquet, Ch. Chauvin and Y. Glaize, Nucl. Instr. Meth. 132 (1976) 129

[ARM76b] P. Armbruster, M. Asghar, J.P. Bocquet, R. Decker, H. Ewald, J. Greif, E. Moll, B. Pfeiffer, H. Schrader, F. Schussler, G. Siegert and H. Wollnik, Nucl. Instr. Meth. 139 (1976) 213

[ARM81] P. Armbruster, U. Quade, K. Rudoph, H.-G. Clerc, M. Mutterer, J. Pannicke, C. Schmitt, J.P. Theobald, W. Engelhardt, F. Gönnenwein and H.. Schrader, CERN 81-09, Vol. II, p 675

[ARM87] P. Armbruster, M. Bernas, J.P. Bocquet, R. Brissot, H.R. Faust and P. Roussel, Europhys. Lett. 4, 793 (1987)

[ARM99] P. Armbruster, Rep. Progr. Phys. 62 (1999) 465

[ASG81] M. Asghar, F. Caitucoli, B. Leroux, P. Perrin and G. Barreau, Nucl. Phys. A 368 (1981) 328

[ASG84] M. Asghar and R.W. Hasse, J. Physique C6, 455 (1984)

[ASG93] M. Ashar, N. Boucheneb, G. Medkour, P. Geltenbort and B. Leroux, Nucl. Phys. A 560 (1993) 677

[AUC71 V. Avrigeanu, A. Florescu, A. Sandulescu and W. Greiner, Phys. Rev. C 52 (1995) R1755

[BAR04] L. Barabanov, W.J. Furman, A.B. Popov, I.S. Guseva and G.A. Petrov, "Interactions of Neutrons with Nuclei XI", Dubna 2004, p 304

[BAR97] A.L. Barabano and W.J. Furman, Z. Phys. A 357, 411 (1997)

[BAU92) W. Baum, PHD thesis, TU Darmstadt, Germany, 1992

[BAU92b] W. Baum, S. Neumaier, U. Nast-Linke, M. Mutterer, J.P. Theobald, C. Ziller, A. Göpfert, G. Barreau, T.P. Doan, B. Leroux, A. Sicre and H. Faust, Inst. Phys. Conf. Ser. 132, 477 (1992)

[BEL83] D. Belhafaf, J-P- Bocquet, R. Brissot, Ch. Ristori, J. Crancon, H. Nifenecker, J. Mougey and V.S. Ramamurthy, Z. Phys. A 309, 253 (1983),

[BEL91] A.V. Belozerov, A.G. Beda, S.I. Burov, G.V. Danilyan, A.N. Martemyanov, V.S. Pavlov, V.A. Shchenev, L.H. Bondarenko, Yu. Mostovoi, P. Geltenbort, J. Last, K. Schreckenbach and F. Gönnenwein, JETP Lett. 54 (1991) 132

[BER79] M. Beranger, A. Gobbi, F. Hanappe, U. Lynen, C. Ngo, A. Olmi, H. Sann, H. Stelzer, H. Richel and M.F. Rivet, Z. Phys. A 291, 133 (1979)

[BER84] J.F. Berger, M. Girod and D. Gogny, Nucl. Phys. A 428 (1984) 23c

[BER89] J.F. Berger, M. Girod and D. Gogny, Nucl. Phys. A 502 (1989) 85c

[BER91] M. Bernas, P. Armbruster, S. Czajkowski, H. Faust, J.P. Bocquet and R. Brissot, Phys. Rev. Lett. 67, 3661 (1991)

[BIR07] E. Birgersson, S. Oberstedt, A. Oberstedt, F.-J. Hambsch, D. Rochman, I. Tsekhanovich and S. Raman, Nucl. Phys. A 791 (2007) 1

[BLE91] R. Blendowske, T. Fliessbach and H. Walliser, Z. Phys. A339, 121 (1991)

[BOC88] J.P. Bocquet, R. Brissot and H.R. Faust, Nucl. Instr. Meth. A 267 (1988)466

[BOC89] J.P. Bocquet and R. Brissot, Nucl. Phys. A 502 (1989) 213c

[BOC90] J.P. Bocquet, R. Brisssot, H.R. Faust, M. Fowler, J. Wilhelmy, M. Asghar and M. Djebara, Z. Phys. A 335, 41 (1990)

[BOH39] N. Bohr and J.A. Wheeler, Phys. Rev. 56, 426 (1939)

[BOH56] A. Bohr, "Peaceful Uses of At. Energy" Geneva 1955, Vol. 2, p 151

[BOH75] A. Bohr and B.R. Mottelson, "Nuclear Structure", W.A. Benjamin, 1969

[BOL72] M.Bolsterli, E.O. Fiset, J.R. Nix and J.L. Norton, Phys. Rev. C 5, 1050 (1972)

[BON86] A. Bonasera, Phys. Rev. C 34, 740 (1986)

[BOU89] N. Boucheneb, P. Geltenbort, M. Asghar, G. Barreau, T.P. Doan, F. Gönnenwwein, B. Leroux, A. Oed and A. Sicre, Nucl. Phys. A 502 (1989) 261c

[BOU91] N. Boucheneb, M. Asghar, G. Barreau, T.P. Doan, B. Leroux, A. Sicre, P. Geltenbort and A. Oed, Nucl. Phys. A 535 (1991) 77

[BRO78] U. Brosa and H.J. Krappe, Z. Phys. A 284, 65 (1978)

[BRO90] U. Brosa, S. Grossmann and A. Müller, Phys. Rep. 197 (1990) 167

[BRO99] U. Brosa and H.-H. Knitter, Phys. Rev. C 59, 767 (1999)

[BRU50] D.C. Brunton and G.C. ;Hanna, Can. J. Research 28 (1950) 190

[BUD88] C. Budtz-Jorgensen and H.-H. Knitter, Nucl. Phys. A 490 (1988) 307

[BUN83] V.E. Bunakov and V.P. Gudkov, Nucl. Phys. A 401 (1983)93

[BUN85] V.E. Bunakov and V.P. Gudkov, Z. Phys. A321, 271 (1985)

[BUN02] V.E. Bunakov and F. Goennenwein, Phys. At. Nucl. 65, 2036 (2002)

[BUN03) V.E. Bunakov and S.G. Kadmensky, Phys. At. Nucl. 66, 1846 (2003)

[BUN08a] V.E. Bunakov and S.G. Kadmensky, "Interactions of Neutrons with Nuclei XVI", Dubna 2008, p 325

[BUN08b] V.E. Bunakov and S.G. Kadmensky, Phys. At. Nucl. 71, 1887 (2008)

[CAR80] N .Carjan and B. Leroux, Phys. Rev. 22, 2008 (1980)

[CAR07] N. Carjan, P. Talou and O. Serot, Nucl. Phys. A 792 (2007) 102

[CHE72] E. Cheifetz, B.Eylon, E. Fraenkel and A. Gavron, Phys. Rev. Lett. 29, 805 (1972)

[CLE75a] H.-G- Clerc, K.-H. Schmidt, H. Wohlfarth, W. Lang, H. Schrader, K.E.

Pferdekämper,
R. Jungmann, M. Asghar, J.-P. Bocquet and G. Siegert,
Nucl. Instr. Meth. 124 (1975) 607

[CLE75b] H.-G. Clerc, W. Lang, H. Wohlfarth, K.-H. Schmidt, H. Schrader, K.E. Pferdekämper and R. Jungmann, Z. Phys. A 274, 203 (1975)

[CLE86] H.-G-Clerc, w. Lang, M. Mutterer, C. Schmitt, J-P. Theobald,U.Quade, K.Rudolph, P. Armbruster, F. Gönnenwein, H.Schrader and D. Engelhardt, Nucl. Phys. A 452 (1986) 277

[CLE99] H.-G. Clerc, "Heavy Elements and Related Phenomena", World Scientific Vol. I, 451(1999)

[CRÖ98] M. Cröni, A. Möller, A. Kötzle, F. Gönnenwein, A. Gagarski and G. Petrov, "Fission and properties of neutron-rich nuclei", World Scientific, 1997, 109

[DAB65] J.W.T. Dabbs, F.J. Walter and G.W. Parker, IAEA Conf. "Phys. and Chem. of Fission" 1965, Vol. I, 39

[DAN77] G.V. Danilyan, B.D. Vodennikov, V.P. Dronyaev, V.V. Novitskii, V.S. Pavlov and S.P. Borovlev, JETP Lett. 26, 197 (1977)

[DAN06] G.V. Danilyan, A.V. Fedoroc, V.A. Krakhotin, V.S. Pavlov, E.V. Brakhman, I.L. Karpikhin, E.I. Korobkina. R. Golub, T. Wilper and M.V. Russina, Phys. At. Nucl. 69, 1158 (2006)

[DAN09] G.V. Danilyan, Ṕ.Granz, V.A. Krakhotin, F. Mezei, V.V.Novitskii, V.S. Pavlov, M. Russina, P.B. Shatlov and T. Wilpert, Phys. Lett. B679 (2009) 25

[DAV98] M. Davi, H.O. Denschlag, H.R. Faust, F. Gönnenwein, S. Oberstedt, I. Tsekhanovich and M. Wöstheinrich, AIP Conf. Proc. 447, 1998, 239

[DEE88] A. D`Eer, C. Wagemans, Nève de Mèvergnies, F. Gönnenwein, P. Geltenbort, M.S. Moore and J. Pauwels, Phys. Rev. C 38, 1270 (1988)

[DEM97] L. Demattè, C. Wagemans, R. Barthélémy, P. D`Hondt and A. Deruytter, Nucl. Phys. A 617 (1997) 331

[DEN00] H.O. Denschlag, I. Tsekhanovich, M. Davi, F. Gönnenwein, M. Wöstheinrich, H. R. Faust and S. Oberstedt, "Sem. On Fission" Pont d`Oye IV, World Scienttific, 2000, 5

[DHO84] P. D`Hondt, C. Wagemans, A. Emsallem and R. Brissot, Ann. Nucl. Energy 11, 485 (1984)

[DIT91] W. Ditz, U. Güttler, R. Hentzschel, P. Stumpf, H.O. Denschlag and H.R. Faust, "Sem. on Fission" Pont d´Oye II, C. Wagemans ed., 1991, p 114

[DJE89] M. Djebara, M. Asghar, J.P. Bocquet, R.Brissot, J, Cran□on, Ch. Ristori, E. Aker, D. Engelhardt, J. Gindler, B.D. Wilkins, U. Quade and K. Rudolph, Nucl. Phys. A 496 (1989) 346

[DUB08] N. Dubray, H. Goutte and J.-P. Delaroche, Phys. Rev. C 77, 014310 (2008)

[ENG95] Ch. Engelmann, F. Ameil, P. Armbruster, M. Bernas, S. Czajkowski, Ph. Dessagne, C. Donzaud, H. Geissel, A. Heinz, Z. Janas, C. Kozhuharov, Ch. Miehè, G. Münzenberger, M. Pfützner, C. Röhl, W. Schwab, C. Stéphan, K. Sümmerer, L. Tassan-Got and B. Voss, Z Phys. A 352, 151 (1995)

(EWA64] H. Ewald, E. Konecny, H. Opower and H. Rösler, Z. Naturforschg. 19 a,194 (1964)

[DUR94] I. Düring, U. Jahnke and H. Märten,

"Nucl. Fiss. and Fiss.-Prod. Spectroscopy" ILL report 94 FAO5T, 1994, p 202
[FAI62] H. Faissner and K. Wildermuth, Phys. Lett. 2, 212 (1962)
[FAR47] G. Farwell, E. Segré and C. Wiegand, Phys. Rev. 71, 327(1947)
[FAU81] H.R. Faust, P. Geltenbort, F. Gönnenwein and A. Oed, ILL report 81 FA45S
[FAU95] H. Faust and Z. Bao, Sem. Fission Pont d`Oye III, C. Wagemans ed., EUR 16295 EN, 1995, 220
[FAU04] H.R. Faust and Z. Bao, Nucl. Phys. A 736 (2004) 55
[FIO93] G. Fioni, H.R. Faust, M. Gross, M. Hesse, P. Armbruster , F. Gönnenwein and G. Münzenberger, Nucl. Instr. Meth. A 332 (1993) 175
[FIO94] G. Fioni and H.R. Faust, "Dynamical Aspects of Nucl. Fission", Dubna 1994, p 147
[FLA80] V.V. Flambaum and O.P. Sushkov, Phys. Lett. 94 B, 277 (1980)
[FLE40] G.N. Flerov and K.A. Petrzak, J. of Physics (USSR) III, 275 (1940)
[FLO93] A. Florescu, A. Sandulescu, C. Cioaca and W. Greiner, J. Phys. G 19 (1993) 669
[FOM97] A.S. Fomichev, I. David, M.P. Ivanov and Yu.G. Sobolev, Nucl. Instr. Meth. A 384 (1977) 519
[FRI98] T. Friedrichs, H. Faust, G. Fioni, M. Gross, U. Köster, F. Münnich and S. Oberstedt, AIP Conf. Proc. 447, 231 (1998)
[FUL62] R.W. Fuller, Phys. Rev. 126, 684 (1962)
[FUR10] W. Furman, "Sem. on Fission" Gent, Belgium, World Scientific 2010, p 53
[GAG98] A.M. Gagarski, I.A. Krasnoschokova, G.A. Petrov, V.I. Petrova, Yu.S. Pleva, V.E. Sokolov, F. Gönnenwein, A. Kötzle, P. Jesinger, K. Schmidt, G.V. Danilyan, V.S. Pavlov, V.B. Chvatchkin, M. Mutterer, S.R. Neumaier, P. Geltenbort, V. Nesvizhevsky, O. Zimmer, S.M. Soloviev and V.Ya. Vasiliev, "Interactions of Neutrons with Nuclei VI", Dubna 1998, p 279
[GAG03] A. Gagarski, private communication
[GAG04] A. Gagarsski, G. Petrov, T. Zavarukhina, F. Gönnenwein, P. Jesinger, M. Mutterer, J. von Kalben, W.H. Trzaska, S. Khlebnikov, G. Tyurin, T. Kuzmina, S. Soloviev, V. Nesvizhevsky, A. Petukhov and E. Lelievre-Berna, "Interactions of Neutrons with Nuclei XII", Dubna, 2004, p 255
[GAG06] A.M. Gagarski, I.S. Guseva, G.A. Petrov, V.E. Sokolov, O.A. Scherbakov, G.V. Valski, A. S. Vorobiev and F. Gönnenwein, "Russian Reactor Conf.", Moscow 2006, p 112
[GAG09a] A. Gagarski, G. Petrov, I. Guseva, T. Zavarukhina, F. Gönnenwein, M. Mutterer, J. von Kalben, W. Trzaska, M. Sillanpää, Yu. Kopatch, G. Tiourine, T. Soldner and V. Nesvizhevsky, "Interactions of Neutrons with Nuclei XVI", Dubna 2009, p 356
[GAG09b] A. Gagarski, G. Petrov, I.Guseva, T. Zavarukhina, F. Gönnenwein, M. Mutter, J. von Kalben, Yu. Kopatch, G. Tiourine, W. Trzaska, M. Sillanpää, T. Soldner and V. Nesvizhevsky, AIP Conf. Proc. 1175, 2009, p 323
[GAG10] A. Gagarski, F. Gönnenwein, I. Guseva, Yu. Kopatch, M. Mutterer, G. Petrov, G. Tiourine, T. Kuzmina, V. Nesvizhevsky and T. Soldner, "Interactions of Neutrons with Nuclei XVII", Dubna 2010, p 17
[GAG10] A. Gagarsky, private communication
[GEL84] P. Geltenbort, PHD thesis, University of Tübingen, 1984
[GEL86a] P. Geltenbort, F. Gönnenwein and A. Oed, Rad. Effects 93, 57 (1986)

[GEL86b] P. Geltenbort, F. Gönnenwein, A. Oed and P. Perrin, Rad. Effects 93, 325 (1986)

[GIN82] J.E. Gindler, L.E. Glendenin, B.D. Wilkins, B.B. Back, H.-G. Clerc and B.G. Glagola "Lecture Notes in Physics" 158, Springer 1982, p 145

[GOE91] F. Gönnenwein and B. Börsig, Nucl. Phys. A 530 (1991) 27

[GOE01] F. Gönnenwein, P. Jesinger, M. Mutterer, A.M. Gagarski, G.A. Petrov, W. H. Trzaska, V. Nesvishevsky and O. Zimmer, "Fission Dynamics of Atomic Clusters and Nuclei", World Scientific 2001, p 232

[GOE04] F. Gönnenwein, Nucl. Phys. A 734 (2004) 213

[GOE05] F. Gönnenwein, A. Gagarski, P. Jesinger, G. Petrov, M. Mutterer, W. Trzaska, V. Nevsvizhevsky, E. Lelievre-Berna and V. Bunakov, "Nucl. Data for Science and Techn.", AIP Conf. Proc. 769 (2005) 870

[GOE07] F. Goennenwein, M. Mutterer, A. Gagarski, I. Guseva, G. Petrov, V. Sokolov, T. Zavarukhina, Yu. Gusev, J. von Kalben, V. Nesvizhevsky and T. Soldner, Phys. Lett. B 652 (2007) 13

[GOE09] F. Gönnenwein, A. Gagarski, G. Petrov, I. Guseva, T. Zavarukhina, M. Mutterer, J. von Kalben, Yu. Kopatch, G. Tiourine, W. Trzaska, M. Sillanpää, T. Soldner and V. Nesvizhevsky, "Exotic Nuclei", AIP Conf. Proc. 1224 (2009) 338

[GOE10] F. Gönnenwein, "Seminar on Fission", World Scientific 2010, p 3

[GON91] F. Gönnenwein in "The Nucl. Fiss. Process", C. Wagemans, CRC Press, 1991, p 287

[GON92] F. Gönnenwein, Nucl. Instr. Meth. A 316 (1992) 405

[GON94] F. Gönnenwein, A.V. Belozerov, A.G. Beda, S.I. Burov, G.V. Danilyan, A.N. Martemyanov, L.N. Bondarenko, Yu.A. Mostovoi, P.Gelternbort, J. Last and K. Schreckenbach, Nucl. Phys. A 567 (1994) 303

[GON99a] F. Gönnenwein, M. Wöstheinrich, M. Hesse, H. Faust, G. Fioni and S. Oberstedt, "Seminar on Fission" Pont d`Oye IV, World Scientific, 2000, p 59

[GON99b] F. Gönnenwein, M. Hesse, M. Wöstheinrich, M. Mutterer, H. Faust, G. Fioni and S. Oberstedt, "Perspectives in Nuclear Physics", World Scientific, 1999, p 293

[GON05] F. Gönnenwein, M. Mutterer and Yu. Kopatch, europhysicsnews 36, 11 (2005)

[GOU05] H. Goutte, J.F. Berger and D. Gogny, Phys. Rev. C 71, 024316 (2005)

[GOU06] H. Goutte, J.F. Berger and D. Gogny, J. Mod. Phys. 15 (2006) 292

[GRA95] U. Graf, F. Gönnenwein, P. Geltenbort and K. Schreckenbach,, Z. Phys. A 351, 281 (1995)

[GRE00] W. Greiner, "Nuclear Shells-50 Years" World Scientific 1999, p 1

[GRU82] C.R. Gruhn, M. Binimi, R. Legrain, R. Loveman, W. Pang, M. Roach, D.K. Scott, A. Shotter, T. J. Symons, J. Wouters, M. Zisman, R. Devries, Y.C. Peng and W. Sondheim, Nucl. Instr. Meth. 196 (1982) 33

[GUE79] C. Guet, C. Signarbieux, P. Perrin, H. Nifenecker, M. Asghar, R. Caitucolli and B. Leroux, Nucl. Phys. A 314 (1979) 1

[GUS07] I.S. Guseva and Yu.I. Gusev, "Interact. of Neutr. with Nucl. 14", Dubna, 2007, p 101

[GUS09] I. Guseva and Yu. Gusev, AIP Conf. Proc. 1175, 2009, p 355

[GUS10] I.S. Guseva, "Interact. of Neutrons with Nucl. 18", Dubna, 2010, to be published

[GUT91] U. Güttler, PHD thesis, University of Mainz, Germany, 1991

[HAD88] M. Haddad, J. Crancon, G. Lhospice and M. Asghar, Nucl. Phys. A 481 (1988) 333

[HAH39] O. Hahn and F. Strassmann, Naturwissenschaften 27 (1939) 11
[HAL65] I. Halpern, Phys. and Chem. of Fission" IAEA, 1965, p 369
[HAL71] I. Halpern, Ann. Rev. Nucl. Sci. 21 (1971) 245
[HAM93] F.-J. Hambsch, H.-H. Knitter and C. Budtz-Jorgensen, Nucl. Phys. A554 (1993)209
[HAM03] F.-J. Hambsch, S. Oberstedt, A. Tudora, G. Vladuca and I. Ruskov, Nucl. Phys. A 726 (2003) 248
[HAN89] H. Hongyin, H. Shengnian,, M.Jiangchen,, B. Zongyu and Ye. Zongyuan, "Fifty Years with Nucl. Fiss.", Am. Nucl. Soc. 1989, p 684
[HAS88] R.W. Hasse and W.D. Myers, "Geometric Relationships of Macroscopic Nuclear Physics", Springer 1988
[HEE89a] P. Heeg, M. Mutterer, J. Pannicke, P. Schall, J.-P. Theobald and K. Weingärtner, "Fifty Years with Nucl. Fiss.", Am. Nucl. Soc. 1989, p 299
[HEE89b] P. Heeg, J. Pannicke, M. Mutterer, P. Schall, J.P. Theobald, K. Weingärtner, K.F. Hoffmann, K. Scheele, P. Zöllner, G. Barreau, B. Leroux and F. Gönnenwein, Nucl. Inst. Meth. A 278 (1989) 452
[HEN81] H. Henschel, A. Kohnle, H. Hipp and F. Gönnenwein, Nucl,. Instr. Meth. 190 (1981) 125
[HEN92] H. Hentzschel, PHD thesis, University of Mainz, 1992
[HEN94] R. Hentzschel, H.R. Faust, H.-O. Denschlag, B.D. Wilkins and J. Gindler, Nucl. Phys. A 571 (1994) 427
[HER79] B. Hering, PHD thesis, University of Tübingen, Germany, 1979
[HES96] M. Hesse, M. Wöstheinrich, F. Gönnenwein and H. Faust, "Dynamical Aspects of Nuclear Fission", Dubna, 1996, p 238
[HER81] E.S. Hernandez, W.D. Myers, J. Randrup and B. Remaud, Nucl. Phys. A 361 (1981) 483
[HOL77] W. Holubarsch, PHD thesis, University of Tübingen, 1977
[HOL88] N.E. Holden and M.S. Zucker, "Nucl. Data for Science and Technology", 1988, p 795
[HUH97] M. Huhta, P. Dendooven, A. Honkanan, A. Jokinen, G. Lhersonneau, M. Oinonen, H. Penttilä, K. Peräjärvi, V.A. Rubchenya and J. Äystö, Phys. Lett. B 405 (1997) 230
[HUL94] E.K. Hulet, Phys. At. Nucl. 57 (1994) 1099
[ITK09] M.G. Itkis, G.N. Knyazheva, E.M. Kozulin and F. Goennenwein, AIP Conf. Proc. 1175, 2009, p 126
[JEN42] W. Jentschke and F. Prankl, Z. Phys. 119 (1942) 696
[JES00] P. Jesinger, A. Kötzle, A.M. Gagarski, F. Gönnenwein, G. Danilyan, V.S. Pavlov, V.B. Chvatchkin, M. Mutterer, S.R. Neumaier, G.A. Petrov, V.I. Petrova, V. Nesvizhevsky, O. Zimmer, P. Geltenbort, K. Schmidt and K. Korobkina, Nucl. Instr. Meth. A 440 (2000) 618
[JES01] P. Jesinger, PHD thesis, University of Tübingen, 2001
[JES02] P. Jesinger, A. Kötzle, F. Gönnenwein, M. Mutterer, J. von Kalben, G.V. Danilyan, V.S. Pavlov, G.A. Petrov, A.M. Gagarski, W.H. Trzaska, S. Soloviev, V.V. Nesvizhevski and O. Zimmer, Phys. At. Nucl. 65 (2002) 630
[JES05] P. Jesinger, Yu.N. Kopatch, M. Mutterer, F. Gönnenwein, A.M. Gagarski, J. von Kalben, V. Nesvizhevsky, G.A. Petrov, W. H. Trzaska and H.-J. Wollerstein, Eur. Phys. J. A 24, 379 (2005)

[KAL00] V.A. Kalinin, V.N. Dushin, B.F. Petrov, V.A. Jakovlev, A.S. Vorobyev, A.B. Laptev, G.A. Petrov, Y.S. Pleva, O.A. Shcherbakov, V.E. Sokolov and F.-J. Hambsch, "Interactions of Neutrons with Nuclei 8", Dubna, 2000, p 261

[KAU91] J. Kaufmann, N. Boucheneb, G. Medkour, M. Asghar, F. Gönnenwein, W. Mollenkopf, G. Barreau, T.P. Doan, B. Leroux, A. Sicre and P. Geltenbort, "Nuclear Data for Science and Technology", Springer, 1991, p 131

[KAU92] J. Kaufmann, W. Mollenkopf, F. Gönnenwein. P. Geltenbort and A. Oed, Z. Phys. A 341, 319 (1992)

[KAP72] S.S. Kapoor, S.K. Choudhury, S.K. Kataria, S.R.S. Murthy and V.S. Ramamurthy, "Nucl. Phys. and Solid State Phys. Symp." Chandigarh, India, 1972, Vol. 15b, p 107

[KAT73] S.K. Kataria, E. Nardi and S.G. Thompson, "Phys. and Chem. of Fission", IAEA, 1973, 389

[KEY73] G.A. Keyworth, C.E. Olsen, F.T. Seibel, J.W.T. Dabbs and N.W. Hill, Phys. Rev. Lett. 31, 1077 (1973)

[KIE91] J. Kiesewetter, S. Okretic, F.M. Baumann, K.-Th. Brinkmann, H. Freiesleben, H. Gassel and R. Opara, Nucl. Instr Meth. A 314 (1992) 125

[KNI87] H.-H. Knitter, F.-J. Hambsch , C. Budtz-Jorgensen and J.P. Theobald, Z. Naturforsch. 42a, 786 (1987)

[KNI91] H.-H. Knitter, U. Brosa and C. Budtz-Jorgensen, "The Nucl. Fiss. Process", C. Wagemans ed., CRC Press, 1991, p 498

[KNI92] H.-H. Knitter, F.-J. Hambsch and C. Budtz-Jorgensen, Nucl. Phys. A 536 (1992) 221

[KOE00a] U. Köster, PHD thesis, TU Munich, Germany, 2000

[KOE00b] U. Köster, H. Faust, T. Friedrichs, S. Oberstedt, G. Fioni, M. Gross and I. Ahmad, "Sem. on Fission", World Scientific, 1999, p 77

[KOE00c] A. Kötzle, P. Jesinger, F. Gönnenwein, G.A. Petrov, V.I. Petrova, A.M. Gagarski, G. Danilyan, O. Zimmer and V. Nesvizhevsky, Nucl. Instr. Meth. A 440 (2000) 750

[KOP99] Yu.N. Kopatch, A.B. Popov, V.I. Furman, N.N. Gonin, L.K. Kozlovky, D.I. Tambovtsev and J. Kliman, Phys. At. Nucl. 62 (1999)840

[KOP02] Yu.N. Kopatch, M. Mutterer, D. Schwalm, P. Thirolf and F. Gönnenwein, Phys. Rev. C 65, 044614 (2002)

[KOP05] Yu.N. Kopatch, V. Tishchenko, M. Speransky, M. Mutterer, F. Gönnenwein, P. Jesinger, A.M. Gagarski, J. von Kalben, I. Kojouharov, E. Lubkiewics, Z. Mezentseva, V. Nesvizhevsky, G.A. Petrov, H. Schaffner, H. Scharma, W.H. Trzaska and H.-J. Wollersheim, "Nucl. Fission and Fission Product Spectroscopy" AIP Conf Proc. 798, 2005, p 115

[KRA69] H.J. Krappe and U. Wille, Nucl. Phys. A 124 (1969) 641

[LAN 80] W. Lang, H.-G. Clerc, H. Wohlfarth, H. Schrader and K.-H. Schmidt, Nucl. Phys. A 345 (1980) 34

[LAN81] L.D. Landau and E.M. Lifshitz, "Quantum Mechanics", Butterworth, 1981

[MAE84] H. Märten and D. Seeliger, J. Phys. G. 10 (1984) 349

[MAE91] H. Märten, "Sem. on Fission", C. Wagemans ed., Pont D`Oye, 1991, p 15

[MAE92] H. Märten, "Dyn. Aspects of Nucl. Fission", Dubna, 1992, p 32

[MAR72] J. Maruhn and W. Greiner, Z. Physik 251, 431 (1972)

[MAR90] G. Martinez, G. Barreau, A. Sicre, T.P. Doan, P. Audouard, B. Leroux, W. Arafa, R. Brissot, J.P. Bocquet, H. Faust, P. Koczon, M. Mutterer, F. Gönnenwein, M. Asghar, U. Quade, K. Rudolph, D. Engelhardt and E. Piasecki, Nucl. Phys. A515 (1999) 433

[MAS09] V. Maslov, private communication 2009

[MAY48] M.G. Mayer, Phys. Rev. 74, 235 (1948)

[MED97] G. Medkour, M. Asghar, M. Djebara and B. Bouzid, J. Phys. G. 23 (1997) 103

[MEI39] L. Meitner and O.R. Frisch, Nature 3615, 239 (1939)

[MEI50] L. Meitner, Nature 4197, 561 (1950)

[MER81] A.C. Merchant and W. Nörenberg, Phys. Lett. 104B, 15 (1981)

[MIL58] J.C.D. Milton and J.S. Fraser, Phys. Rev. 111, 877 (1958)

[MIL61] J.C.D. Milton and J.S. Fraser, Phys. Rev. Lett. 7, 67 (1961)

[MIS02] S. Misicu and W. Greiner, J Phys. G 28 (2002) 2861

[MOE70] P. Möller and S.G. Nilsson, Phys. Lett. 31B, 283 (1970)

[MOE95b] A. Möller, F. Gönnenwein, J. Kaufmann, J. Lemli, G. Petrov and C. Signarbieux, "Sem. on Fission", C. Wagemans ed., Pont d`Oye, 1995, p 76

[MOE96] A. Möller, M. Cröni, F. Gönnenwein and G. Petrov, "Large Scale Coll. Motion of At. Nucl.", World Scientific, 1996, p 203

[MOL75] E. Moll, H. Schrader, G. Siegert, M. Asghar, J.P. Bocquet, G. Bailleul, J.P. Gautherin, J. Greif, G.I. Crawford, C. Chauvin, H. Ewald, H. Wollnik, P. Armbruster, G. Fiebig, H. Lawin and K. Sistemich, Nucl. Inst. Meth. 123 (1975) 615

[MOL81] P. Möller and J.R. Nix, At. Data and Nucl. Data Tables 26, 165 (1981

[MOL91] W. Mollenkopf, PHD thesis, University of Tübingen, 1991

[MOL92] W. Mollenkopf, J. Kaufmann, F. Gönnenwein, P. Geltenbort and A. Oed, J. Phys. G 18 (1992) L203

[MOL95] P. Möller, J.R. Nix, W.D. Myers and W.J. Swiatecki, At. Data and Nucl. Data Tables 59, 185 (1995)

[MOL09] P. Möller, A.J. Sierk, T. Ichikawa, A. Iwamoto, R. Bengtsson, H. Uhrenholdt and S. Aberg, Phys. Rev. C 79, 064304 (2009)

[MOO78] M.S. Moore, J.D. Moses, G.A. Keyworth, J.W.T. Dabbs and N.W. Hill, Phys. Rev. C 18, 1328 (1978)

[MUE84] R. Müller, A.A. Naqvi, F. Käppeler and F. Dickmann, Phys. Rev. C 29, 885 (1984)

[MUL99] S.I. Mulgin, V.N. Okolovich and S.V. Zhdanova, Phys. Lett. B 462 (1999) 29

[MUS73] M.G. Mustafa, U. Mosel and H.W. Schmitt, Phys. Rev. C 7, 1519 (1973)

[MUT96a] M. Mutterer, P. Singer, Yu. Kopatch, M. Klemens, A. Hotzel, D. Schwalm, P. Thirolf and M. Hesse, "Dynamical Aspects of Nuclear Fission", Dubna 1996, p 250

[MUT96b] M. Mutterer and J.P. Theobald, in "Nuclear Decay Modes", IOP Publishing, 1996, p 487

[MUT98] M. Mutterer, P. Singer, M. Klemens, Yu.N. Kopatch, D. Schwalm, P. Thirolf, A. Hotzel, M. Hesse and F. Gönnenwein, "Fission and Properties of Neutron-rich Nuclei", World Scientific 1997, p 119

[MUT00] M. Mutterer, W. H. Trzaska, G.P. Tyurin, A.V. Evsenin, J. von Kalben, J. Kemmer, M. Kapusta, V.G. Lyapin and S.V. Khlebnikov, Trans. Nucl. Sci. 47 (2000) 756

[MUT02] M. Mutterer, Yu.N. Kopatch, P. Jesinger and F. Gönnenwein,

"Dynamical Aspects of Nuclear Fission", World Scientific, 2002, p 326
[MUT04] M. Mutterer, Yu.N. Kopatch, P. Jesinger, A.M. Gagarski, F. Gönnenwein, J. von Kalben, S.G. Khlebnikov, I. Kojouharov, E. Lubkiewics, Z. Mezentseva, V. Nesvizhevsky, G.A. Petrov, H. Schaffner, H. Scharma, D. Schwalm, P. Thirolf, W.H. Trzaska, G.P. Tyurin and H.-J Wollersheim, Nucl. Phys. A 738 (2004) 122
[MUT08] M. Mutterer, Yu. N. Kopatch, S.R. Yamaledtinov, V.G. Lyapin, J. von Kalben, S.V. Khlebnikov, M. Sillanpää, G.P. Tyurin and W.H. Trzaska, Phys. Rev. C 78, 064616 (2008)
[MYE81] W.D. Myers, G. Mantzouranis and J. Randrup, Phys. Lett. B 98 (1981) 1
[NAG96] Y. Nagame, I.Nishinaka, K. Tsudada, Y. Oura, S. Ichkawa. H. Ikezoe, Y.L. Zhao, K. Sueki, H. Nakahara, M.Tanikawa, T. Ohtsuki, H. Kudo, Y. Hamajima, K. Takamiya and A.H. Cunk, Phys. Lett. B 387 (1996) 26
[NAI04] H. Naik, R.J. Singh and R.H. Iyer, J. Phys. G 38 (2004) 107
[NAI07] H. Naik, S.P. Dange and A.V.R. Reddy, Nucl. Phys. A 781 (2007) 1
[NET91] D.R. Nethaway, private communication
[NIF73] H. Nifenecker, C. Signarbieux, R. Babinet and J.Poitou, "Phys. and Chem. of Fission", IAEA, 1973, Vol. II, p 117
[NIF80] H. Nifenecker, J. de Physique 41 (1980) 47
[NIF82] H. Nifenecker, G. Mariolopoulos, J.P. Bocquet, R. Brissot, Ch. Hamelin, J. Cran□on and Ch. Ristori, Z. Phys. A 308, 39 (1982)
[NIS98] K. Nishio, Y. Nakagome, H. Yamamoto and I. Kimura, Nucl. Phys. A 632 (1998) 540
[NOW82] I. Nowicki, E. Piasecki, J. Sobolewski, A. Kordyasz, M. Kisielinski, W. Czarancki, H. Karwoski and P. Koczon, Nucl. Phys. A 375 (1982) 187
[OED81] A.Oed, G. Barreau, F. Gönnenweinn, P. Perrin, C. Ristori and P. Geltenbort, Nucl. Instr Meth. 179 (1981) 265
[OED83a] A. Oed, P. Geltenbort and F. Gönnenwein, Nucl. Instr. Meth. 205 (1983) 451
[OED83b] A. Oed, P. Geltenbort, F. Gönnenwein, T. Manning and D. Souque, Nucl. Instr. Meth 205 (1983) 455
[OED84] A. Oed, P. Geltenbort, R.Brissot, F. Gönnenwein, P. Perrin, E. Aker and D. Engelhardt, Nucl. Instr. Meth. 219 (1984) 569
[OED86] A. Oed, P. Geltenbort, F. Gönnenwein and B. Stütz, Radiation Effects 96 (1986) 53
[PAN86] J. Pannicke, "IXe Journées d`études sur la fission nucléaire", Centre d`études nucléaires de Bordaux-Gradignan, 1986, D-13
[PAS71] V.V. Pashkevich, Nucl. Phys. A 169 (1971) 275
[PAS95] V.V. Pashkevich, "Low Energy Nuclear Dynamics", World Scientific 1995, p 161 and private communication
[PAT71] N.J. Pattenden and H. Postma, Nucl. Phys. A 167 (1971) 225
[PER97a] K. Persyn, E. Jacobs, S. Pommé, D. De Frenne, K. Govaert and M.L. Yoneama, Nucl. Phys. A 615 (1997) 198
[PER97b] K. Persyn, E. Jacobs, S. Pommé, D. De Frenne, K. Govaert and M.L. Yoneama, Nucl. Phys. A 620 (1997) 171
[PET89] G.A. Petrov, G.V. Valskii, A.K. Petukhov, A.Ye. Alexandrovich, Yu.S. Pleva, V.E. Sokolov, A.B. Laptev and O.A. Scherbakov, Nucl. Phys. A 502(1989)297c

[PFE70] E. Pfeiffer, Z. Physik 240, 403 (1970)
[PIA70] E. Piascki, M. Dakowski, T. Krogulski, J. Tys and J. Chwaszcewska, Phys. Lett. B 33 (1970) 568
[PIA73] E. Piascki and J. Blocki, Nucl. Phys. A 208 (1973) 381
[PIK94] G.A. Pik-Pichak, Phys. At. Nucl. 57 (1994) 906
[PLE72] F. Pleasonton, R.L. Feerguson and H.W. Schmitt, Phys. Rev. C 6, 1023(1972)
[POE99] D.N. Poenaru, W. Greiner, J.H. Hamilton, A.V. Ramayya, E. Hourany and R.A. Gherghescu, Phys. Rev. C 59(1999)3457
[POE05] D.N. Poenaru, R.A. Gherghescu and w. Greiner, Nucl. Phys. A 747(2005)182
[POM93] S. Pommé, E. Jacobs, K. Persyn, D. De Frenne and K. Govaert, Nucl. Phys. A 560(1993)689
[PRA79] M. Pralash, V.S. Ramamurthy and S.S. Kapoor, "Phys. and Chem. of Fission", IAEA, 1979, 353
[PYA10] Yu. V. Pyatkoc, d.V. KÖamanin, W. von Oertzen, A.A. Alexandrov, I.A. Alexandrova, O.V. Falomkina, N.A. Kondratjev, Yu.N. Kopatch, E.A. Kuznetsov, Yu.E. Lavrova, A.N. Tyukavkin, W.Traska and V.E. Zhuhcko, Eur. Phys. J. A 45(2010)29
[QUA79] U. Quade, K. Rudolph and G. Siegert, Nucl. Instr. Meth 164(1979)435
[QUA82] U. Quade, K. Rudolph, P. Armbruster, H.-G. Clerc, W. Lang, M. Mutterer, J. Pannicke, C. Schmitt, J.P. Theobald, F. Gönnenwein und H. Schrader, Lecture Notes in Physics, Springer, 158(1982)40
[QUA88] U. Quade, K. Rudolph, S. Skorka, P. Armbruster, H.-G. Clerc, W. Lang, M. Mutterer, C. Schmitt, J.P. Theobald, F. Gönnenwein, J. Pannnicke, H. Schrader, G. Siegert and D. Engelhardt, Nucl. Phys. A 487(1988)1
[RAO74] V.K. Rao, V.K. Bhargava, S.G. Marathe, S.M. Sahkundu and R.H. Iyer, Phys. Rev. C 9, 1506(1974)
[RAS69] J.O. Rasmussen, W. Nörenberg and H.J. Mang, Nucl. Phys. A 136(1969)465
[REY00] F. Reymund, A.V. Ignatyuk, A.R. Junghans and K.-H. Schmidt, Nucl. Phys. A 678(2000)215
[ROC02] D. Rochman, H. Faust, I. Tsekhanovich, F. Gönnenwein, E. Storrer, S. Oberstedt and V. Sokolov, Nucl. Phys. A 710(2002)3
[ROC04] D. Rochman, I. Tsekhanovich, F. Gönnenwein, V. Sokolov, F. Storrer. G. Simpson and O. Serot, Nucl. Phys. A 735(1904)3
[RUB82] V.A. Rubchenya, Sov. J. Nucl. Phys. 35, 334(1982)
[RUB88] V.A. Rubchenya and S.G. Yavshits, Z. Phys. A 329, 217(1988)
[RUB91] A. Ruben and H. Märten, "Dynamical Aspects of Nuclear Fission", Dubna 1991, 246
[RUB01] V.A. Rubchenya, W.H. Trzaska, D.N. Vahtin, J. Äystö, P. Dendooven, S. Hankonen, A. Jokinen, Z. Radivojevich, J.C. Wang, I.D. Alkhazov, A.V. Evsenin, S.V. Khlebnikov, A.V. Kuznetsov, V.G. Lyapin, I. Osetrov, G.P. Tiourin, A.A. Aleksandrov and Yu.E. Penionzkhevich, Nucl. Instr and Meth. A 463(2001)653
[SAN89] A. Sandulescu, A. Florescu and w. Greiner, J. Phys. G 15(1989)1815
[SCH65] H.W. Schmitt, W.E .Kiker and C.W. Williams, Phys. Rev. 137, B837(1965)
[SCH66] H.W. Schmitt, J.H. Neiler and F.J. Walter, Phys. Rev. 141, 1146(1966)
[SCH78] H. Schultheis and R. Schultheis, Phys. Rev. C 18, 1317(1978)

[SCH81] W.U. Schröder, J.R. H;uizenga and J. Randrup, Phys. Lett. 98B, 355(1981)
[SCH84] C. Schmitt, A. Guessous, J.P. Bocquet, H.-G. Clerc, R. Brissot, D. Engelhardt, H. R. Faust, F. Gönnenwein, M. Mutterer, H. Nifenecker, J. Pannicke, Ch. Ristsori and J.P. Theobald, Nucl. Phys. A 430(1984)21
[SCH88] K.Schreckenbach, ILL report 88SCO9T, 1988
[SCH92] A. Schuber, J. Hutsch, K. Möller, W. Neubert, W. Pilz, G. Schmidt, M. Adler and H. Märten, Z. Phys. A 341, 481(1992)
[SCH94a] W. Schwab, H.-G. Clerc, M. Mutterer, J.P. Theobald and H. Faust, Nucl. Phys. A 577(1994)674
[SCH94b] P. Schillebeeckx, C. Wagemans, P. Geltenbort, F. Gönnenwein and A.Oed, Nucl. Phys. A 580(1994)15
[SCH98] K.-H. Schmidt, A.R. Junghans, J. Benlliure, C. Böckstiegel, H.- G. Clerc, A. Grewe, A. Heinz, A.V. Ignatyik, M. de Jong, G.A. Kudyaev, H. Müller, M. Pfitzner and S. Steinhäuser, Nucl. Phys. A 630(1998)208c
[SCH00] K.-H. Schmidt, S. Steinhäuser, C. Böckstiegel, A. Grewe, A. Heinz, A.R. Junghans, J. Benlliure, H.-G.Clerc, M. de Jong, J. Müller, M. Pfützner and B. Voss, Nucl. Phys. A 665(2000)221
[SER00] O. Serot, N. Carjan and C. Wagemans, Eur. Phys. J. A 8, 187(2000)
[SER98] O. Serot and C. Wagemans, Nucl. Phys. A 641(1998)34
[SER05] O.Serot, C. Wagemans, S. Vermote, J. Heyse, T. Soldner and P. Geltenbort, "Nucl. Fission and Fission-Prod. Spectroscopy AIP 798, 2005(2005)
[SHE95] R.A. Shekhmametev, Yu.V. Pyatkov, F. Gönnenwein and P. Geltenbort, "XIII Meet. Phys. of Nucl. Fission", Obninsk, 1995, 242
[SIC86] A. Sicre, G. Barreau, A. Boukellal, F. Caitucoli, T. P. Doan, B. Lerroux, P. Geltenbort, F. Gönnenwein, A. Oed and M. Asghar, Radiation Effects 93, 401(1986)
[SID89] J.L. Sida, P. Armbruster, M. Bernas, J.P. Bocquet, R.Brissot and H.R. Faust, Nucl. Phys, A 502(1989)233c
[SIE76] G. Siegert, H. Wollnik, J. Greif, R. Decker, G. Fiedler and B. Pfeiffer, Phys. Rev. 14, 1864((1976)
[SIE92] P. Siegler, T. Schunk, M. Mutterer, W. Schwab, J.P. Theobald, U. Quade and H. Faust, "Nuclear Data for Science and Technology", Springer 1992, 128
[SIG73] C. Signarbieux, R. Babinet, H. Nifenecker and J. Poitou, "Physics and Chemistry of Fission", IAEA, 1973, 179
[SIG81] C. Signarbieux, M. Montoya, M. Ribrag, C. Mazur, C.Guet, P. Perri and M. Maurel, J. Physique Lett. 42(1981)L-437
[SIG91] C. Signarbieux, private communication
[SIM90] G. Simon, J. Trochon, F. Brisard and C. Signarbieux, Nucl. Instr. Meth. A 286(1990)220
And G. Simon, PHD Thesis, Université Paris-Sud, 1990
[SIN89] A.K. Sinha, D.M. Nadkarni and G.K. Mehta, Pramana-J. Phys. 33(1989)85
[SOK04] V.E. Sokolov, A.M. Gagarski, G.A.Petrov, V.I. Petrova, Yu.S. Pleva, Yu.P. Rudnev, G.V. Valsky, T.A. Zavarukhina and V.I. Zherebchevsky, "Interaction of Neutrons with Nuclei XII", Dubna 2004, 25

[SOK05] V.E. Sokolov, A.M. Gagarski, I.S. Guseva, S.P. Golosovskaya, I.S. Krasnoshchokova, G.A. Petrov, V.I.Petrova, A.K. Petukhov, Yu. S. Pleva, V.P. Alfimenkov, A.N. Chernikov, L. Lason, Yu D. Mareev, V.V. N ovitski, L.B. Pikelner, T.L. Pikelner and M.I. Tsulaya, "Nuclear Data for Science and Technology", AIP 769, 2005, 708

[SPE04] M. Speranky, V. Tishchenko, Yu. Kopatch, A. Gagarski, M. Mutterer, F. Gönnenwein, W. Trzaska, h.-Wollersheim, J. von Kalben and V. Nesvizhevky, "Interaction of Neutrons with Nuclei XII", Dubna 2004, 430

[STE98] S. Steinhäuer, J. Benlliure, C. Böckstiegel, H.-G. Clerc, A. Heinz, A. Grewe, M. de Jong, A.R. Junghans. J. Müller, M. Pfützner, K.-H-Schmidt, Nucl. Phys. A 634(1998)89

[STR87] Ch. Straede, C. Budtz-Jorgensen and H.-H. Knitter, Nucl. Phys. A 462(1987)85

[SUS81] O.P. Sushkov and V.V. Flambaum, Sov. J. Nucl. Phys. 33(1981)31

[STU95] P. Stumpf, H.O. Denschlag and H.R. Faust, "Seminar on Fission", C. Wagemans ed., Pont d`Oye III, !995, 196

[TER62] J. Terrell, Phys. Rev. 127, 880(1962)

[THE86] J.P. Theobald, "Seminar on Fission", C. Wagemans ed., Pont d`Oye I, 1986, 467

[THE89] J.P.Theobald, P. Heeg and M. Mutterer, Nucl. Phys. A 502(1989)343c

[TIS06] V.G. Tishchenko, Yu.N. Kopatch, M. Mutterer, F. Gönnenwein, A.M. Gagarski, P. Jesinger, J. von Kalben I. Kojouharov, E. Lubkiewics, Zh. Mezentseva, V. Neviszhevsky, G.A. Petrov, H. Schaffner, H. Schara, M. Speransky, W. Trzaska and H.-J. Wollersheim, "Interaction of Neutrons with Nuclei XIII", Dubna 2006, 342

[TRA72] B.L. Tracy, J, Chaumont, R. Klapisch, J.M. Nitschcke, A.M. Poskanzer, E. Roeckl and C. Thibaud, Phys. Rev. C 5, 222(1972)

[TRO89] J. Trochon, G. Simon and C. Signarbieux, "50 Years with Nuclear Fission.", Am. Nucl. Soc., 1989, 313

[TSE99] I. Tsekhanovich, H.O. Denschlag, M. Davi, Z. Büyükmumcu, M. Wöstheinrich, F. Gönnenwein, S. Oberstedt and H.R. Faust, Nucl. Phys. A 658(1999)217

[TSE01] I. Tsekhanovich, H.O. Denschlag, M. Davi, Z.Büjükmumcu, F. Gönnenwein, S. Oberstedt and H.R. Faust, Nucl. Phys. A 688(2001)633

[TSE03] I. Tsekhanovich, Z. Büjümumcu, M. Davi, H.O. Denschlag, F. Gönnenwein and S.F. Boulyga, Phys,.Rev. C 67, 934610(2003)

[TSE04a] I. Tsekhanovich, N. Varapai, V. Rubchenya, D. Rochman, G.S. Simpson, V. Sokolov, G. Fioni and I. AlMahamid, Phys. Rev. C 70, 044610(2004)

[TSE04b] I. Tsekhanovich, G. Simpson, R. Orlandi, A. Scherillo, D. Rochman and V. Sokolov, "Seminar on Fission", editor C. Wagemans, World Scientific 2004, 55

[TSE08] I. Tsekhanovich, J.A. Dare, A.G. Smith, B. Varley, D. Cullen, N. Lumley, T. Materna, U. Köster and G.S. Simpson, "Seminar on Fission", editor C. Wagemans, World Scientific 2008, 189

[TSI47] T. San-Tsiang, H. Zar-Wei, R. Chaptel and L. Vigneron, Phys. Rev. 71, 382(1947)

[TUR51] A. Turkevich and J.B. Niday, Phys. Rev. 84, 52(1951)

[VAL76] G.V. Valskii, Sov. J. Phys. 24, 140(1976)

[VAN73] R. Vandenbosch and J.R. Huizenga, "Nuclear Fission",Acadmic Press 1973

[VER10] S. Vermote, C. Wagemans, O. Serot, J. Heyse, J. vam Gils, T. Soldner, P. Geltenbort, I. AlMahamic, G. Tian and L. Rao, Nucl. Phys. A 837(2010)176

[VIO85] V.E. Viola, K. Kwiatkowski and M. Walker, Phys. Rev. C 31, 1550(1985)
[VOR04] A.S. Vorobyev, V.N. Dushin, F.-J. Hambsch, V.A. Jakovle, V.A. Kalinin, A.B. Laptev, B.F. Petrov and O.A. Shcherbakov, "Nucl. Data for Science and Technology" AIP Conf. Proc. 769, 2004 613
[VOR69] A.A. Vorobiev, D.M. Seliverstov, V.T. Grachov, I.A. Kondurov, A.M. Nikitin, A.L. Yegorov and Yu.K. Zalite, Phys. Lett. 30B, 332(1969)
[VOR72] A.A. Vorobiev, D.M. Seliverstov, V.T. Grachov, I.A. Kondurov, A.M. Nikitin, N.N. Smirnov and Yu. K. Zalite, Phys. Lett. 40B, 102(1972)
[VOR75] A.A. Vorobiev, V.T. Grachev, I.A. Kondurov, Yu.A. Miroshnichenko, A.M. Nikitin, D.M. Seliverstov and N.N. Smirnov, Sov. J. Nucl. Phys. 20, 248(1975)
[WAG81] C. Wagemans, E. Allaert, F. Caitucoli, P. D`Hondt, G. Barreau and P. Perrin, Nucl. Phys. A 369(1981)1
[WAG89] C. Wagemans, P. Schillebeeckx and J.P. Bocquet, Nucl Sci. Engin. 101, 239(1989)
[WAG91] C. Wagemans ed., "The Nuclear Fission Process" CRC Press, 1991
[WAG98] C. Wagemans, J. Wagemans, P. Geltenbort, O. Zimmer and F. Gönnenwein, "Nuclear Fission and Fission-Product Spectroscopy", AIP Conf. Proc. 447, 1998, 262
[WAG04] C. Wagemans, J. Heyse, P. Janssens, O. Serot and P. Geltenbort, Nucl. Phys. A 742(2004)291
[WAG08] C. Wagemans, S. Vermote and O. Serot, "Sem. on Fiss.", World Scientific, 2008, 117
[WAH62] A.C. Wahl, R.L. Ferguson, D.R. Nethaway, D.E. Troutner and K. Wolfsberg, Phys. Rev. 126 1112(1962)
[WAH88] A.C. Wahl, At. Data and Nucl. Data Tables 39(1988)1
[WAH02] A.C. Wahl, Los Alamos Report, LA-13928, 2002
[WEI86] E. Weissenberger, P. Geltenbort, A. Oed, F. Gönnenwein and H. Faust Nucl. Instr. Meth. A 248(1986)506
[WIL85] J.F. Wild, P.A. Baisden, R.J. Dougan, E.K. Hulet, R.W. Lougheed and J.H. Landrum, Phys. Rev. C32, 488(1985)
[WIL76] B.D. Wilkins, E.P. Steinberg and R.R. Chasman, Phys. Rev. C 14, 1832(1976)
[WOE96] M. Wöstheinrich, M. Hesse and F. Gönnenwein, "Dynamcal Aspects of Nuclear Fission" Dubna 1996, 231
[WOE99] M.Wöstheinrich, R.Pfister, F. Gönnenwein, H.O. Denschlag, H. Faust and S. Oberstedt, Acta Physica Slovaca 49, 117(1999)
[WOH76] H. Wohlfarth, W. Lang, H.-G. Clerc, H. Schrader, K.-H Schmidt and H. Dahn, Phys. Lett, 63B, 275(1976)
[ZIE74] M. Zielinska-Pfabé and K. Dietrich, Phys. Lett. 49B, 123(1974)
[ZIE08] J.F. Ziegler, J.P. Biersack and M.D. Ziegler, "The Stopping and Range of Ions in Matter", internet: www.SRIM.org